World Trade since 1431

World Trade since 1431

Geography, Technology, and Capitalism

Peter J. Hugill

The Johns Hopkins University Press
Baltimore and London

Published in cooperation with the Center for American Places,
Harrisonburg, Virginia

The Johns Hopkins University Press
2715 North Charles Street
Baltimore, Maryland 21218-4319
The Johns Hopkins Press Ltd., London

Library of Congress Cataloging-in-Publication Data

Hugill, Peter J.
World trade since 1431 : geography, technology, and capitalism / Peter J.
Hugill.
p.   cm.
"Published in cooperation with the Center for American Places,
Harrisonburg, Virginia"—T.p. verso.
Includes bibliographical references and index.
ISBN 0-8018-4241-7
1. Economic history.   2. Capitalism—History.   3. International trade—
History.   4. Technological innovations—History.   I. Title.
HC21.H84   1992
382'.09—dc20   92-6765

A catalog record for this book is available from the British Library.

To Judy, Lacey, and Laura, with all my love

Whoso commands the sea commands the trade of the world; whoso commands the trade of the world commands the riches of the world.

Sir Walter Raleigh, 1608

Contents

List of Illustrations

MAPS

List of
Illustrations

List of Tables

From my perspective as a geographer human interaction with the environment is always mediated by technology. Even the most primitive human groups use tools; more advanced groups manipulate the organic environment by growing food rather than gathering it, and still more advanced ones transform the physical environment by directing flows of inorganic energy against raw materials to transform them into increasingly sophisticated manufactures. Most social scientists treat technology as irrelevant, as a dependent variable (in the Marxist tradition), or at best, as a "black box" to describe changes they cannot otherwise account for. I have no intention of arguing for a narrow technological determinism, although unwary readers may accuse me of it. I do insist that we raise technologies to the status of independent variable and accept that they may be causative, in particular in those (numerous) cases when they are not indigenous to a given culture but have been acquired by diffusion. As a geographer I am also concerned at the appalling ignorance of the environment displayed by most academics, probably because they operate on a too restricted time line, paying little or no attention to events more than a few generations away. They treat the environment, like technology, as a dependent variable or a "black box," despite evidence that earlier civilization collapsed from problems caused by environmental change. The "party line" in modern Western civilization declares that our technology is powerful enough to repair any damage we do, yet defines technology as a dependent variable. We have come to believe that, whatever mess we get ourselves into, we will triumph by sheer will. I find it disconcerting that, in a society that bets heavily on a market economy, we should have come to the contradictory belief that we can have a command technology.

Capitalism was invented once on this planet, in the Protestant polities of western Europe. It diffused, slowly at first, then at increasing speed, to the rest of the planet. These two sentences are at once historical and geographical, economic, social, and political statements. They contain all the keys to my inquiry and even suggest the frustration of the social scientist at the word *once*. We can make no tests of whatever hypothesis we advance to explain the genesis of capitalism. At best our account can be a logical, well-argued historical narrative (hence my dissatisfaction with Marxist claims to propound the laws of history). We may construct thought experiments along the lines of alternative histories, but they will never be hypotheses that can be tested. We can surmise what might have happened if, for example, we had never "opened" Japan to the West in the mid-1800s. Or, if an isolated Japan had developed capitalism indigenously, we would have two cases to analyze and thus a stronger possibility of formulating a hypothesis. Had we yet another isolated civilization on which we might experiment, we could imagine testing our hypothesis by secretly manipulating that civilization.

Below the level of the question of how capitalism began is the more easily answered question of how it has progressed. I choose the word *progressed* with care, to avoid the suggestion that activity must always intensify, which would have been inherent in the word *developed*. Equally I imply by *progress* that I believe in the liberal agenda of Adam Smith. If life does not get better for people, if the "wealth of nations" does not increase under capitalism, then the capitalist experiment in organizing human societies will have been a failure and will deserve to end. Thus far capitalism, for all its manifest faults, has given us both progress and development, though we too often confuse the two. We have a roughly five-hundred-year narrative by which we may guide ourselves toward continued progress. Regularities in that narrative are evident in the form of long cycles, although there is debate about just how regular they are, what may cause them, how they may be measured, and even whether they truly exist at all or are merely epiphenomena. Such historical regularities allow us to do what we cannot do for the origins of capitalism: formulate a hypothesis that may eventually be tested so that we may arrive at a theory of progress within the capitalist system.

Self-reflection is most important here. Keynes could not have removed us from the worst of the constant depressions of the business cycle without a societal understanding of what business cycles were and what Keynes was about. However much Japan was forced into opening to foreign trade in the mid-1800s, it was Japanese self-reflection that pushed the Japanese into acquiring Western technologies. Further self-reflection made the Japanese realize, sometimes dangerously slowly, that Western technology works best with the Western social, economic, and political structures that initially nurtured those technologies—in short, democratic capitalism. Capitalism is astoundingly flexible: many forms of it have existed through space and over time; many forms can coexist in a given place at a given time. In its purest theoretical form, as envisaged by Smith, it maximizes the number of economic decision makers and assigns all decision making to the impersonal forces of the market. In practice, varying levels of restraint are placed on the market to resist or encourage the creation of oligopolies or monopolies at any given level of organization, be that organization technical, social, economic, or political. To take one of many examples, we have long had the notion of "natural" monopolies in which the market cannot work efficiently, such as in the provision of electrical power. Such "natural" monopolies must be regulated for safety and to resist excess profit taking unless the technology shifts, as it did in the case of the telephone, in such a way as to allow market forces to operate. Our openness to nonmarket forces operating in areas of capitalist enterprise is a direct result of self-reflection and the accumulation of experience: societies, like individuals, are foolish if they repeat mistakes.

The implication of this is simple. Once we recognize, as a society, that something is a real phenomenon, we have much more power to affect and change it. I argue that the origin and much of the progress of the capitalist system has been achieved without self-reflection. As it becomes clearer how progress occurs, self-reflection will allow societies to progress more efficiently, although such progress may be a two-edged sword. I am reminded of the science fiction story in which a human researcher deduces that joke telling is a trait planted in humans by undefined aliens for whom humans are experimental animals. Immediately the deduction is made, the experiment ends; jokes are no longer funny, and the human researcher recognizes that some other experiment will be forthcoming. The complexity of capitalist systems and the lack of obvious and simple causation ensures, however, that self-reflection will not result in easy manipulation.

As with any work of synthesis, the work before you has many origins. I have long been disturbed by the unwillingness of modern non-Marxist social scientists to address large-scale questions in the history of our culture, in particular that of the origins and progress of capitalism. Equally disturbing has been the willingness of Marxists to persuade themselves that they alone could successfully answer that particular question and the quite remarkable unwillingness of historians to engage in anything approaching a theoretical debate. Even in disciplines where theory is important, many theories have come to be regarded as unsound scholarship instead of being subjected to at least occasional reexamination. Many well-meaning people in my own discipline—geography—have, for example, long refused to pay serious attention to the possibility that certain types of physical environment might inhibit or encourage certain types of human activity, on the grounds that such models would be determinist. The better practitioners of such environmental determinism from early in this century, however, tended more to the view that environmental conditions influence a society's range of choices. Given the ability of technology to override certain environmental elements, this is a view I endorse.

I am also convinced that change occurs first and fastest, if not always for the best, in the borderlands. Experiments occur best away from core influences, which tend always to the solution that causes the least change. This inertia ensures that the core eventually pays "the penalty of the commanding lead." The core also tends to be overpopulated, making it hard for bright persons with new ideas to make themselves heard. Scale is as severe a detriment to innovation as inertia. Many experiments are necessary to find answers to problems old and new, and many experiments fail, some dismally. In the core they may be lost to sight because of scale and inertia. In the borderlands those that succeed succeed spectacularly because they have no inertia to overcome and have fewer barriers imposed by scale.

The title of this book refers to 1431, the probable year in which the Portuguese navigator Velho first successfully sailed out into the Atlantic to the Azores and back to Portugal again. Many other dates would have been reasonable. A.D. 400 marks the approximate beginning of the use in northwestern Europe of the moldboard plow, a piece of technology that allowed remarkable production increases by bringing vast new acreages of river valley clay soil under crop, but data from this vital historical period are sorely lacking. A.D. 800 marks the increased productivity wrought by replacing two-field with three-field agriculture, but my theme is that the capitalist world has prospered most by simple, cheap geographic expansion. The year 1347 marks the arrival in Europe of the Black Death, but my theme is expansion of production, not reduction of consumption, however important reduced consumption may often be. Fourteen ninety-two is always a popular choice, but however vital they are to the success of European geographic expansion, the Americas are not the whole story: 1492 marked the culmination of a long expansion out of Europe, and in any case, Columbus was trying to get to the Indies. The year 1431 marks the true, if halting, return by Europeans to geographic expansion, an expansion that would quickly increase production, set Europe on the road (more properly the trade-wind) to capitalism, and ensure a sustained five hundred years and more of European-led progress.

Throughout this text there are numerous references to places, many of which are well known, some of which are not. Where events happened is at least as important as how and why they happened; often, geographical location can help to explain causes and processes. Although the ideal situation would be to have an atlas as a companion while reading, that is not always practical. Consequently, the set of reference maps, titled "World Maps," at the end of the book is provided to assist the reader who may not have ready access to an atlas or other geographical resource.

Acknowledgments

My acknowledgments are many, but the main one is to the several thousand students of human geography at Texas A&M University on whom I have tried my ideas over the past twelve years, as well as the students in my graduate course, "Historical Geography of the World-System." I also acknowledge the role of my mother's mother, whose tales of my grandfather, of imperial wars, and of her life as a soldier's wife in South Africa enthralled me as a child. Not until I had acquired a formal education and a sense of history did I understand the implications of what she told me.

At Leeds University I was lucky enough to have S. R. Eyre as my first undergraduate tutor. From him I came to understand the importance of technology in mediating human interaction with the physical environment. Also at Leeds Maurice Beresford began for me a long-standing interest in economic history. My master's supervisor at Simon Fraser University, Phil Wagner, introduced me to U.S. cultural geography and the work of Carl Sauer. Wagner continued to expand my interests in technology and gave me a much keener sense of the interplay of technology with culture. He also deserves my thanks for introducing me to the work of historian William H. McNeill, by whose thinking I have been greatly influenced, and that of the anarchist Peter Kropotkin. Kropotkin's work convinced me that scale is of central importance in human activities.

The debt I owe David Sopher, my dissertation supervisor at Syracuse University, can never be repaid, nor can that to Don Meinig, who came to be the de facto second chair of my committee. Sopher, who died tragically some years back, made me understand the overarching need for theory, something I had hitherto resisted. Meinig's exemplary account of macro themes in human geography and his powerful understanding of geopolitics are as central to my work as Sopher's concern with theory. Eli Gerson, then in the Department of Sociology at Syracuse, introduced me to the work of Mead and the American pragmatists, and thus to the notion of societies as self-reflecting systems that may change themselves over time. John Agnew deserves my thanks for resisting my tendency to liberal beliefs in the certainty of progress.

Here at Texas A&M I owe thanks to my first department head, Campbell Pennington, for creating an academic climate in which I had

**Acknowledg-
ments**

time for reflection and interaction with others on campus. My first dean here, Earl Cook, dead these several years, shaped more of this monograph than it might seem. Not only do I thank him for his *Man, Energy, Society,* but also for steering me to Albion's classic *Forests and Seapower.* Brian Blouet greatly clarified my understanding of the works of Mackinder and Mahan. Jonathan Smith helped me understand some of the more arcane meanings of my own text.

Almost as soon as I came to Texas A&M in 1978, I joined an interdisciplinary seminar known as the NAR (on the grounds of the refusal of its founder, Steve Picou of Sociology, to be involved in any semi-nar). Over twelve years of Friday afternoon meetings in various homes and bars, at which we abused the many authors of the books we selected for dissection, I have acquired a strong liking for interdisciplinary arguments and Shiner Bock beer. I particularly thank (in alphabetical order) Art DiQuattro of Political Science (resident Marxist), Bruce Dickson of Archaeology (resident Realist), Alex MacIntosh of Sociology (resident Parsonian), Wes Peterson of Agricultural Economics (ex–resident market economist and Beaujolais nouveau importer), Ed Portis of Political Science (resident Weberian), Jane Sell of Sociology (resident rat experimenter), and Paul Thompson of Philosophy (resident philosopher and grantsman)—all of whom would like you to know that NAR franchises are still available at very reasonable rates. They may not recognize the arguments here, but they have assuredly influenced them. They will, I have no doubt, trash them as enthusiastically as they have the ideas of all the other authors whose work we have dissected over the years. Thanks, guys (in the nonsexist Canadian sense; I have not yet brought myself to the southern "y'all"). The NAR would like to remind university administrators everywhere that, under its unwritten constitution, committee or faculty meetings after 4 P.M. Friday are illegal. NAR franchisees should also note that they are not allowed to consider any books with the words "policy" or "dancing" in the title, or any books on French structuralism suggested by me.

Thanks are also due to the many people who helped bring the manuscript to fruition. Fraser Hart at Minnesota encouraged me to seek out George Thompson, president of the Center for American Places and a publishing consultant to the Johns Hopkins University Press, a move I have never regretted. I should also like to thank Carville Earle and Peter Hall, who read earlier versions of the manuscript. This thanks extends beyond the mere fact of their function as reviewers. Their published works have, in each case, influenced my thinking on the topics covered here well beyond my ability to reference that influence. The editorial staff of Johns Hopkins also deserve praise, as does the meticulous copy editing of Terry Schutz. Without the constant help of my graduate assistant, Mary Ann Kniseley, I would never have been able to cope with the varied and hectic demands of academic life while writing this book. Mary Ann also helped collect and manipulate significant amounts of the more arcane data, coping with my odd requests with good grace and proving eminently adept at extracting data where there seemed to be none. Sandy McConnell helped cheerfully with the typing and with tables that sometimes turned out to be far too complex. Tim Davis and John Maslonka of the Cartographics Laboratory at Texas A&M did a fine job on the maps and illustrations. The National Railway Museum at York and the Henry Ford Museum at Dearborn deserve rounds of applause for simply existing: their collections are unique. Many other museums gave me the chance to see what it was that I wanted to write about, as well as considerable pleasure: among the most important have been the Royal Navy Museum at Portsmouth (HMS *Victory*), the USF *Constellation* Foundation in Baltimore, the SS *Great Britain* Trading Company in Bristol, the National Maritime Museum in San Francisco, the Baltimore & Ohio Railroad Museum in Baltimore, the National Air and Space Museum in Washington, D.C., the Deutsches Museum in Munich, the Musée de l'Air in Paris, the Imperial War Museum's aviation collection at Duxford, the Royal Air Force Museum at Hendon, the United States Air Force Museum at Dayton, and the Canadian National Aeronautical Collection in Ottawa.

The staff of the Public Records Office in London were unfailingly helpful in tracking down the information I needed (like all its users, I could wish for longer hours). The library of the Commonwealth Society, also in London, was of particular use on the problems encountered by Britain early in this century, especially in coping with the "cotton famine." The Texas A&M library has unsuspected strengths in the history of engineering. Several of my engineering colleagues at Texas A&M have contributed to the development of my thought. Howard Wolko and Bill Harris, however, stand out as my ideal engineers, able to see through the immediate nature of a technical problem to its full range of implications for our society. Howard Wolko is at the National Air and Space Museum and deserves thanks for sharing his insights into the development of airframe technology, for shepherding me through NASM's substantial document collection, and for being kind enough to read chapter 6 with a critical eye. I thank Bill Harris, who holds the Snead Chair of Transportation Engineering at Texas A&M, for being good enough to put his unmatched knowledge of the development of the railroad at my disposal and for reading through that section of chapter 4. Any sins of omission or commission in these chapters are mine and not theirs. Informal conversations with members of the Institute for Nautical Archeology here at Texas A&M, notably Dick Steffy and George Bass, have greatly helped, as has service on the graduate committees of their students.

Last, but very much not least, I thank my family and friends. I could never have finished this book without the unfailing support of my wife, Judy Warren, and our two daughters, Lacey and Laura. They have been dragged, usually willingly, to all manner of strange places for me to track down technologies and documentation. A transatlantic marriage, with parents, cousins, and friends on both sides, has been an immense help in providing both simple access to needed materials and understanding of how global systems must work.

Chapter One

Geographic Reality in the Development of Capitalism

Till the end of the fifteenth century Europe was dominated by her position as part of the continental triad of the Eastern Hemisphere. In spite of her long, indented coastline and considerable maritime development, hers was mainly coastwise navigation, not oceanic . . . Herself a country of the temperate zone she sought the tropical products of the Orient. Thither the way was easy, for the Indian Ocean stretches out two welcoming arms, the Red Sea and the Persian Gulf, towards the Mediterranean.
But the day came when [the routes to the East] were held by the Ottoman Turks, a power with whom it was fruitless either to fight or to reason. Europe gazed longingly, despairingly towards the old avenues to the Orient, . . . then reluctantly, at first almost hopelessly, she looked towards the Atlantic.

Ellen Churchill Semple
American History and Its Geographic Conditions

Although much has been written over a very long time on the development of capitalism, it has often been far too abstract and theoretical. We have learned to tie the success of capitalism to the tremendous increase in world trade that began in the 1400s as Europe expanded overseas, a process driven by vastly increased geographic knowledge and technical skill in ship construction and handling. Any adequate understanding of the development of capitalism must therefore, of necessity, be grounded in the geographical reality of that expansion. At relatively low levels of technology, the costs of overcoming distance and the problems posed by the environment were much greater than they are for us. Because geographical reality is not a constant, all theory must be accompanied by a historical narrative.

THE DEVELOPMENT OF HISTORICAL SOCIAL SCIENCE

Since the publication of Wallerstein's *The Modern World-System* in 1974, social scientists have accelerated the return to macro theory tentatively begun by Rostow in 1962. Three of the great founders of modern social science, Adam Smith, Karl Marx, and Max Weber, were macro theorists, seeking in particular to explain the origins of capitalism in western Europe. In being able to respond to Smith, Marx was able to come up with an explicitly theoretical rather than a mostly descriptive model. Put simply, Marx's historicist model saw capitalism as the second inevitable stage in historical and economic development, a stage that followed feudalism but preceded the dictatorship of the proletariat (stage 3) and true communism (stage 4). Weber was able to respond to Marx as well as Smith and took issue with Marx, claiming that capitalism was a product of an extremely specific force in European history: the emergence of radical Protestant Christianity in the ideas of John Calvin. Calvinism took hold most efficiently in Holland and England, propelling those polities to the forefront of European history after about 1600. Calvinism's advantage was that, in a society just changing from agrarian to industrial, it decreased religious consumption and increased the savings rate, thus providing more capital for investment and accelerating industrialization.

In very broad terms these macro theories were displaced in Western social science by micro theory after the events of the 1930s. In part this was because social science emulated natural science, physics in particular, with its radically reductionist approach. In part it was because of the success of Keynesian economics in the late 1930s and, most spectacularly, in the period from the end of World War II to the economic crisis of 1973, which was brought on by the OPEC-induced oil price rise. Keynes's theories were short run, with investment discounted over thirty years, about one human generation. When asked why, Keynes made the famous reply, "In the long-run we're all dead."

Avoidance of long-run or macro theory by social scientists was compounded by features other than physics envy or the desire to emulate the success of microeconomics. The failure of geographers to rework environmental determinism into a viable macro theory in the 1930s ensured that ecological and environmental themes would be ignored by the academic community (Eyre, 1978, 149). Conservationist work focused on environmental themes (Cook, 1976) has usually been dismissed as the work of Cassandras working against progress. In actuality the critics of the Cassandras have usually confused progress with development, the intensification of which may well result in the reversal of progress.

In a more academic vein, long-run historical data of the type desired by social scientists have been hard to come by until relatively recently. Historians traditionally are more concerned with politics and wars than with economic and social forces operating at the level of the masses. After World War II, however, a few historians began to break new ground and a group of historical social scientists developed; they attempted to apply social science techniques to laboriously accumulated historical data of the appropriate type. Three efforts stand out.

The French *Annales* School of Social History

The *Annales* school had its origins in the work of the French regional geographers and historians early this century, in particular Lucien Febvre and Paul Vidal de la Blache. Such scholars emphasized the need for an intense understanding of a small part of the earth's surface. They focused historical inquiry on the influence of soil and crop types, labor systems, and demography, instead of just war and politics. Marc Bloch, the founder of the *Annales* school, named for the French journal *Annales d'histoire sociale*, pursued such themes strongly before World War II in his landmark works *French Rural History* (1966) and *Feudal Society* (1961). Bloch, a Jew, was shot by the Gestapo in 1944 for his part in the French Resistance. He heavily influenced Fernand Braudel, whose landmark study of the Mediterranean, *Capitalism and Material Life, 1400–1800*, was published in French in 1949 and appeared in English translation only in 1973. On the basis of this work, he has since produced his monumental three-volume work, *Civilization & Capitalism, 15th–18th Century* (1981, 1982, 1984). Immanuel Wallerstein has done a great deal to bring these ideas into American scholarship. His work derives directly from that of Braudel, and the great influence of the *Annales* school clearly derives from the explicitly theoretical stance taken by Wallerstein.

The Cambridge School(s) of Historical Social Science

The Cambridge geographer, H. Clifford Darby, pioneered the intensive study of the Domesday Book, the great Norman census of England after the conquest of 1066. Domesday is a relatively firm data base that covers an entire major European region at a date well before any comparable resource. Some of this shows up in a pioneering collection of essays that Darby published in 1936, in particular his own contribution, "The Economic Geography of England, A.D. 1000–1250." The influence of the French social geographers and their intense concern with "man in rapport with the natural environment through his technology" is clear (Brunhes, quoted in Darby, 1936, vii, my translation). The focus of the Cambridge school is clearer in Darby's more recent collection (1973) and in the remarkable body of work represented by the papers published in the *Journal of Historical Geography*. Although the journal is by no means limited to the Cambridge school, most of the best practice of that school appears there, much as the best French social history appears in *Annales*. Darby also recognized a strong debt to the noted Cambridge economic historian, Sir John Clapham (1949).

The second Cambridge school of note, which is by no means linked to the Darby school of historical geography, is the Cambridge Group for the History of Population and Social Structure, a unit of Britain's Social Science Research Council. One of the origin points of this group was the publication by Peter Laslett in 1965 of *The World We Have Lost: England before the Industrial Age*, now in its third edition (1984). England is unique in Europe in the historical depth of its demographic data at the local level, and the Cambridge historical demographers have exploited this depth brilliantly. At the establishment of the Church of England, King Henry VIII gave parish priests the power to tax the local population by recording all births, deaths, and marriages. Laslett, not least through his innovative use of radio programs, was able to organize community members interested in their local history to transliterate local parish records for use by the Cambridge historical demographers. This remarkable effort has paid off in works by Laslett and Wall (1972) and Wall, Robin, and Laslett (1983), but most specifically in Wrigley and Schofield's landmark *The Population History of England, 1541–1871: A Reconstruction* (1981).

These two Cambridge schools are very strong empirically and in technique, but they tend toward micro rather than macro theoretical perspectives. Neither has been concerned with the origins of capitalism, and both have tended to focus heavily on preindustrial European society. Although this society was on the way to capitalism, and a thorough analysis of it is necessary to understand the transition, neither Cambridge school has focused on that transition. A related effort in America has been that of the urban social historians on the East Coast, in particular at Harvard. A number of excellent, hard-data-oriented monographs has emerged from this school, but the only highly organized major research effort is focused on Philadelphia. The Philadelphia Social History project uses data from censuses and city directories as well as other sources to work toward an interdisciplinary history of the city (Hershberg, 1981).

The Berkeley School of Cultural Geography

The third effort is more diffuse; it is based around the Berkeley school of cultural geography developed under Carl Sauer in the 1920s and 1930s. Sauer's interest was in preindustrial agrarian societies, in particular the traditional societies of the New World. His concern, which was for the enduring relationship between people and their environment, is exemplified in such titles as *Land and Life* (1963) and *Man in Nature* (1939), a quite remarkable book written for high school use. Sauer's interests were close to those of the French social geographers, who influenced him heavily. He was far more concerned with the natural world than were the European geographers and historians. Unlike the Europeans, who came from a strong liberal tradition, Sauer was a natural conservative. These are far from political terms. Liberals, at least until World War I, believed in progress, the steady improvement of human existence through technology, which gives humans dominion over nature. Sauer's conservatism led him to see human beings as dangerously destructive of a fragile ecosystem, at least if they are given the technology to dominate nature. Sauer's contribution, therefore, is a concern with the long-term consequences to the ecosystem of human technology; it is a contribution to macro theory that is still unrecognized. Even such explicitly conservationist writers as Eyre (1978) fail to acknowledge this debt. Sauer was concerned with the destructive ecological practices of European commercial agriculture in the New World and its displacement of the much stabler ecology of the American Indians.

Sauer's own major contribution to the study of human technology was his pioneering and controversial work on the origins of agriculture (1952), a contribution continued by some of his students (Carter, 1977). His understanding of the role of transport in preindustrial societies was also strong. In an essay on the Caribbean he pointed out that the Caribbean resembles the Mediterranean to the extent that it is a natural highway for humans using simple boats (1962). It was therefore the logical entry point to the New World for Europeans (1966). What is in many ways the clearest current statement of Sauer's perspective comes from Donald W. Meinig, who, though he is not one of Sauer's students, clearly owes a considerable intellectual debt to Sauer. Meinig adds an explicitly geopolitical and theoretical component to Sauer's concern with ocean travel and the development by Europeans of the Atlantic economy from the 1430s onward (1986). He retains much of Sauer's conservative attitude about the danger of human disruption of the ecosystem.

The work of Sauer and his students is also explicitly diffusionist, a view I strongly support in this book. Even on foot, and certainly in simple boats, humankind has always been mobile. At low levels of transportation technology, humans may not be able to move much in the way of material goods, but they can assuredly move ideas. Archaeologists, not without good reason, have resisted the idea of diffusion in simple preagricultural societies, but the evidence pro and con is fragmentary (Hugill & Dickson, 1988, xiv–xix). In commercial and industrial societies the evidence for diffusion is overwhelming, although national pride and commercial rivalry have attempted to hide it. Some historians have looked to diffusion for a better explanation of events; most notable among them is William H. McNeill (1963), one of the foremost practicing American historians. As a group, business historians have also been active in the formulation of diffusionist explanations of why particular economies develop in particular ways at particular periods in time (Hugill & Dickson, 1988, xix–xx).

These three efforts at historical social science can be summarized as follows:

1. a search, grounded in French social history but explicitly seeking theory, for alternatives to Marxist historicism and Weberian idealism

2. an empirical and technique-oriented "new history" examining nontraditional economic and social records, in particular census and demographic materials

3. a concern with both ecosystem survivability and the spatial patterning of human activity in the preindustrial and commercial phases of European expansion into the Atlantic

For my purposes here these efforts, though significant, lack structure. In specific terms they also show too little concern with the industrial world that has developed in the last two hundred years. They tend to view industrial societies as separate from agrarian societies. From many perspectives, they are; but from the perspective of world system theory, the distinctions blur. Wallerstein has properly concerned himself with the origins of the world system in capitalist (as opposed to precapitalist and subsistence) agriculture. Capitalist industry is a special case of capitalist agriculture, at least in its first incarnation in the late eighteenth century, when industry still largely depended for raw materials on capitalist agriculture. Since capitalist industry has been, from its inception, a truly global system, the transportation of goods and people is a significant constraint on its operation. More recently, especially since the late nineteenth century, it has become less dependent on the organic world for raw materials and increasingly dependent on the inorganic world. The British geographer E. A. Wrigley points out in a fine recent essay that Britain was the first country to go from an advanced organic economy to a mineral-based economy (1983).

Any understanding of the world system must thus be tempered through two further understandings. The first is of the technologies of transportation and production, whether agricultural or industrial. The second is of macroeconomic theorizing, most specifically the development of long-wave theory.

THE ROLE OF TECHNOLOGY

The best framework for the historical analysis of technology is that of Lewis Mumford, who, following the Scottish urban planner Patrick Geddes, first proposed a three-part division of human technics in 1934. To Geddes's Paleo- and Neotechnic eras Mumford added the Eotechnic phase of cultural preparation (Mumford, 1934, 109). Stated briefly, though with additions of my own for which Mumford and Geddes should not be held to account, the Mumford/Geddes scheme is as follows.

The Eotechnic

Although industrial, this technical phase was still highly agricultural. Raw materials for industry had to be grown, as did fuel. Because industry was predominately concerned with textiles, fiber was the main raw material. We can thus characterize such an economy as concerned primarily with growing the three Fs—food, fiber, and fuel.

The Eotechnic had its beginnings in the increase of interregional trade that characterizes the economy of northwestern Europe after about A.D. 800. This was a trade in bulky, low-value commodities as opposed to compact, high-value preciosities. As such, it required efficient bulk movers. Given the technology available, that meant wind-driven ships made out of wood. Such industrial processing as occurred was local and only became capital- and labor-intensive toward the end of the Eotechnic period, in the late eighteenth century, with the emergence of the factory system. Although inanimate power was used in industrial production as well as in transportation, it was limited in amount and geographic availability. Eotechnic mills, built of wood and powered for the most part by falling water, some by wind, could develop only small amounts of power at sites determined by geography, not human choice. Only a few horsepower were generated at each site, the usual calculation for a wooden water wheel being 1 horsepower per meter of wheel diameter. Mills larger than 5 horsepower were unusual. No marked regional concentration of power was possible, except in a small number of geographically favored areas, where the climate was temperate and numerous streams were available. Northern England and New England are good examples of such regions.

The main structural material of the Eotechnic was wood, which was also the main fuel. Since wood competed for acreage with food and fiber, which until the very early 1800s was wool, its importance imposed severe ecological restrictions. It is thus easy to remember that, in Eotechnic northwestern Europe, the three Fs were also the three Ws—wheat, wool, and wood—though in some regions depending upon climate, access to soil type, and transportation, other crops were occasionally substituted. The availability of wood imposed structural limits on such things as ship and building size, water wheel diameter, bridge spans, and the like. Except where enough capital existed to build canals, and where political conditions were stable enough to make such long-term investments profitable, inland transport was animate: humans or pack animals carried penny quantities of low-bulk, high-value goods.

The Paleotechnic

The first stage of capitalist industry was the Paleotechnic. The switch from falling water to steam power allowed large concentrations of inanimate power in a region, thus large concentrations of labor to process large amounts of agricultural raw materials. Steam-powered industrial cities replaced water-powered industrial villages. Such concentrations of humanity, without the leavening effects of technology to remove human wastes, produced a rapid decline in living standards for average people. Alone, these conditions might not have produced the beginnings of organized political action by labor, but they were coupled with a hangover from radical Protestantism that defined as immoral those who could not, as well as those who would not, work. Marx, among many other socialist, anarchist, Christian, and utopian thinkers, suggested social and political solutions to problems brought on by technical change in this specific economic and social climate.

Iron, and later steel, replaced wood as the primary structural material. The substitution of coal for wood in the iron-smelting process brought a great increase in the capacity of Paleotechnic society. It also removed most wood from the three Fs. Steam was also applied to long-distance movement of people and goods, both on sea and by land. It regularized transoceanic travel and, more importantly, allowed the first true penetration of the continental interiors. Continental and even transoceanic polities under relatively democratic governance became possible for the first time. Rail travel opened up vast virgin tracts to capitalist agriculture. Raw materials for industry, as well as food for industrial labor forces, became plentiful. One significant new crop was added to the three Fs; cotton. The crop was climatically limited to the subtropics, but steamboats and railroads made it possible to move cotton from subtropical growing regions to manufacturing regions. Cotton became an engine for Paleotechnic economic growth, first in Britain, then in the northern United States, industrial Europe, and Japan.

The Neotechnic

The Neotechnic is the second stage of capitalist industry. Whereas the Eotechnic was marked by extremely limited inland transportation and the Paleotechnic was restricted to fixed lines of transport that operated poorly for short hauls, the Neotechnic radically evened out geographic access. Electrically driven streetcars and interurbans, bicycles, and automobiles, buses, and trucks with internal combustion engines made short-haul transport economically viable and filled in the broad interstices between Paleotechnic transportation lines. Since land represents a major fixed cost of production, access to large quantities of "new" land radically reduced fixed costs. More rapid movement of goods allowed drastic reduction in inventory costs and lowered the percentage of perishable goods lost to spoilage. Commercial aviation, especially jet aviation, has permitted long-distance, high-speed movement of mail, people, and high-value goods. In turn, the integration of continental and transoceanic economies at the managerial level holds out the promise of a truly global economy.

A steady shift away from totally natural ecologies began during World War I, when German explosives manufacturers substituted nitrates synthesized from the atmosphere for nitrates mined from bird droppings. The same nitrates also formed the basis for production of nitrogenous fertilizers, which, though they have increased agricultural productivity immensely, are consuming the finite supply of hydrocarbon fuels in greater and greater amounts. Petroleum-fueled tractors increased agricultural productivity at similar cost.

A capitalist agriculture based on extensive use of land with labor costs controlled by a variety of devices (slavery, sharecropping, the family farm, and primitive mechanization) began to give way to one based on intensive land use with heavy application of inorganic fertilizers and sophisticated mechanization. Capitalist agriculture is now a subset of capitalist industry. The same technology that made it possible to synthesize nitrates made it possible to synthesize hydrocarbon alternatives to organic raw materials. Synthetic fibers began to displace wool and cotton, and synthetic rubbers the natural variety, in the late 1930s. Since

World War II capitalist industry has moved to heavy dependence on liquid hydrocarbons in parts of the food production process, for industrial raw materials, and for transportation. Compared to the Eo- and even the Paleotechnic ecology, this is an unstable situation, heavily dependent on both nonrenewable resources and an increasing level of technical skill. Electronic communications promise a reduced need to move people physically in advanced Neotechnic societies, but hydrocarbon consumption in agricultural or raw material production would remain the same.

Long-Wave Theory

The second tempering of world system theory comes from the recent revival in macroeconomics of *long-wave theory*. In the 1930s the Russian economist Kondratiev questioned Marx's assumption that capitalism was doomed to failure because of the law of the tendency of the rate of profit to fall (Marx, 1959, 211–66). Marx's law is simple. Capitalist production calls for free entry to the marketplace of would-be producers. With free entry, new producers accept lowered profits as a cost of entry, and older producers lower profits to compete. As long as free entry goes on, profits tend to fall. One way out of this trap was evident to Marx. Monopolistic or oligopolistic production would allow one firm, or a small number of firms organized through a cartel, to fix prices. Marx, however, underestimated the strength of belief in free enterprise within the capitalist system. Fully developed capitalist polities have taken legal action to reserve monopoly power to the polity itself. They have permitted monopolies only when a particular technological level has seemed to warrant the absence of competition, as was the case in telephonic communications before the advent of microwave linkages and microelectronics. Polities that permitted oligopolies, such as imperial and Nazi Germany and imperial Japan, were not fully capitalist states in this definition. They made that transition only after World War II.

In the late nineteenth and early twentieth centuries a second way out of the trap postulated by Marx was deduced. Captive markets abroad could buy industrial products at rates high enough to compensate for low profits on the domestic market. Hobson first propounded this idea in his 1902 defense of the "new Imperialism" of the late nineteenth century.

Overproduction in the sense of an excessive manufacturing plant, and surplus capital which could not find sound investments within the country, forced Great Britain, Germany, Holland, France to place larger and larger portions of their economic resources outside the area of their present political domain, and then stimulate a policy of political expansion so as to take in the new areas. The economic sources of this movement are laid bare by periodic trade-depressions due to an inability of producers to find adequate and profitable markets for what they can produce. (Hobson, 1902, 80)

The great British geopolitician Sir Halford Mackinder strongly pre-
figured Hobson in his address to the Institute of Bankers in 1899 (1900).
His focus was more on the export of capital than of product, but both
are tied to Marx's law. The rate of profit can be raised by sending capital
where returns are higher, just as it can be raised by sending products to a
captive imperial market. When capital is exported, imperialism seems
even more necessary to ensure that the political and legal system does
not work against the interest of the foreign owners of capital. Without
imperial oversight, debtor nations can repudiate debts, if necessary.

Despite the pioneering statements of Hobson and Mackinder, it was
the work of Lenin that tied imperialism to Marxism and made the latter
Marxist Leninism. Lenin was at fault in his assumption that monopoly
power would continue, but in the first two decades of this century the
power of imperial Germany's cartels was not yet broken, nor had U.S.
antitrust legislation shown that monopoly power was curbable in polities with a genuine commitment to capitalism. Lenin saw imperialism
as a step beyond monopoly power in terms of maximizing profits. In
1916 he completed his famous pamphlet, "Imperialism, the Highest
Stage of Capitalism" (1939), in which he drew heavily and approvingly
on Hobson. Lenin followed Hobson in suggesting that the driving force
behind late nineteenth-century imperialism was higher return on capital
invested in secure imperial possessions (Lenin, 1939, 100).

The Austrian economist, Joseph Schumpeter, took issue with Lenin's
notion in the 1930s. In fully capitalist societies, he argued, "the competitive system absorbs the full energies of most of the people at all
economic levels . . . In a purely capitalist world what was once energy
for war becomes simply energy for labor of every kind . . . A purely
capitalist world therefore can offer no fertile soil to imperialist im-
pulses" (Schumpeter, 1955, 69). Schumpeter also pointed out that "the
balance sheet of export monopolism is anything but a brilliant success"
(1955, 87). He saw the wellspring of imperialism in the creation of tariffs
behind which the interests of regional ruling classes could most profitably operate. For Schumpeter imperialism was rooted in the autocratic
class system of feudal Europe with its numerous regional polities. Where
autocratic classes were displaced by mercantile ones, as in Holland by
the 1580s and Britain by the 1680s, free trade became the dominant
principle (Schumpeter, 1955, 89–91). In this analysis imperialism is not
the "highest stage of capitalism," merely a bizarre hangover from feudal, precapitalist Europe.

Schumpeter recast Marx's law as the "theory of vanishing investment
opportunity" (Schumpeter, 1950, 112). In this Schumpeter saw the
nineteenth and early twentieth centuries as a special case of windfall
profits brought about by rapidly rising populations and geographical
expansion. He saw investment opportunities "vanishing" in the twentieth century because demographic change was reducing effective demand as family size fell. The end of geographic expansion driven by
improved land transportation had a similar result. Unlike most of his
contemporaries in economics, who believed that "but minor achieve-

ments remain[ed]" after the wave of technical innovation that characterized the early Neotechnic, Schumpeter identified a pattern of recurring innovation (Schumpeter, 1950, 117).

> We are just now in the down grade of a wave of enterprise that created the electrical power plant, the electrical industry, the electrical farm and home and the motorcar. We find all that very marvelous, and we cannot for our lives see where opportunities of comparable importance are to come from. As a matter of fact, however, the promise held out by the chemical industry alone is much greater than what it was possible to anticipate in, say, 1800 . . . Technological possibilities are an uncharted sea. (Schumpeter, 1950, 117–18)

Two important ideas suffuse this quote. The first is that there are "waves of enterprise," the second that such waves are driven by technology. In an earlier footnote Schumpeter commented that the capitalist economy shows signs of Kondratiev "long waves," cycles of about fifty-five years each (Schumpeter, 1950, 68n).

I suggest that charting the sea is not that difficult. Transportation technology is crucial, as are the technologies of agricultural and industrial production. Technology can also be better understood if the terms *hardware* and *software* are borrowed from the computer world. Hardware is the machinery itself: prime movers, vehicles, farm and factory machines that materially increase production. Software describes the technologies of utilization of human labor, usually in interaction with hardware. Banking and insurance are classic software. Arkwright's factory system was also largely a software innovation, as were the U.S. system of mass production introduced in the 1840s and the rational scientific U.S. managerial technology introduced by Taylor and Gilbreth in the early twentieth century.

Schumpeter's ideas were later picked up by the German economist, Gerhard Mensch (1979), and have been recently elaborated by Berry (1991). Mensch points out that these "waves of enterprise" are more than mere extensions of the business cycle. They seem, in fact, to be central to the whole development of capitalist economies. He also points out that capital surges into new technologies about every three human generations, or about every hundred years (Mensch, 1979, 124). This puts Marx's law of the tendency of the rate of profit to fall in its place as applying only to two Kondratiev cycles. As the rate of profit in old technologies falls, capital in a truly capitalist economy moves to new, more profitable technologies. In polities that accept monopoly or oligopoly power, or that believe in imperialism, capital may elect to remain in the old technology.

As Goldstein has recently pointed out, a variety of long waves seem to be compounded by other cyclical phenomena operating at different periodicities. He suggests three basic worldviews—liberal, revolutionary, and conservative—and six current schools of long-cycle research, two from each camp (Goldstein, 1988, 9, 151). I find myself merging the revolutionary world system school with the liberal innovation and leadership cycle schools and at the same time adding a truly conservative concern with the need for long-term balance between production and consumption.

In economic terms we have not only the seven- to ten-year business cycle driven by short-term fluctuations in investment, but twenty-five-year Kuznets cycles and fifty- to fifty-five-year Kondratiev cycles driven by the complex pattern of changing investment opportunities (Berry, 1991). Two Kondratiev cycles merge into a roughly one hundred–year world leadership cycle of the type developed by Modelski and Thompson (1988). Berry notes that one Kondratiev cycle peaks as new technologies emerge, usually in the eighties and nineties of each century, and as large surplus profits are available from the previous cycle. Because the rate of profit has fallen, such capital has been held liquid and moves freely into new technologies. The other cycle peaks in the thirties or forties of every century, when much capital is tied up in investment in the technologies developed in the eighties and nineties. Very little capital is liquid in the situation typical of the thirties and forties since the rate of profit in investments made in the teens and twenties is still high. As the rate of profit starts to fall, usually in the late twenties, it precipitates a major depression that may be deepened if it coincides with a business cycle depression or if governments do not act in a Keynesian fashion and increase public spending. After the depression, investors tend to seek higher rates of return in the next wave of technical innovation. Since the previous technologies have turned unprofitable, the higher rates of risk in the newer technologies are more acceptable than they would have been before the depression.

Long-wave theory does not just apply to economic events. Goldstein notes that an adequate theory of long cycles must include the role of war (1988, 278). Modelski and Thompson count the number of capital ships in the world's navies in the five hundred years after 1494, thus empirically demonstrating clear swings in the balance of power (1988). Periods in which a single polity has hegemonous power give way to a multipolar world in which several states, usually including the past hegemon, vie for succession. A new hegemon emerges, usually after a period of warfare between the major contenders. This warfare is as much economic as military. Modelski and Thompson point out that a polity achieves hegemony only by having a sounder economy than its competitors. Such an economy usually, but not necessarily, includes relatively low public-spending levels, in particular on the military, low wages, and high levels of saving and investment in long-term infrastructure. Once hegemony is achieved, however, the tendency is to divert public monies to military spending in order to maintain hegemony. This process has been variously described, most recently by the historian Paul Kennedy as "imperial overstretch" (1987).

It is, however, the empirical proof of long-wave theory that makes Modelski and Thompson's work significant. They demonstrate five clear hegemons, one in each of the five pairs of Kondratiev or world leadership cycles experienced in the past five hundred years: Portugal, Holland, England, Britain, and the United States.

1. Portugal achieved hegemony by virtue of the Treaty of Tordesillas of 1494, which divided the non-Christian world between Portugal and Spain. Portugal got by far the more productive hemisphere, 60 degrees west longitude to 120 degrees east. This hemisphere encompassed modern Brazil, all of Africa, nearly all of Mackinder's "world island" of Eurasia, and half of the Indonesian archipelago. Spain's hemisphere included North America and the western half of South America, the eastern fringe of Eurasia, the eastern half of the Indonesian archipelago, and had it been known, nearly all of Australia. Much of Spain's acquisition was the Pacific Ocean, and to extract value from western South America its ships had to cross the Pacific to its western outpost, the Philippines, before coming back to Spain through Portuguese waters around the tip of Africa.

2. In the period of multipolarity that followed Portuguese hegemony, Portugal, Spain, England (not yet Britain since Scotland had not been pressed into union with England), and Holland competed for hegemony. Portugal and Spain, now allied by royal marriage as well as geography, attempted to break the English challenge in 1588, when they sent the Armada against England. Had they succeeded they might then have crushed Holland, where they had already established considerable land forces in their attempt to drive out the Calvinist heresy. They failed in both endeavors because of the ability of the English fleet to deflect them, but England was unable to use its naval superiority to gain hegemony. Holland's superior mercantile economy gave it the edge over England, and the Dutch rose to hegemony.

3. By the 1660s England was able to challenge Holland, although France continually sought to challenge both. England succeeded Holland as hegemon, although three closely fought Anglo-Dutch wars did not settle the issue. England captured many Dutch merchant ships in the first war, but later lost them again. Increasingly sophisticated British banking, initially along Dutch lines, and the foundation of the Bank of England in 1694 attracted much capital from Amsterdam to London late in the seventeenth century. Dutch fear of France, Catholic attempts to suppress Protestantism, and the accession of the Dutch monarchs William and Mary to the throne of England in 1688 sealed England's emergence as hegemon. In 1707 England consolidated her state power over the British Isles with the Act of Union with Scotland. This removed the threat of a hostile, French-backed Catholic polity to the north and allowed the development of a British polity. England was not as clear a hegemon in the early 1700s as Holland had been in the early to mid-1600s. Part of the

problem in an organic economy was England's poor resource base. England was relatively easily challenged by France, a polity with a very much broader organic economic base, despite the problem of integrating that economy with the minimal land transportation of the period. France challenged English and British power throughout the eighteenth century, but the only major French success came in the 1770s, when France allied with the American revolutionaries against Britain (Stinchcombe, 1969). The alliance between France and America was short lived but it produced a multipolar world again, this time with three powers struggling for hegemony: Britain, the French and Spanish in the Bourbon alliance and, less obviously, the new American republic.

4. The resulting Franco-British war was not over until 1815 and only after the rest of the European polities had been dragged into it. Britain emerged as unchallenged hegemon at the end of the Napoleonic Wars, as much because of the superior productivity of British industry, which was substantially enhanced by the factory system, as because of superior military performance. Britain also made the transition to a mineral-based, Paleotechnic economy in the late 1700s, just in time to defeat France, whose economy retained an advanced organic basis just too long (Wallerstein, 1989).

5. British economic power declined again after the 1860s, and two competing polities attempted succession. Imperial Germany at first seemed more likely to succeed, but Britain attempted to retain power with a complex series of alliances that brought the United States center stage. Both Germany and the United States invested much more successfully than did Britain in the Neotechnic technologies of electricity and the internal combustion engine when they became available in the 1870s and 1880s. Considerable economic growth attended U.S. involvement in World War I, and it seemed that the United States was poised for military as well as economic hegemony in 1918. A failure of political will ensued, and the United States retreated into isolationism, to be forced out of it not only by a resurgent Germany but also by a Japan that stepped into the power vacuum the United States had created. The United States achieved military hegemony late, in 1945 rather than in 1919.

Modelski and Thompson thus provide a useful empirical basis for the application of long-wave theory. Yet their work is remarkable for its omission of any concern for what drives these long waves and transitions of hegemony. They concern themselves with naval technology and the strength of the economy in the polities involved, but they suggest no mechanism by which the economy gains or retains strength or which accounts for superiority in naval technology (whether hardware or software).

In this respect Mumford's sociotechnical classification can be usefully expanded to account for the mechanisms missing in the long-wave and world system theories. The immediately obvious problem is that, although the Paleo- and Neotechnic are both mineral-based economic phases, one based on coal and one on oil, that match nicely with the fourth and fifth world leadership cycles and the hegemonies of Britain and the United States, the Eotechnic must account for three hegemonies. In this shortcoming Mumford's Eotechnic is merely simplistic, failing to capture the complexity of what was a long, slow, but steady period of change in an advanced organic economy.

At least one major division in the Eotechnic came in the 1400s, when Europe broke out of its historic geographic bounds and began the process of overseas expansion that brought on the capitalist world system of Wallersteinian theory. The origins of this division, which are complex, are rooted in the development of superior ships to exploit Europe's particular geography. The early Eotechnic, from about A.D. 800 to the early 1400s, was an experimental period in which the necessary cultural, commercial, political, and technical agencies of interregional commerce were established. The early 1400s to the late 1700s represent the fully commercial phase of the Eotechnic.

The Early Eotechnic, ca. A.D. 800 to ca. 1430

As Eric Jones has reminded us, Europe is a "peninsula of peninsulas" (1981, 90). Only an archipelago could have better waterborne connections than those made possible in Europe by a combination of seas and rivers. Jones also argues persuasively that Europe shifted to trade because no one polity was ever able to grow large enough to dominate the others. States had to trade the necessities of life to survive, but no overarching polity guaranteed security. Regional rulers had to be favorable to trade, to tax rather than preempt it, and to provide the military means for its defense. Since, however, no one polity policed all Europe, merchants and merchant ships went armed. The "defensible" ship had clearly emerged in Europe by the mid-1200s, with bowmen positioned fore and aft in castles (Hugill, 1988c). The early Eotechnic was thus taken up by experimentation as rulers learned how to maximize their income from trade, traders learned how to exploit Europe's geography to achieve complementarities of production, and sailors learned how to navigate and, most of all, defend their ships.

As a consequence ship design changed rapidly. The clinker-built Viking ship had castles added in the thirteenth century. In the early fourteenth century clinker construction gave way to the cheaper, flat-bottomed plank construction of the cog, which was also equipped with castles. Most of these ships were operated by traders, not polities. Indeed, from its formation in 1256 until its defeat by the Dutch in the Baltic in 1441, the Hanseatic League, an apolitical organization of traders based primarily in the Baltic, carried most of Europe's trade. Baltic trade was critical to northern Europe because it was the major source of wood for ships and a not unimportant source of wheat. The Dutch polity could never have become a major power after 1441 without ships built from Baltic lumber and domestic populations fed increasingly on Polish wheat.

Yet the level of trade in the early Eotechnic, however important, was finite, limited by European geography. Trade allowed regions to concentrate on what they produced best and to trade that for the surplus of other regions. Thus were achieved the efficiencies of specialization that the economist David Ricardo later described in his theory of economic rent, where rent was "that portion of the earth, which is paid to the landlord for the use of the original and indestructible powers of the soil" (Sraffa & Dobb, 1951, I, 65). Ricardo believed soil productivity was fixed by nature, a not unreasonable supposition in the eighteenth century. Productivity gains would thus only be possible from regional specialization and free trade. Ironically Ricardo, writing at a time when agricultural tariffs were normal, sought unknowingly to return to the early Eotechnic, when the Hansa practiced free trade throughout the Baltic and North seas (Hugill, 1988c).

The Commercial Eotechnic, ca. 1430 to ca. 1783

The expansion of Europe outside its historic bounds, which began in the 1430s with Portuguese adventuring into the Atlantic, temporarily removed the Ricardian constraint: Europe's finite ecology, in which productivity gains had to be won through managerial and technical innovation, was replaced by what seemed the infinite ecology of the globe.

In regard to the technology that brought on the second phase of the Eotechnic, the Portuguese cannon-armed caravel of the mid-1400s was the first ship that could sail anywhere on this planet, defend itself against piracy, pose a threat to those unwilling to trade, and carry enough trade goods to be profitable. The range, security, and profitability of shipping are the subsequent keys to European success and to the hegemony of whichever polity found the best combination of the three. If long-wave theory is correct, the commercial Eotechnic should thus have three subdivisions to explain the successive hegemonies of Portugal, Holland, and England in the first, second, and third world leadership cycles. The key to the Eotechnic after the mid-1400s is clearly both increased production through geographic expansion outside western Europe and increased productivity through long-distance seaborne commerce and regional specialization. Wallerstein well recognized the importance of increased effective land area in his first volume (1974) when he analyzed the growth of the Polish wheat frontier in the 1500s. Such an increase in land providing resources to Europe has been characterized as "ghost acreage" (Borgstrom, quoted in Jones, 1981, 83). Jones emphasized that this acreage would, if statistics were available, equal "the tilled land that would be needed to supply, with given techniques, food of equivalent value to that brought from outside into the system. It may be subdivided into fish acreage, that required to raise a supply of animal protein equivalent to that derived from the fisheries; and trade acreage, that required to supply the equivalent of the net import of grown foodstuffs" (Jones, 1981, 83).

Regrettably Jones equates this land with the simple accession of acreage after 1500, quoting Walter Prescott Webb's figures of 26.7 persons per square mile for western Europe plus the "Great Frontier" by 1500, 4.8 by 1650, and 9.0 by 1800 (Jones, 1981, 83). Given Eotechnic transport, only coastal areas and such archipelagos as the Caribbean and the Indonesian chain were accessible to long-distance trade. The really large accession of "ghost acreage" comes with steam-powered riverboats after 1815 and railroads after about 1840 which allowed expansion into the continental interiors. The truly massive impact of geographic expansion was thus felt in the Paleotechnic rather than the Eotechnic.

We may thus characterize at least two Eotechnics. From about 800 to the early 1400s, Europe lived on its own organic base, producing the three Fs on its finite acreage. This period of more than six hundred years was long enough that differentiation occurred and was critical to the success of regional polities, but it was change in a finite universe. Only drastic exogenous variables, such as the demographic collapse of the Black Death, had much impact on production per capita. Better agricultural technology, better systems of labor organization, and better systems of retaining surplus product caused shifts toward higher productivity (White, 1962). Nevertheless, such shifts were slow and, however important at the time, slight enough to be difficult to measure retrospectively.

After 1430 production per capita rose reasonably rapidly because of the "ghost acreage," which, however, could only be affordably exploited with controlled-cost labor systems. Slavery became common where new populations could be imported, such as in a New World demographically shattered by European disease. Serfdom and sharecropping became normal in European-influenced areas where populations resisted European disease, mostly on the Baltic and eastern Mediterranean fringes of western Europe. Wallerstein's work is central to an understanding of the importance of controlled-cost labor in the periphery of capitalist agriculture (1974). After the cannon-armed ship, the next most important technology for profit maximization was a labor system that controlled production costs. Slavery was not an innovation, merely a diffusion of European technique, and one already in wide use on the eastern frontier of Europe.

Some parts of the world were able to resist this commercial phase of the Eotechnic. Areas resistant to European disease, with high population densities and strong polities, such as China and Japan, kept Europeans out until internal political weakness, superior Paleotechnic technology, or both, allowed either Europeans or European ideas to muscle their way in. India lacked political cohesion and was eventually fragmented, although not until near the end of the commercial Eotechnic. Africa had environmentally vectored diseases that kept Europeans out of permanent residence until tropical medicine developed in the late

Paleotechnic. Russia was a land-based polity that offered little to the western European polities once they had elected to build their wealth on seaborne trade. Only the Paleotechnic development of land transportation allowed massive European penetration of Russia, and then it was a penetration by European ideas, not Europeans themselves. Russia's common boundary with western Europe had, in any case, resulted in considerable if selective Europeanization before the late nineteenth century. Russia also benefited from a landborne expansion based on the trade of fur, one of the few agricultural products light and valuable enough to stand the high cost of overland, largely riverine, transport in the Eotechnic (Gibson, 1969, 6–7).

The three phases of the commercial Eotechnic thus require explanation. In this seemingly homogeneous technical period, how did hegemony pass, sequentially, from Portugal to Holland to England, and how did Portugal acquire hegemony in the first place? Schumpeter argued that demography, geography (in the sense of land available for agricultural production), and technology are the key variables to profit (1950, 115–18). Wallerstein has argued that hegemony is achieved first through efficiencies in the agroindustrial arena (1984, 40).

World Leadership Cycle 1: Portugal as Hegemon

Portugal's accession of the "ghost acreage" represented by the coasts of Brazil, Africa, and India, let alone the Indonesian archipelago and even (against the Treaty of Tordesillas) Japan, was clearly spectacular. Nevertheless it would have been impossible without the technologies of the cannon-armed ship and open-ocean navigation, and without a European state system that allowed Portugal almost total autonomy within the confines of Christian Europe.

Portugal began Atlantic expansion westward in the 1430s at the same time as the Turks were laying siege to Constantinople in the East. Constantinople fell in 1453, and the Orthodox Christian church was displaced by Islam. Such a crisis would have had a very different outcome had it occurred in a centralized polity rather than one that operated as a politically independent unit within the loose framework of a western Europe united only by a common elite language (Latin) and a common ideology (Christianity) (Dawson, 1932, 218–38). On the other side of Eurasia an analogous situation did occur. Until 1433 the Chinese were systematically expanding westward from their trade links with the Malabar coast of India, formed in the early thirteenth century. They crossed the Indian Ocean and explored the coast of Africa past Madagascar. Had they continued they would have almost certainly entered the Atlantic long before the Portuguese expanded around Africa into the Indian Ocean. Yet the Chinese pulled back sharply in 1433, partly because of internal disagreement over expansionism, partly because of the Mongol threat to their northern frontiers. The Ming dynasty simply redirected resources from external expansion to internal security (McNeill, 1982, 45–46). Portugal suffered no such constraints when the Ottoman Turks were beating down the gates of Constantinople, not only because the fate of Orthodox Christianity was of little concern to Roman Christians, but also because the European state system recognized no overarching political power.

However important this precondition was, and it was clearly vital, the Portuguese expansion was still technological. The Portuguese developed reliable techniques for open-ocean navigation, whereas it is not clear that the Chinese did. By the 1430s Portuguese ships were highly seaworthy and, armed with cannon, could go on the offensive or the defensive at will. Chinese junks of the 1430s might well have been their equal in a show of force, but the test was never made. The Portuguese, substituting inanimate power in the form of gunpowder for force of human arms, also got by with far fewer crew than the Chinese. The Chinese could easily have done the same, but their voyaging was officially sanctioned state adventuring for prestige. The Portuguese were interested in trade and profits and reserved far more space for goods. How much the Portuguese relied on technology diffused from China is unknown, and will remain so until the nautical archaeologists exhume a Chinese ship of the pre-1433 period of Indian Ocean voyaging. Certainly the Portuguese learned much from the Islamic world in both navigation and sail design.

Other factors in the success of western European expansion deserve mention. McNeill put it well in his magisterial *Rise of the West* when he described the "three talismans of power" possessed by the Europeans of the Atlantic seaboard as

1. a deep-rooted pugnacity and recklessness operating by means of
2. a complex military technology, most notably in naval matters, and
3. a population inured to a variety of diseases which had long been endemic throughout the Old World ecumene. (McNeill, 1963, 569)

The disease advantage has already been mentioned, although its principal early impact was on the Americas. In Mesoamerica, for example, some 94 percent of the indigenous population was swept away by about 1650 in the "great dying" (Wolf, 1982, 133–34). But western Europeans, almost all of whom came from Viking or seminomadic stock that settled down to farm from about A.D. 400 on, certainly possessed "a deep-rooted pugnacity and recklessness." Some of that drove the mere fact of expansion, some of it drove the technical search for ever more vicious engines of destruction and domination of non-European folk.

The Portuguese were thus the first of a series of European polities to exploit the general advantages of the European state system, as well as European economic and social structure, over the much weaker systems of the rest of the world. Once the Portuguese had shown what was possible, the rapid diffusion of Portuguese technologies to the rest of Europe allowed other polities to compete in the new global arena. Other polities had better resource bases, in particular for shipbuilding, or rulers more amenable to trade or social structures that allowed more aggressive persons to come forward, whatever their social rank. All or some of these forces allowed other Europeans to come up with a better balance of range, security, and profit in their seaborne commerce.

Portugal held hegemony only briefly. The Portuguese resource base was too slight to build an adequate supply of ships, Portuguese social structure favored land ownership over commerce as a mark of status, and demographically Portugal was simply too weak to supply even administrators for a vast global enterprise. Power abhors a vacuum, and four polities attempted to step into the breach: Spain, England, Holland, and France. Spain, seemingly locked out of the land hemisphere by the Treaty of Tordesillas, used claims of marriage to take the Portuguese throne by force in 1580. Spain had much greater resources for shipbuilding than Portugal, although by the early 1600s Spanish shipbuilding was showing signs of severe strain caused by resource depletion (Phillips, 1986, 19–24). The Spanish elite also tended to favor land ownership over trade, and this preference did not encourage maritime enterprise. Finally the Spanish overspent on their military without gaining any real benefits in the way of new industry, industry being worse than trade in Spanish status ranking.

The Portuguese success must therefore be described as essentially driven by innovations in hardware. It was the three-masted ship, reliably navigated with good maps and navigational instruments and armed with cannon, that made Portugal's hegemony possible.

World Leadership Cycle 2: The Success of Holland

In the multipolar world that emerged at the end of the first world leadership cycle, England seemed a more likely successor state to Portugal than Spain combined with Portugal. France concerned itself with the consolidation of a large, land-based polity and paid too little attention to seaborne expansion. Holland seemed an unlikely challenger. England had modernized its economy under Henry VIII and was undergoing an impressive global expansion under Elizabeth I. The English pioneered much better ships, in particular the "low-charged" galleon (Hugill, 1988c), and much cheaper iron cannon (Cipolla, 1965, 37–41). The "low-charged" galleon had little in the way of the high superstructure that characterized Iberian ships. On a wind-driven ship, superstructure acts as a fixed sail and makes maneuvering very difficult. By 1588, when the Armada sailed against England, the Spanish were horrified to encounter English ships that literally ran rings round them (Howarth, 1981, 93–94). The English had hoped to use this superior maneuverability to stay out of the way of the heavy but short-ranged Spanish guns and, with their own longer-ranged guns, to batter the Iberian fleet into submission. This tactic failed, and the Armada was defeated more by the chaos that followed the unleashing of fireships at the battle of Gravelines off the Flemish coast. Even then, most of the Armada escaped the fireships, but the Spanish were forced to sail north around Scotland to get back into the Atlantic because the English blocked the Channel and the persistent westerly winds told against the less easily maneuvered Spanish ships. The huge loss of Spanish ships to bad weather serves to emphasize the superior construction of northern European ships by the late sixteenth century. Iberian ships were more lightly built, both to conserve resources and to meet the less taxing environment of the Mediterranean and South Atlantic.

Despite the success of English arms in 1588, England failed to achieve hegemony. In a commercial society the size and ability of the commercial fleet is far more significant in the long haul than military competence. Holland simply had a much bigger and better commercial fleet than England in the late sixteenth century. In part such a large fleet was necessitated by the bulk trade with the Baltic, in part by the Dutch dominance of the North Sea herring fishery. It was also made possible by Holland's three-way geographic position between the Baltic bulk trades in wheat and wood, the Iberian and French trades in cheap sea salt and wine that developed in the 1500s, and the entrepot trade with the lands along the Rhine (Israel, 1989, 18). The large number of ships all this commercial activity necessitated was made possible by Holland's access to Baltic resources, but success was primarily a result of the particular turn taken by the Protestant Reformation in the Dutch republic.

Weber long ago advanced the thesis that Dutch success was wrought by radical Protestantism (1904–5). In the notion that money should be put to work to glorify God just as humans should, the ideas of John Calvin clearly contributed to increasing the velocity of money. Since lenders and borrowers shared a common ideology, the risk to lenders dropped markedly compared to the old system of Jewish moneylender and Christian borrower. With the risk removed of a borrower simply repudiating a debt on the grounds of usury, interest rates fell and the Dutch "cheap-money" policy became possible. Calvin's ideas also promoted efficient production, hence the amount of labor-saving machinery in Dutch shipyards. High wages had the same effect, and high wages were produced in Holland by the numerous job opportunities on the large merchant fleet. Finally, although the Dutch did not work out a system of insurance, they managed the risk inherent in using wind-driven ships by spreading ownership thinly. Israel points out that few merchants owned ships outright, preferring shares as small as a sixteenth, a thirty-second, and even a sixty-fourth (1989, 21).

The Dutch evolved a manufacturing system for ships which was unexcelled for centuries afterward; it was driven by a "cheap money" policy and a sophisticated banking system. Shipyards could borrow to build with no firm orders, and buyers could borrow on the basis of ships in stock. In the yards themselves "wind-powered sawmills, powered feeders for saws, block and tackles, great cranes to move heavy timbers" supplanted human labor on a scale unknown elsewhere (Wallerstein, 1980, 43). Shipyards standardized on one basic design, the *fluyt*. This was a three-masted ship with no superstructure and a boxy, capacious hull that was ideal for cargo. It was built in large numbers, and since most labor on a sailing ship goes to raising and lowering sails, its short masts and numerous pulleys radically reduced onboard labor requirements. Short masts limited the distances sails had to be moved, and pulleys geared down and thus multiplied the power of human muscles (Unger, 1978, 37). As early as 1605 Sir Walter Raleigh complained that Dutch ships carried 20 tons per crew member, whereas English ships carried only 7 (quoted in McGowan, 1981, 53). Israel implies that this disparity in crewing was simply because the Dutch had so many ships in bulk trade with the Baltic, and bulk carriers were "designed for minimal crews and maximum economy" (1989, 21). However accurate this information may be, an experienced sailor like Raleigh would not have complained so vociferously had the disparity in crewing been simply explained.

The Dutch thus had a huge number of cheap-to-buy, cheap-to-crew ships compared to their competitors. As early as 1532 they had about 400 seagoing ships, and by 1636 at least 2,500 (Unger, 1978, 11). In 1629 the English, their chief commercial rivals, had 180 (Davis, 1962, 15). Hegemony was based on trade, and the taxable profits from trade paid for military protection. For naval ships the Dutch merely copied the "low-charged" English galleons and bought English iron cannon technology to build their military capacity equal to that of England after England's victory over the Armada (Cipolla, 1965, 48–60). Using "cheap money" and their banking advantages, the Dutch then turned to Sweden, northern Germany, and even Russia for cannon production. Such Baltic areas had the plentiful timber needed for charcoal to smelt iron.

In land warfare the Dutch were more innovatory. Prince Maurice of Nassau reached back into Roman history to produce the first "modern" army with organized drill and a sophisticated division of labor (McNeill, 1982, 131–39). The new military system spread rapidly through Europe, but not rapidly enough to reach the Spanish before their military power was destroyed. Their attempt to hold the increasingly unruly heretics of Holland, first with the Inquisition, then with military power, fell before Dutch prowess on land and English and Dutch naval strength.

Despite Dutch innovations in hardware, however, it was clearly in software areas that they were most successful. It was an ideological innovation (Calvinism) that allowed the Dutch to lower the cost of financing ships and to develop the first reasonably modern banking system. They improved risk management by taking shares in ships rather than owning them outright. Finally, better management of their land-based troops sped their victory over Spain.

World Leadership Cycle 3: Commercial England

England was not content to allow Holland to steal hegemony from under its nose, but England lacked a merchant fleet and the economic base to challenge Holland. These England acquired in the three Anglo-Dutch Wars, the third of which actually was more between France and Holland. The first, of 1652–54, was precipitated by the English Navigation Act of 1651, requiring that trade with English colonies be carried on in English ships. Since much was carried in cheaper, more efficient Dutch ships, the Dutch resisted. Militarily the war was a marginal English victory, but the English took as prizes some thousand to seventeen hundred merchant ships. This increased the English mercantile fleet to about the same size as the Dutch (Davis, 1962, 51). English shipyards began to copy Dutch designs and techniques. Although one English gain in the first Anglo-Dutch War was improved access to the Baltic and its lumber, it was the new England across the Atlantic that began to tip the balance of shipbuilding power toward England. This trend was helped in the second Anglo-Dutch War of 1664–67 when the English took the New Netherlands Colony, later the American states of New York and New Jersey. The rest of the second war was less favorable to England, with France entering in support of Holland. In the third Dutch War of 1672–78 England and France allied against Holland. England was quickly knocked out of the battle, but its absence proved advantageous. Holland fought a long and expensive war with France and in doing so lost much of her banking to London.

The loss of hegemony by the Dutch in the late seventeenth century is, however, far from clear. Israel points out that the Dutch were at the peak of their success in the late 1600s, but that "by 1670 the European powers were no longer prepared to tolerate it" (1989, 292). The two logical successors in the multipolar world of the 1680s were England and France, but neither gained the upper hand with ease. The transition lacks the clarity of the Dutch accession in the second world leadership cycle and of the transitions to Britain and the United States in the fourth and fifth world leadership cycles.

England had no clear technical advantage in the late 1600s, although the Royal Navy had developed line-of-battle software technology for naval warfare in 1653 for the first Anglo-Dutch War (Modelski & Thompson, 1988, 25). The Dutch adopted this technology quickly and proved just as adept at it as the English. Although the new tactic brought easy victory to England in the battle off the Gabbard in June 1653, by August at Scheveningen both fleets fought in the line of battle and the English victory was hard won and bloody (Padfield, 1979, 225–32).

Other events are clearly better explanations than such an easily copied software innovation for which both sides were well prepared. The advantage England came to enjoy may have lain more with internal social structure than with technical systems. England proved more adept at alienating labor from the land in the later seventeenth century. This lowered the cost of labor for fledgling industries by creating a pool of landless laborers. In contrast, labor was expensive in Holland because of full employment in mercantile activities. Running a global commercial system with a relatively small population base made labor relatively scarce in Holland. In the other would-be hegemon, France, the population base was relatively large. In France, however, magistrates refused to abrogate the remnants of feudal law and tended to confirm the right of tenants to stay on land they had been on for generations, however small and inefficient their holdings. No large pool of landless laborers was created, and landlords were denied opportunities to improve productivity through larger landholdings and economies of scale. Landlords also found it hard to introduce more efficient crops when tenants looked to their own needs first (Wolf, 1982, 115–23, 268–69).

English policy ran risks that the more conservative French approach did not. Landless laborers were potential revolutionaries, and at midseventeenth century such groups as the Levellers and Diggers seriously disrupted English society. Although both groups were Puritans, they were put down by Cromwell's Puritan government, bloodily in the case of the Levellers. Cromwell was himself a large landowner. It has been estimated that by the late 1600s, as a result of the willingness of local magistrates to alienate labor from the land, as much as 40 percent of the English population had left agricultural employment (Wolf, 1982, 269).

Holland avoided ignominious defeat in the third Anglo-Dutch War by the exertions of William of Orange, who took over military command in the first year of the war. His competence later encouraged the English invitation of 1688 to ascend the throne of England. The combination of the loss of fiscal authority from Amsterdam to London, which modeled its "cheap money" policy on that of Amsterdam, the sheer costs to Holland of the war with France, the removal of William to England, England's creation of a cheap labor force, and English innovations in sharing the risk of trading with wind-driven ships through insurance, all combined to ensure the transition of power from Holland to England. England's lower military costs must also have helped. Unlike Holland, England had no need of a standing army. A competent navy and the defensive moat of the English Channel sufficed.

England under William then put down the last serious threat to disunity within the British Isles by bloodily emptying the seven provinces of northern Ireland of Catholic supporters of James II, whom Parliament dethroned before it invited William and his wife Mary to jointly assume the English throne. The northern Irish provinces were then resettled with Protestant Lowland Scots loyal to the Protestant monarchy of England. William's successor, Queen Anne, peacefully settled the Scottish problem by the Act of Union of 1707, which allowed the British polity to emerge. The Act of Union did not entirely remove the threat of Catholic disruption in Scotland, as the events of 1715 and 1745 demonstrated. However, only the first of these two Highland Scottish uprisings threatened the British state. After the second one, the Highland "clearances" treated Highland Scotland almost as bloodily as William of Orange had handled the Catholics of northern Ireland.

Once British internal security had been guaranteed by the events of 1715, Britain was free to develop as a hegemonous power. British commerce boomed, helped by the Navigation Acts, but also by New World ship production and new industrial technologies at home. Britain was the first polity to break away from the ecological bottleneck of all agrarian societies: production is limited to the area of land, types of climate, and quality of soil available, agricultural technology being held constant.

This bottleneck was always present in the Eotechnic, but geographic expansion after 1430 reduced its significance. Then other factors intervened to restore it to importance. Early in the eighteenth century Johann Heinrich von Thunen pointed out that agricultural activity, *pace* Ricardo, is constrained far more by distance to market than by soil quality (Chisholm, 1962, 21–35). Land rent was thus redefined as a combination of the costs of production and transport. Von Thunen pointed out that two factors most controlled transport cost: bulk relative to value, and perishability. Perishable and high-bulk/low-value crops had to be grown near the market. Less perishable and low-bulk/higher-value crops could be grown further away.

Von Thunen's model describes a homogeneous soil environment in a region with no topographic barriers and only animate transport. He understood that it would be modified by changes in any of those factors. Since transport was a critical factor, von Thunen recognized that much cheaper water transport would radically distort his model. Local fabrication of surplus was also a distortion if that fabrication served a distant market. The classic example has always been the distillation of local surplus grain into alcohol, as in the production of Scotch whisky from surplus Scottish barley. The resultant product was low in bulk and high enough in value to stand considerable transport costs. The zone furthest from the market in von Thunen's model produced only livestock. Because livestock could be driven to market, transport costs were less relevant. On the other hand, milk and fresh vegetables, which perished quickly before canning and refrigeration, had to be grown near the market. Wood also had to be grown close to the market because it was needed in large amounts at a low price for fuel and structural material.

One use of wood in Eotechnic society does not fit this model. Ships, like distilled alcohol, represented considerable value added by local labor, and they could be sailed to market as an animal could be driven. They could thus use resources on the fringe when ecological pressure was high near the market and transport costs were low. Albion's classic work (1926) on the use of lumber in shipbuilding is instructive here. The Dutch brought lumber the relatively short distance from the Baltic and turned it into ships in Dutch shipyards. The English, with poorer access to Baltic resources, not least because of constant Dutch interference, came to depend on American lumber once New England was established. The costs of shipping lumber across the Atlantic were five to six times higher than shipping it from the Baltic, and much wood was lost in transit. Americans cut wood in the winter so that they could haul it from the interior over frozen ground to local rivers, which, swelled by the spring melt, carried it to the coast. Because the best season for sailing ships in the Atlantic was the summer, lumber was loaded wet from its spring trip and crossed the Atlantic in the heat of summer. Dry-rot spores proliferate in such hot, damp conditions, and they had plenty of time to take hold on an Atlantic crossing. Similar conditions in the Baltic trade had less severe results. The Dutch stockpiled lumber in the Baltic, giving it time to dry out, and had an excellent quality control system and a shorter, cooler trip to the shipyards. Dry rot was rare in

Baltic lumber but common in American lumber once it reached England. The logical solution was thus to fabricate ships in America. By the time of the American Revolution of 1776 a full third of the British merchant fleet was registered in American ports, and an unknown but presumably large number of ships were built in America for British customers. Perhaps as much as half the British merchant fleet was thus built of American resources.

The two greatest consumers of wood were the household and the iron industry. The latter had deforested much of southern England by the early 1600s in its search for hardwood to make charcoal to smelt iron. Some of this industry was also exported to the hardwood forests of the Atlantic coast of North America. By 1775 America accounted for "about 15 percent (30,000 tons) of world output" of iron (Lemon, 1987, 141). Although much was for local consumption, exports were common. Local iron nails facilitated the cheap, rapid construction of wooden ships when British shipwrights were still constrained by guild laws to use wooden dowels. No ship lasted long in the Atlantic, and iron nails were not the drag on longevity they might otherwise have been.

Despite the ability to use American resources, which freed soil within the British Isles for food and fiber production, English demand for fuel and construction lumber seemed insatiable in the early 1700s. Some relief was had by the burning of coal in domestic fires, especially in cities like London with good sea access to the coastal coalfield around Newcastle in northern England. The iron industry made a slow but extremely significant transition to coal once Abraham Darby had demonstrated that coal, like wood, could be heated to make coke, which burned, like charcoal, at high enough temperatures to smelt iron. Darby's discovery, as Jones has pointed out, extended Europe's "resource frontier vertically downwards by mining coal" (1981, 84). Jones further characterizes this vertical frontier as "an unprecedented ecological windfall." Its true importance, however, was that it broke the bottleneck of agricultural production in a society in which technological improvements in agricultural productivity were slow in coming. Coal's first great achievement was, in the late 1700s, to increasingly free agricultural land from fuel in favor of food or fiber.

World Leadership Cycle 4: Industrial Britain and the Paleotechnic

In an insightful recent essay the English geographer Wrigley pointed out that Britain was first to switch from an advanced organic society to a mineral-based economy by expanding along the vertical frontier. In the third world leadership cycle, English hegemony had its roots in a more efficient system for acquiring cheap labor, but its continued success was in breaking the critical ecological bottleneck by making the transition from wood to coal as the primary domestic and industrial fuel. In the third world leadership cycle, industry did not supplant, but merely supplemented, long-distance trade as the primary source of English wealth. In the multipolar world at the end of the third world leadership cycle, however, the advantages of the British Isles were eroding as two challenger economies developed: France, which was military and commercial; and the United States, which was simply commercial.

On the face of it England seemed bound to lose hegemony in the late 1700s. France, using a fine national system of canals, had successfully integrated the first major European land-based polity by the mid-1700s. A relatively competent and sophisticated bureaucracy seemed to maximize local production and distribution. French naval technology was the best in the world. British naval captains regarded the French seventy-four-gun third-class warship as more than a match for any British copy and went to great lengths to have themselves appointed to captured French ships. Despite a series of losses, in particular in India and Canada in the 1760s, the French seemed embarked on a winning streak after their successful support of the American Revolution. Internal disruptions notwithstanding, the French Revolution held the promise of improved social conditions for production at home. Napoleon's rise to power in the 1790s produced land-based military might to challenge all of Europe, let alone any other single European polity.

The case for America as challenger is less obvious. Yet by the 1770s, America was a commercial power to be feared. A third of British merchant ships sailed under the flags of colonial American ports, and even more were built in America for British merchants. America's gross product in 1775 was about 40 percent that of Britain, compared with only about 4 percent in 1700 (Lemon, 1987, 143–44). Wealth, although concentrated in the colonial elite, was not so politically powerful as in Europe. America was by no means a classless society, but upward social mobility was far easier than in Britain or France. The social forces of production were thus good, although the cost of labor was far higher than in Europe, population being scarce and wages very high. Military spending was nonexistent, since as long as America was British, Britain provided the military. Even after the founding of the American republic, military spending remained extremely low. Americans were secure behind their Atlantic wall, much as Britain had been behind the English Channel. When a navy became a necessity to protect American merchant ships withdrawn from the umbrella of Britain's fleet, Americans proved even more adept at ship construction than the French. Such big frigates as the USS *Constitution* and *Constellation* were considered by contemporaries to be superior to more heavily gunned British and French ships. In individual ship-to-ship fights they proved themselves so (Sternlicht & Jameson, 1981, 76–79).

America also had considerable untapped resources in the late eighteenth century. The streams of the Green and White mountains, of the Berkshires, and of the Blue Ridge were as easily dammed for water power to run factories as were the streams of northern England. Large stands of uncut hardwood forest remained available for fuel, building, or ship construction. The construction of an excellent canal network opened substantial acreages of land for agricultural development across the Lake Ontario shore plain and down the Great Valley of Virginia south and west from the Shenandoah River. From a resource perspective, had the Eotechnic continued, America would have been a logical hegemon.

Nor did Americans lack commercial will. In 1783 the United States was still primarily a commercial nation, producing agricultural surplus for the Atlantic trade. Americans did not turn their interest to their own great interior until after the Revolution had shut off their trade with Britain and British possessions. John Quincy Adams borrowed (and revised) Bishop Berkeley's famous line to read, "Westward the star of empire takes its way," for an oration delivered at Plymouth, Massachusetts, in 1802. Berkeley wrote his line about a century earlier at the time of England's Atlantic hegemony to praise Britain's westward expansion. By 1802, Britain had severely restricted American trade. Only vessels licensed by Britain could trade with British possessions.

Although licenses were generously granted during the long Napoleonic Wars, especially for trade with the West Indies, it was not trade at the scale to which Americans had been accustomed as part of Britain's empire. The four great colonial cities of America's northern Atlantic seaboard—Boston, New York, Philadelphia, and Baltimore—as well as a host of lesser cities, were Atlantic trading cities. All were badly placed for entry to America's interior, and all had eventually to build expensive canals or railroads to tap that interior trade (Vance, 1990, 267–86). New York earned its tag, the Empire State, because the state-funded Erie Canal created the first good access to the western inland empire Adams dreamed of.

In the early republican era Americans built up Eotechnic industry. Technology, particularly as it concerned factory organization and cotton- and textile-spinning machinery, was vastly diffused across the Atlantic (Jeremy, 1981). Samuel Slater came to America in 1789. He was apprenticed to Strutt, the original partner of Arkwright, the inventor of both the factory system and much cotton textile machinery. Slater passed himself off as a farm laborer to emigrate from Britain in the face of strict laws preventing emigration of skilled mechanics and the export of drawings, patterns, or models of textile machinery, all of which Slater had memorized. Once in America he quickly entered into partnership with Moses Brown in Providence, Rhode Island, and the buildup of the great northeastern cotton textile industry began (Bagnall, 1893, 145–52).

Nor was the young republic lacking in political will. Revolutionary British were quickly transformed into Americans. Even Jefferson, feared by the New England traders for his links to the populist West, showed remarkable political acumen in his dealings with France and his proposed dealings with Britain. The United States was clearly willing to play great power politics. Jefferson sent James Monroe to France in 1803 to buy New Orleans. The Spanish governor had withdrawn the right of transit from U.S. traders in 1802, throttling the Atlantic trade of the lands along the Mississippi and Ohio rivers. If the French refused to deal, Monroe's instructions were to make a deal with Britain. In Jefferson's words, "The day that France takes possession of New Orleans . . .

we must marry ourselves to the British fleet and nation" (Morison, 1965, 364–66). The upshot, of course, was the Louisiana Purchase and the sustained continental development of the United States, a development that still rested on Atlantic trade.

France and the United States, however, failed to make a transition to hegemony in the fourth world leadership cycle because Britain changed from a primarily commercial power into a primarily industrial one. For all Britain's military victories over the French, the cost of the long set of wars that ended in 1815 came close to beggaring the country. Only the remarkable set of production increases made possible by expansion along the vertical frontier and the productivity increases made possible in the domestic economy by the factory system allowed Britain to recover economic hegemony after 1815.

This hegemony was maintained, even strengthened, by successive waves of innovations in steam-powered transportation, which allowed a return to the easy production gains to be won from geographic expansion. Steam-powered riverboats and steam railroads allowed the continental interiors to be opened up for the first time; they were a vast new "ghost acreage" attuned first and foremost to the British market. Only after the U.S. Civil War were Americans able to develop their industry behind a wall of tariffs and to begin to stem the outflow of U.S. food and fiber—wheat and cotton—for British workers and factories. British-funded telegraphs and submarine cables provided better global information flow, large banks were formed for the first time to capitalize the growing trade, and steam-powered ships began to regularize ocean transport for high-value mails and elite passengers. Cheap steamshipping and low fares for all came only at the very end of this fourth world leadership cycle.

As Dodgson has pointed out, by the early 1800s the world system theorized by Wallerstein was in transition between the "vast but weak" and essentially coastal system that had originated with the fifteenth-century expansions and a much more intense, commodity-oriented system. By the early 1800s the spatial structure of the world system was becoming global in response to the forces of "the self-regulating market system acting via an emergent city system to allocate world-wide resources in a manner consistent with a world scaled von Thunen-type model" (Dodgson, 1977, 17). The metropolitan core of this global von Thunen model was the industrial Britain of the fourth world leadership cycle, made wealthy by higher levels of production and productivity than any other contemporary society. The ability to import basic foodstuffs very long distances, not only from islands and the coastal littorals of continents, but also from the continental interiors, caused a vast increase in urbanization in Britain. Between 1801 and 1901 the population of Britain (excluding Ireland) jumped from 10.5 million to 37 million. Urbanization rates are hard to measure for 1801, when a lower limit was accepted as defining an urban place. According to the higher limits prevailing in 1901, however, only 21 percent of Britain's 1801 population was urban, compared with nearly 80 percent by 1901 (Morris, 1986) More significant was the marked growth of the new cities located on the vertical frontier, the coalfields of the British Midlands and North.

London lost its historic dominance, or at least some of it: "In 1811 London was twelve times [as large as] its nearest rival, in 1861 five and a half times, with a slight recovery to six times in 1911" (Morris, 1986, 164). The maps of urbanization show a spectacular increase in the size of these Paleotechnic industrial cities over the century 1811–1911, well in excess of the natural population increase (Morris, 1986, 165, 167). Agricultural employment declined steadily after the Enclosure Act of 1830 and the repeal of the Corn Laws, a tariff on imported wheat, in 1846. Rural population densities peaked in 1851, then declined (Lawton, 1986, 10) as labor-intensive agriculture at home was replaced by land-extensive agriculture abroad.

Such urbanization and industrialization, together with such an increase in reliance on distant commodity producers, required excellent, cheap ocean as well as inland transportation. Iron, then steel-hulled, sailing ships gave that, and a remarkable advance in cheap steamshipping followed in the 1880s. So efficient was the British merchant steamship of the 1890s, with its triple-expansion engine and assembly line construction, that it could carry a ton of cargo per knot (a distance of 1.852 kilometers covered in an hour) on "scarcely more than the energy released by the burning of a couple of sheets of writing paper" (Craig, 1980, 14).

The high point of the British world system was thus reached in the late nineteenth century. The British global von Thunen system was united by sea commerce using steamships and a global web of coaling stations and telegraphs to efficiently provide fuel, cargo, and sailing orders. This system was ably summarized at the time. George Parkin's 1894 article, "The Geographical Unity of the British Empire," provides a valuable array of maps and a rather jingoistic interpretation. Parkin pointed out that the sea not only stretches around the globe but that, through such inland waterways as the American Great Lakes, it also reaches to the very edge of Canadian wheat country. "From the Straits of Belle Isle, where a steamship crossing the Atlantic first enters the inland water of Canada, to the head of Lake Superior is a distance of 2384 miles [requiring] . . . only seventy miles of canal . . . to make this system complete" (Parkin, 1894, 233).

Parkin drew an organic analogy. The waterways were the arteries of the body of empire. Submarine cables represented the nervous system, and out of a global network of 152,936 miles, Britain had laid 102,656 (Parkin, 1894, 236). Coal was the fuel that operated the system, and the flow of agricultural raw materials, food, finished products, and people were the oxygen flowing to the various specialized living cells. Parkin further theorized that the vigor of the British Empire was in large part a product of the frontier, a variant of the theory that it is border states, states on the frontiers of civilization, that rise to dominate civilizations that have softened.

Britain itself is cramped in size; it has a climate good for rearing men, but not favorable to the easy enjoyment of life; it is the center of an advanced and, therefore, conventional civilization. Men go abroad to get elbow-room; to find a sunny and exhilarating climate; to escape the shackles of conventionality . . . That the change is a healthy one no man can doubt; it gives free play to that Berserker energy of our Norse blood which still clings to us; it is what unlimited oxygen is to lungs long shut up in crowded rooms (Parkin, 1894, 239).

Parkin's circulatory analogy led him further, however.

Sated with freedom, space, movement, action . . . he comes back as opportunity permits; he brings or sends back his children to come under the influence of art, taste, culture, refinement, historic surroundings; . . . a new value is put upon them from the comparative roughness of colonial life . . . This movement [is] . . . a safeguard against that erosion of wealth, luxury, power, and over-civilization which has overthrown the greatest nations of the past. (Parkin, 1894, 239–40)

World Leadership Cycle 5: The Slow Emergence of America and the Neotechnic

It was not, however, decadence that ended British hegemony so much as loss of industrial superiority. This was made worse by failure to adopt the new technologies of the late nineteenth century, by the export to the geographic frontier of large amounts of capital that might otherwise have gone into such new technologies, and by a social system that made it hard for bright persons to succeed without emigrating. The central technical theme was the transition from a mineral-based economy using coal first to one using the electrical distribution of energy generated largely by the consumption of coal, then to one in which petroleum became the important mineral. The second phase amounted to the opening of a new and immensely productive vertical frontier. In the 1880s, however, coal production was still the preeminent measure of the strength of the economy. Only the American economy ever surpassed that of Britain, and then only in the teens and twenties of this century. Coal was the primary source of electricity production, and coal and electricity were the primary sources of the revolutionary shifts that were beginning in chemical production toward fertilizers and synthesized raw materials for industry.

The second phase of the Neotechnic was the development of petroleum production as a new vertical frontier. Petroleum fueled a revolution in transportation and a further round of geographic expansion. As early as the teen years of this century petroleum-fueled automobiles, trucks, and tractors were proving efficient workhorses even when built to much smaller scales than those possible with steam-powered machinery. These new vehicles greatly reduced the importance of fixed lines of transportation and allowed powered vehicles to extract surplus agricultural production from areas between the more capital-intensive fixed transport lines.

As more efficient or powerful engines were developed, diesel-engined ships lowered the costs of ship transport in the 1920s and aircraft began to move people relatively quickly in the 1930s. New weapons—the diesel-electric submersible boat, the aircraft carrier, the strategic bomber, and the combination of tactical aircraft with the tank—radically altered the technical conditions of warfare and thus the balance of power in the multipolar world that had emerged in the 1880s.

Measured by economic strength, Britain lost hegemony first to imperial Germany in the 1880s. Germany, fattened by reparations paid by France after the Franco-Prussian War of 1870–71, achieved political and full economic unification in 1871. With slightly more than 41 million people, Germany became a substantially larger market than Britain (26 million) or France (36 million), and slightly larger than a United States just embarking on mass immigration and industrial growth (just under 41 million). Imperial Germany dreamed of an ever greater level of economic integration in "Mitteleuropa," a German-dominated customs union reaching from the Pyrenees to the river Elbe. Mitteleuropa, embracing Austria, Belgium, Denmark, France, Germany, Holland, and Luxembourg, held 149 million people by 1911. The United States's population approached only 94 million in that year.

The German economy grew very quickly indeed in this period. From 1880 to 1900 industrial output grew an average of 5.3 percent each year. In the same period Britain's growth rate was 1.7 percent (Trebilcock, 1981, 432). Germany reinvested at a rate one and a half to two times that of Britain, and even slightly more than the U.S. rate. Capital formation ran 21 percent of gross national product from 1871 to 1890 in Germany, compared to 20.6 percent from 1879 to 1888 in the United States and only 12.4 percent from 1880 to 1889 in Britain (Trebilcock, 1981, 436). Trebilcock comments that German "investment was heavily concentrated in industrial and infrastructural fixed capital (non-residential construction + canals, roads, railways + equipment)" (1981, 437).

These remarkable achievements occurred in a country driving hard toward self-sufficiency through high technology. Trebilcock's measured account of the rise of imperial Germany emphasizes "two, technologically distinct, surges of expansion." The first was pushed by Paleotechnic railroad construction through about 1870, the second by Neotechnics, in particular electrical and chemical engineering, through 1914 (Trebilcock, 1981, 46–48). These surges were clearly linked with the remarkable surge in agricultural productivity that occurred after 1850. Prussians had long been attuned to the notion of self-sufficiency. They had encouraged productivity increases with a complex system of agricultural credit extended to landowners as early as the 1780s (Trebilcock, 1981, 25). Regional self-sufficiency, however, was useless if surpluses could not be moved from one region to another to cover local shortfalls. Such shortfalls were common through the 1850s and were resolved only by better inland transportation on canals and railroads.

After 1870 German attention seems to have turned from a concern with Ricardian efficiencies in internal free trade to attempts to modify the productive powers of the soil. After a less than 20 percent increase in the productivity of wheat per hectare between 1850 and 1870, productivity shot up by slightly less than 50 percent between 1870 and 1900. U.S. productivity did not increase at all in that same thirty years (Hugill, 1988b, 119). Such an increase was vital if Germany was to achieve preeminence in the agroindustrial arena. Germany had a large population to be fed at home and lacked access to low-cost agricultural production from the geographic frontier. German strategic thinkers of the nineteenth century stated that Germans must acquire such a frontier in the form of an empire or be reduced to the status of a backward agrarian nation—to "carriers of water and hewers of wood" in the memorable phrase of Frederick List (quoted in Trebilcock, 1981, 40). Shut out from the geographic frontier by British and U.S. maneuvering, Germany had to pursue alternative solutions. These included aggressive expansion on the vertical frontier, productivity increases in domestic agriculture to ensure the food supply, and a search for substitutes for the principal tropical raw material for industry, cotton fiber. Given the mercantilist theories propounded by late nineteenth-century German economists, a secure market was sought in the form of "Mitteleuropa," which became a major German war aim in World War I (Fischer, 1967, 202).

Once embarked on the course of Mitteleuropa, Germany vitally needed productivity gains. By 1910 German farmers were applying 50 kilograms of chemical fertilizer per hectare of arable land. No other nation of comparable size came even close: Britain applied 28 kilograms per hectare, France only 20. All these figures were heavily augmented by natural fertilizer. In countries with a medium livestock density, such as Britain, France, and Germany, about another 80 kilograms per hectare of pure nutrient content would be available. Americans applied more chemical fertilizer than Germans, about 60 kilograms per hectare in 1910 (U.S. Bureau of the Census, 1975), but, having a much lower livestock density, they used substantially less animal fertilizer. Certainly the figures for wheat productivity do not imply much U.S. use of any sort of fertilizers, and the U.S. farm labor supply was far too small to be able to transfer large amounts of animal manure from stable to field. Earle and Hoffman have demonstrated that the supply of agricultural labor supply in the U.S. North was small except at harvest time (1980). Only in the cotton South was labor tied to the land all year long, and there labor was needed to weed in the subtropical climate. European farms used their labor much more intensively, and in Germany agriculture accounted for a much larger segment of the work force, even compared with such a major agricultural exporter as the United States. In 1881, 11 percent of Britain's gross national product was accounted for by agriculture, 36 percent of Germany's, and 27 percent of the United States's. By 1907 the figures stood at 6, 25, and 17 percent, respectively. The United States entered the realm of single-digit figures in the early

1930s, Germany not until the mid-1950s (Hugill, 1988b, 121).

Germany thus led in the preeminent thrust of Neotechnics: the development of a technically intensive world system dependent upon intensification of domestic food agriculture and the synthesis of substitutes for organic raw materials for industry. It shifted away from the focus on an ecologically diverse geographic frontier that characterized the first four hundred years of the world system to an increasingly intense focus on the vertical frontier and a concern with increases in productivity rather than simply in production. The vertical frontier provides not only fuel in the Neotechnic world system, but also much of the fiber and many other industrial raw materials that would otherwise come from the organic world.

Several factors somewhat confuse the emergence of the German Neotechnic world system. Polities with heavy investments in the old, land-extensive world system maneuvered to protect those investments (usually through restrictions on imports from more cost-effective producers). Submarines developed by imperial Germany were capable of disrupting trade with the geographic frontier. German levels of investment in organized research, the primary means of productivity increase in the new, technology intensive world system, were higher than American or British investments. Finally, improvements in energy distribution and transportation capabilities favored Britain and the United States.

Nowhere is the success of the selective blocking and adoption of the Germanic Neotechnic more evident than in the case of the United States. In the late nineteenth century the United States operated a continental version of the British maritime world system. The subtropical South was a major source of fiber for U.S. textile mills as well as those of the rest of the industrialized world. The Great Plains fed American cities as well as many European ones on the abundance of their grain production, albeit at very low outputs per hectare. Vast coal resources fueled industry as efficiently as did those of Europe, and a sophisticated railroad network tied the northern core to the agrarian peripheries of the South and West. Yet the United States did not lag in adopting the technologies of the Neotechnic or the means of improving them. American universities grew rapidly, and land-grant schools focused on the practical problems of agriculture and mechanics. Investment flowed into the electrical industry, petroleum production, and petroleum-fueled personal transport.

This success rather stands Wallerstein and Mumford on their heads. Wallerstein's Marxian reasoning is that the social relations of production are all-important, whereas technical forces have been preeminent in Neotechnic society. Mumford argued that electrical distribution of energy and petroleum-powered transportation would allow renewed expansion along the geographic frontier. I place both sets of forces into categories subsidiary to the chain of technically induced productivity increases and substitutes for organically produced raw materials.

Neotechnics severely destabilized the social relations of production established in the commercial phase of the Eotechnic by radically reducing the demand for agricultural labor in the early Neotechnic and by radically reducing the demand for industrial labor in the robotized factories of the later Neotechnic. By substituting a large, unskilled work force in the growing service sector of the economy for a moderately skilled work force in the declining industrial sector, this change has repolarized society. The expanding research, development, and managerial sector of the labor force is dependent upon an increasingly sophisticated educational system, in particular at the university level. A large amount of low-cost, unskilled labor on the geographic frontier is no longer needed. After about 1900 an expanding number of industrial jobs took up the slack from an agricultural sector made more productive by chemical fertilizers and mechanization. Since about 1970 the increasing use of industrial robots has made much industrial labor redundant. Although there is much talk of a rapidly growing service sector in the capitalist economies, the most rapid growth is in low-skill service jobs, such as fast food and retail sales. Such a workplace is less easily politicized than the old industrial proletariat. Its class consciousness cannot easily be "raised" because it produces nothing; therefore Marxist arguments about surplus production are invalid. It is also geographically diffuse, thus not subject to cost-effective centralized organization into unions and union locals. Although it has elements of a permanent underclass, not all its members are from the lower social and economic classes. Class solidarity is also lacking because the service economy provides temporary, age-related employment to many teenage children of the relatively prosperous classes. Such low-paid workers are now part of the chain of consumption in industrial societies, not the chain of production.

Robotized industrial production is increasingly in the hands of the managerial and research sectors of industrial society. In Marxist terms these groups have no class consciousness, usually believing themselves well rewarded by capitalist society with prestige and income. Like those in the low-paid service sector, they are geographically diffuse and thus hard to organize politically. For all practical purposes, Marx's arguments that the social relations of production are all-important have thus ceased to be valid.

Mumford's argument—that improvements in transportation and power generation would reopen the geographic frontier—has greater validity. Electricity and internal combustion–engined vehicles have opened up new regions, although the principal growth has been in the expansion of existing urban regions. Cheap electricity has meant air conditioning, thus an increased willingness of folk reared in temperate areas or ones with long, acute winters to move to areas with long, acute summers. Mechanical refrigeration was developed for shipping meat in the 1870s, but it was not until the 1950s that cheap electrical energy allowed it to be routinely used to cool living and working spaces. The remarkable growth of American cities in the hot, dry Southwest and the hot, humid Southeast since 1945 has been heavily air conditioned.

Motor vehicles have greatly contributed to a reopening of the geographic frontier. Motor vehicles allow urban growth without tying up large amounts of capital in fixed investments in transportation. Because most motor vehicles are owned by individuals or small companies, capitalization is more at the individual level. The investment risks are spread out rather than concentrated as they are with railroads, for which privately owned (although usually government-subsidized) companies capitalize both vehicles and routes. Motor vehicles represent mobile fixed capital in the transportation infrastructure, capital that can easily be transferred from region to region or town to town.

The greater efficiency of the mobile component of capital invested in transportation in the Neo- as compared to the Paleotechnic has allowed far more flexible geographies to develop. The motor vehicle includes not only automobiles and trucks but tractors and airplanes. Tractors have increased agricultural productivity and also displaced the draught animals that once ate up some 20 percent of the food supply. In combination with trucks, tractors opened to production land previously too far from the railhead to be economically farmed. Trucks have also allowed industrial production to move away from railheads onto cheaper land outside the old cities. Automobiles allow workers to reach such spread-out plants, as well as to live on cheaper land. Airplanes allow managers to be moved quickly from one plant to another so that the "fixed" costs of management in a large bureaucratic structure can be "amortized" over a larger number of factories. Finally, the microgeography of the factory itself has been revolutionized by electricity and motor vehicles. Whereas the flow of work through the factory was once controlled by the flow of energy generated by falling water or steam and transmitted to machines by belts, shafts, and pulleys, electrical transmission of energy allows machines to be put anywhere. In factories where the flow of power was mechanical, the friction of distance had to be minimized. Such factories routinely had several stories. Raw materials entered on the upper story, and the work flow was, at least in part, powered by gravity. Finished goods exited at ground level. Electrical energy removed the need for such compactness, and electrically powered production lines could move raw materials in the process of fabrication from workstation to workstation. Workstations could be put wherever they would enable workers to achieve maximum productivity. Frank Gilbreth (1911) and Frederick W. Taylor (1911) arrived at this understanding from first principles. Henry Ford seems to have arrived at it independently and empirically. Together these three, by publication, consultation, and example, spread a new style of much more efficient management. In the United States the spread was quick because labor was scarce and expensive. Elsewhere the spread was slower, despite rapid translation of relevant publications. Ford made no secret of his methods and by the early 1930s had large factories in England, France, and Germany run on U.S. lines, albeit by local managers. When Europeans did attempt to implement U.S. managerial technology, they usually failed to apply it properly. The need expressed by Taylor to roll at least some part of productivity increases over into higher pay for workers was often ignored. The French auto maker Louis Renault imposed "Taylorism" in the teens without increasing wages, and his labor struck (Laux, 1976, 192). Ford increased wages dramatically in his plants in the teens and had no labor problems. Much of the discontent at Renault was caused by skilled metalworkers, who saw their livelihood threatened by sophisticated machines operated by cheap labor. No such reserve of skilled metalworkers existed in Detroit.

By the late teens of this century it was evident that, in the Neotechnic, success rested on efficiency in production, adequate electricity and petroleum supplies, a reasonably educated work force paid enough to afford the goods they produced, a large domestic or captive market for those goods, and access to adequate investment capital. The best conjuncture of these factors was in the United States. Imperial Germany lagged badly behind the United States on the first two but had adequate internal capital supplies and an excellent educational system. Low wages, however, meant low consumption, as did the failure to achieve a large market in Mitteleuropa. Furthermore, German social structure mitigated against the easy upward social mobility enjoyed by Americans. The German advantage of a superior educational system was thus offset by better social relations of production in the United States.

Three forces created a huge domestic market for U.S. goods. Sheer geographic size was one; although population was relatively low, it was increasing rapidly in the late nineteenth and early twentieth centuries thanks to both high fertility rates and immigration. Second, this population enjoyed high wages, paid for by superior U.S. managerial technology and productivity. Third, the United States enjoyed privileged access to what was at that time the world's wealthiest consumer economy because Britain maintained a policy of free trade.

In Britain, attempts to overhaul the educational system were relatively unsuccessful until 1945, as were attempts to improve the social relations of production. Wages remained low, and the domestic market was small. Relatively free trade meant that the empire preferred to buy its goods from the United States, particularly in the all-important growth industries surrounding electricity and the automobile. Canadians, Australians, South Africans, and the like vastly preferred more durable Fords, Chevrolets, and Buicks, to British Austins, Morrises, and Wolseleys.

The social relations of production did not improve in Britain until a new social compact was forged in the depths of World War II and revolutionary educational changes were brought in by the Labour government in the 1944 Education Act. In the 1920s and 1930s British entrepreneurs seem to have been slightly more willing to adopt U.S. managerial technology, and British access to the Middle Eastern oilfields was far better than that of Germany. Britain also had excellent capital reserves, though it allowed capital to flow freely to wherever the returns were highest, thus often to the United States. With the advantages of hindsight, the weight of U.S. advantage was obvious by the late teens.

The United States's failure to assume hegemony at the close of World War I thus seems to have been a failure of political will rather than of economic competence. As a consequence the chaotic conditions of a multipolar world extended into the late 1930s and allowed a further challenger, Japan, to appear. Like Germany and Britain, Japan by no means lacked the will to succeed, but it lacked productive efficiency, energy resources, a mass educational system, a large domestic market, and capitalization. The application of military force, however much disguised as the "Greater East Asia Co-Prosperity Sphere" and as liberation of Asian peoples from colonial (European and U.S.) oppression, to create both resources and a market was no substitute for economic competence.

In retrospect, the intense concentration by both academic and popular historians on the role of war in the transition from the fourth to the fifth world leadership cycle is understandable. Only by emphasizing military prowess could either Germany or Japan succeed Britain, because the economic strength of both was questionable. The U.S. preference for economic strength and unwillingness to exert military strength combined to make the military option seem even more sensible. Japanese leaders with training in the United States, such as Admiral Yamamoto, recognized the fatal flaw in this military option: once the United States committed its vast economic resources to war production, Japan's advances in the Pacific would be quickly rolled back. Yamamoto predicted a year of advance for Japan and steady defeat thereafter, but his peers ignored his counsel. German ignorance of U.S. productive capacity was almost as profound, Hitler preferring to believe that Americans were a mongrel race of Jews and pacifists.

Whatever the causes, the results were quickly evident after the signing of the armistice that ended World War I. Instead of a U.S. hegemony beginning in 1918, the multipolar world persisted until 1945. What should have been a single war of transition between the Balkan crisis of 1911 and the end of World War I in 1918 stretched out to a second war between 1931, when Japan invaded Manchuria (Mayer, 1984, 92), and 1945. More realistically, one war was interrupted by a long armistice, as Germany claimed at the time.

Contra Wallerstein, I must thus argue against primacy in the agroindustrial arena as a precondition for success in the later stages of the world system. By the later Neotechnic, certainly after 1940, only food production remained in the realm of agriculture. Fuel production moved into the realm of mineral-based production in the early 1700s and has become ever more entrenched there. With the development of hydrocarbon-based synthetic fibers in the 1930s, first in Germany, then in the United States, fiber production has also moved into the realm of mineral and industrial production. Many other industrial raw materials that were once produced organically are now synthesized from mineral resources.

Such technical change has rendered a spatially extensive, land-based world system either unnecessary or, worse, indefensible against challenger economies that have invested in highly technified production systems. In the recent history of the world system we may identify three basic options for achieving and holding hegemony. These are

1. *Extensive production of agricultural surplus in a colonial periphery and intensive industrial production in the domestic core.* This is, in essence, Wallerstein's model. It is, fundamentally, Eotechnic, and with detail variations, it can be applied to Portuguese, Dutch, and English hegemony, as well as to the early phases of British hegemony before the mineral-based system took over. It was also the model followed by Japan. The survival of the hegemon in such a system depends upon seaborne trade and naval power and was first codified by Admiral Mahan (1890). Any disruption of seaborne trade is fatal, as each hegemon and would-be hegemon has, in turn, found out. The combined kingdoms of Portugal and Spain lost control of the seas to the English and Dutch in the late 1500s. The Dutch progressively lost mercantile dominance to England in the late 1600s. British shipbuilding capacity was severely reduced in the late 1700s, when thirteen of the American colonies were detached from the British Empire with French assistance. Finally German diesel-electric submersible boats posed an intense challenge to British hegemony in World Wars I and II, and U.S. submersibles were a major factor in blocking Japan's bid after 1941.

2. *Extensive production of agricultural surplus in a domestic periphery and intensive industrial production in the core.* This option required the emergence of large, continental polities and was only made possible by Paleotechnic improvements in transportation: the steam-powered riverboat and the railroad. France attempted to use canals to integrate a large, land-based polity as early as the seventeenth century (Vance, 1990, 53–65) but had no clear internal periphery to give its commercial and industrial economies the necessary supply of low-cost agricultural goods. The model for this option is the United States, which was able to detach its Afro-Caribbean South (Meinig, 1986, 432) from the periphery of various European nations, mostly Britain, and attach it to that of the commercial and industrial republic of the North. By retaining low-cost labor systems, first slavery, then sharecropping, the U.S. polity gained their economic advantages in a free-trade system of states that was very reminiscent of the competitive European state system of the Middle Ages, but with an overarching political unity derived from the successful Dutch republic. Thus the republic of the North gained the same sort of advantage in the agroindustrial arena that Britain had early had with its global trade. It took, however, civil war and the erection of tariff walls around the nascent industries of the republic of the North before the full advantages of the internal periphery could be realized. It was the U.S. Civil War that finally detached the South from the British world system and locked it to the U.S. one (Agnew, 1987, 45).

3. *Intensive production of agricultural surplus domestically as well as intensive industrial production in the core.* In part, domestic production of agricultural surplus is made possible by substitution of mineral resources for organic ones first in fuel, then in fiber production. Without

the second component in place, France moved somewhat in the first direction under Napoleon's continental system. Britain modified its global, sea-based system to substitute domestic coal for organically produced fuel, laying the basis for its Paleotechnic hegemony. But Britain did not abandon its extensive colonial agriculture. Germany was the first polity to complete the transition. German farmers intensified domestic agricultural production by the addition of large amounts of mineral-based or electrochemically synthesized fertilizers in the late nineteenth and early twentieth centuries (Hugill, 1988b). German domestic coal and electricity production was substantial and, although Germany lacked domestic oil resources, it made systematic attempts to convert coal to synthetic oil, to reduce oil consumption, and to use alternative energy sources. Finally, German scientists synthesized substitutes for the major crops grown for industry, in particular subtropical crops such as cotton fiber. By 1939 Germany was synthesizing as much fiber as German cotton mills had been consuming just a few years earlier (Hugill, 1988b). Germany also pursued a classic Mahanian naval blockade by using diesel-electric submersible boats against Britain. In two prolonged world wars this blockade caused immense strain to a British economy that had been faltering since the 1870s.

The development of the German version of the world system modifies Wallerstein's claim that a nation bent on hegemony must first attain supremacy in agroindustrial production. In a world system based on minerals rather than organically grown materials, energy production and use is more important than agrarian production. Skill in converting energy to raw materials is also a critical necessity, and the Germans were the first to achieve a highly sophisticated technical education. We may thus suggest that, at this stage in the development of the world system, primacy in mineral and technoindustrial production is the vital step toward hegemony.

Unfortunately for Germany the technical developments of German scientists diffused easily to other industrial societies, in particular to the United States. The growth of the Japanese economy after 1945 confirms the ease with which technology can diffuse to a willing receiver. In an organic economy the society with the greatest land resources and the most efficient production usually does best, as the histories of Britain and the United States attest. In a mineral-based economy the availability of minerals, the skill of the labor force, and the productivity of industry seem to be the critically needed resources. Mineral resources are a provision of nature, and those countries without them are at somewhat of a disadvantage if access to the resources of others becomes a political rather than an economic decision. The German technical educational system of the late nineteenth century took slow hold outside Germany, with only the United States copying it to any great extent. Americans pursued mass rather than in-depth education, which served the country adequately in most areas. Early twentieth-century U.S. managerial technology, made possible in part by the success of mass education, took hold poorly outside the United States, at least until would-be receivers realized that it throve best where the social relations of production allowed rapid mobility for the needed new managerial classes.

It was the conjunction in the United States early in this century of adequate domestic reserves of coal and oil, a reasonable education system, and competent managerial technology within a geopolitically secure framework that allowed the country to achieve hegemonic status. That it was delayed from 1918 to 1945 reflects the belief of many Americans that ocean barriers and an excellent land-based transport system not subject to Mahanian commerce raiding meant absolute geopolitical security. As it turned out, aircraft carriers, strategic bombers, nuclear weapons, and intercontinental ballistic missiles have progressively rendered geopolitical security more relative than real. More to the point, other nations have merged German technical with U.S. mass education, improved their social relations of production, and fully adopted and begun to improve on both U.S. managerial technology and German reliance on both highly technified agriculture and synthesized raw materials. The U.S. loss of hegemony in the economic realm is now marked, and it is clear that the 1990s represent a return to the classic multipolar world typical of the thirty years or so around the end of each century over the historical time frame occupied by mercantile and industrial capitalism.

Chapter Two

Technology and Geography in the Elaboration of Capitalism

The only raw materials which Britain herself produces are metallic ores and coal. Wool can be left on one side as it is an agricultural product.

The face of the country has been completely changed by the Industrial Revolution. What was once an obscure, poorly-cultivated bog [south Lancashire] is now a thickly-populated industrial district.

Friedrich Engels
The Condition of the Working Class in England

Most historians and social scientists regard technology as a black box. It is not. The deliberate manipulation of the environment is the common thread of human history. The tools we have employed most in that history are plants. They are so commonplace that we have virtually forgotten they are tools. Yet they convert the energy of the sun into food for us and our domesticated animals, fiber for our clothes, and fuel for cooking and heating. Without using them as tools, we could not have prospered (Sauer, 1952). We also tend to overlook simpler industrial technologies than our own, even when they belong to our own history (Mumford, 1934). The progress of northwestern Europe from about A.D. 800 to 1700 was brought on as much by such technologies as the sailing ship and the water-powered mill as by improved agriculture, religion, democracy, military prowess, great leaders, and all the rest of the usual suspects beloved of historians and social scientists. From 1700 to about 1900 that progress accelerated because of the release of vast amounts of energy made available by the burning of coal. After 1900 the accelerator was the release of even more energy from oil and the better distribution of energy by electricity. Such things are hardware technologies, but software was also important. Even such simple tools as the clock allowed far more structured work practices to develop in medieval

Europe. Arkwright's factory system is as important an item of technology as the mule spinner or the steam engine. In our own time Fordism has had as big an effect on the management of the economy as Ford cars have had on society.

ORIGINS OF THE NORTHWEST EUROPEAN ORGANIC ECONOMY

The Neolithic spread of Mediterranean crops and agricultural technology into northern and western Europe was heavily constrained by geography. Only on the thin, well-drained soils of the chalk uplands could wheat, barley, or oats be grown using the scratch plow of the Mediterranean basin pulled by an ox or two (Curwen & Hatt, 1953, 78). This plow was the basis of Celtic upland agriculture in northern and western Europe. After about the time of Christ the Celts lost power in this region as new folk began to cultivate the fertile, heavily wooded, but poorly drained clay soils of the river valleys. The new people had a furrow-turning plow, which cut into the ground with a mild steel knife, then used a wooden moldboard to turn the furrow through about 120 degrees, stacking successive furrows so that water could drain out of them along the channels dug several inches down (fig. 2.1). The dry soil could then be harrowed into an approximation of the dry, loamy soils of the Mediterranean, which were ideal for the wheat, barley, and oats that had evolved in such semiarid environments.

The Moldboard Plow

Figure 2-1   The moldboard plow. By stacking furrows and allowing wet soil to drain these plows made possible the very first European geographical "expansion," after about A.D. 400, away from the thin chalk soils of the uplands and into the fertile clay soils of the valley bottoms. Redrawn from Burke 1978.

Unlike the soils of such river basins as the Nile, where fertility was annually renewed by a thin film of alluvium spread by annual floods, Europe's clay soils lost fertility if planted year after year in wheat. The problem was reduced by letting land lie fallow one year in two and allowing livestock to graze the stubble. This practice provided both animal manure and a year of earthworm action to return nitrogen to the soil. Although two-field crop rotation left half the cultivated land idle each year, the gain in productivity per hectare from the return of nitrogen lost in plant growth more than repaid production lost by using depleted soils.

The Angles, Saxons, and other Germanic tribes did not so much displace Neolithic and Bronze Age peoples from their upland farming in Europe as outproduce them, both in agriculture and population. Even this process was a slow, basically peaceable one, despite the quasilegendary exploits of the last major Celtic king of Britain, Arthur, appropriately rearmed with a legendary steel rather than bronze sword to resist the steel-wielding Anglo-Saxons (Fleure & Davies, 1970, 128–32; Morris, 1973, 111–12). Fleure and Davies point to the importance of iron in the success of the Anglo-Saxon invasion of the British Isles, quoting a well-known Welsh, thus Celtic, folktale of the tension between the two cultures: "A tall, fair farmer from the valley . . . marries a slight, dark, fairy bride who rises from the lake. If he strikes her with iron she will disappear into the lake, and, after many happy years, a carelessly thrown bridle brings this about" (Fleure & Davies, 1970, 37). The real villain of the piece was, of course, the iron knife that cut into the ground ahead of the wooden moldboard and allowed the tall, fair farmers from the valleys to outproduce the slight, dark upland population.

Two-field crop rotation left much to be desired. It was not highly productive, and it required concentration in the small-grain harvest of labor that was not used adequately the rest of the year. Three-field agriculture was more productive per hectare, since the peas and beans added to the rotation actively fixed nitrogen back into the soil, thus raising small-grain yields. It was also more efficient in its use of labor because peas and beans were planted and picked at different times than small grains.

Because of the high costs of feudal warfare feudalism can be regarded as a construct of three-field agriculture. By A.D. 800 constant pressure on northwestern Europe from aggressive nomadic folk to the east and intermittent pressure from Moslems coming around the Pyrenees had created a demand for a new military technology. The adoption of the stirrup by Charles Martel allowed Frankish cavalry to develop the technique of mounted shock combat. In this system the stirrup braced the mounted warrior so that he and his long lance became one with the horse. Suitably trained horses could be used to charge the enemy line head on, the long lances opening a path for the following archers and infantry. The principal cost of the new system was feeding and breeding heavy war-horses, which were oat- rather than grass-burning machines, grass being insufficiently caloric. War-horses required either diverting food from humans or a more productive agriculture. Three-field rotation was a software solution that increased productivity.

White has noted that three-field rotation gave an immediate 50 percent increase in food production after its adoption in the early 800s (White, 1962, 71–72). He underestimates the productivity gain by not looking at the labor available for harvest. Because three-field agriculture had a spread-out harvest period, the same labor force used in two-field agriculture could be more efficiently employed. If new land was available, as it often was, the labor force could actually produce twice as much food, not even counting the extra productivity gains that would have come from the use of legumes in the rotation to fix nitrogen back into the soil and give a better wheat crop. Medieval accounts of "assarting," the process of taking in new land, are common. Forests were felled, marshes drained, and polders dyked (White, 1962, 72). Certainly the account given by Bennett of the period 1150–1400 suggests that in three-field agriculture the average peasant was kept busy most of the year (1937, 77–86). Marc Bloch goes so far as to refer to the period beginning in 1050 as the Second Feudal Age, characterized by both considerable assarting "in the heart of the old territories [and by] the colonization of the Iberian plateaux and of the great plain beyond the Elbe" (Bloch, 1961, 69).

White thus introduces two of the main problems that must be faced by any agricultural production system: the temporality of labor supply and demand, and the necessity of balancing human- with nonhuman-oriented land uses. Only toward the end of the Eotechnic, and then not everywhere, did an agriculture emerge which used labor on the same year-round basis as later the Paleotechnic factory system did. As Earle and Hoffman have shown, midnineteenth century agriculture in Britain employed labor year round. In the same period in the United States the greatest labor demand was at harvest time, and the rest of the year agricultural labor was chronically underemployed (1980). Two-field agriculture produced chronic underemployment away from the harvest season. The feudal innovation of three-field reduced much of that underemployment.

As horse breeding expanded to meet the demands of mounted shock combat, horses not suited to combat became available for agriculture. Although horses ate oats rather than grass, they could plow as much as 50 percent faster than oxen and could work longer days (White, 1962, 62). The number of frost-free days available each year between spring and fall drops steadily to the north and east in Europe (fig. 2-2). In the frost-free "window" the ground must be plowed, the wheat must be planted, the crop must have time to grow, and the harvest must be finished. Since the time taken for crop growth is fixed by the genetic structure of the wheat, the only way wheat could be pushed north and east was by reducing the time needed for ground preparation, or for fall harvest time, or both. Horses sped plowing, allowing expansion of three-field agriculture and feudalism east of the Elbe.

Number of Frost-Free Days in Europe

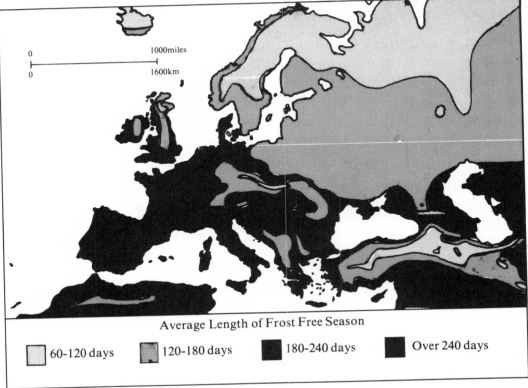

Average Length of Frost Free Season

60-120 days   120-180 days   180-240 days   Over 240 days

Figure 2-2   Number of frost-free days in Europe. By plowing faster horses made expansion of small-grain agriculture north and east possible after about A.D. 1000. Redrawn from International Association of Agricultural Economists 1969.

Decentralization and Feudalism

The geographical, technical, and social forces necessary for capitalism thus first came together in northwestern Europe after Charles Martel's defeat of the Islamic invaders at Tours in 732. Historians' stress on feudalism as a social and political system has obscured its central importance as a system for managing the technological, economic, and geopolitical structure of small polities. Before feudalism Christian Europe was subject to constant assault from three sources: Islam to the west, spilling over the Pyrenees from the Iberian peninsula; steppe nomads to the east along a broad line roughly from the mouth of the Elbe to the Adriatic; and Danish and Viking sea nomads from the Baltic. Feudal warfare removed the first two of these and somewhat constrained and eventually diverted the third. Feudalism replaced the central authority of a monarch with the authority of local lords owing nominal allegiance to a monarch. This spatial decentralization of power is confused for historians by the success of Charlemagne (768–814) in unifying the Frankish lords with those of Saxony and Lombardy (fig. 2-3). With

The Carolingian Polity

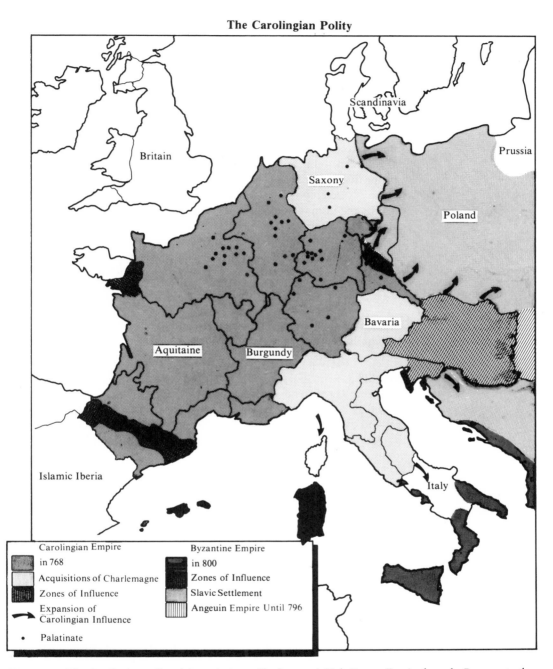

Britain

Scandinavia

Prussia

Saxony

Poland

Bavaria

Aquitaine

Burgundy

Islamic Iberia

Italy

Carolingian Empire
in 768
Acquisitions of Charlemagne
Zones of Influence
Expansion of
Carolingian Influence
• Palatinate

Byzantine Empire
in 800
Zones of Influence
Slavic Settlement
Angeuin Empire Until 796

Figure 2-3   The Carolingian polity of the early 800s: Charlemagne's Holy Roman Empire from the Pyrenees to the river Elbe and Europe's first "superstate." It was dangerously vulnerable to incursions of nomads from the east across the northern German plain and to the consequent loss of laboriously accumulated economic surplus. Redrawn from Westermann 1978.

Charlemagne's death feudal fragmentation reasserted itself, and the Carolingian polity shattered. Fragmentation made considerable geopolitical sense. Small polities made the administrative expense of nomadic conquest too high. A small number of nomads might be able to take over one polity, as they did frequently in China, or even a small number of polities, but they could never conquer and administer the totality of polities that made up feudal Europe. West of a line corresponding roughly to the Maas-Saone-Rhone, secure behind a barrier of small states, moderate-sized polities reemerged quite quickly after feudal fragmentation, most notably in France and England. The map of Europe in the sixteenth century (fig. 2-4) shows this fragmentation at its height. It also emphasizes the eastward progression of the border states, pushing the nomads further and further back into the steppe, a process culminating in the creation of the greatest of all border states, Russia. Fragmentation meant that the western European polities were able slowly to accumulate economic surplus without losing much of it to raiding by nomads or Muslims, and that they did not have to bear high protection costs to hold back the nomads.

Eotechnics and Trade: Specialization and Spatial Extension through Shipping

The cost of fragmentation was relative inefficiency in macro administration. The principal weakness lay in organizing the exchange of economic surplus between unlike geographic regions. States had to specialize because they rarely occupied more than a limited number of soil and climatic regions. Mercantile activity therefore became of vital importance, and the onset of the feudal system around 800 led to a commercial revolution (Lopez, 1971). The nature of the state itself was radically altered. From occasional appropriation of the small amounts of high-value goods traded by outlanders, it went to taxing bulky, thus easily tracked and measured, but low-value goods from neighboring states. The taxation principle required that rulers actively foster trade to maximize their tax revenue.

Bulk goods could only be moved efficiently by water. As Adam Smith so cogently put it,

> as by means of water-carriage a more extensive market is opened to every sort of industry than what land-carriage alone can afford it, so it is upon the seacoast, and along the banks of navigable rivers, that industry of every kind naturally begins to subdivide and improve itself, and it is frequently not till a long time after that those improvements extend themselves to the inland part of the country. A broad-wheeled waggon, attended by two men, and drawn by eight horses, in about six weeks time carries and brings back between London and Edinburgh [608 kilometers] near four ton weight of goods. In about the same time a ship navigated by six or eight men, and sailing between the ports of London and Leith, frequently carries and brings back two hundred ton weight of goods. (Smith, 1776, 1:18)

Europe In The Sixteenth Century

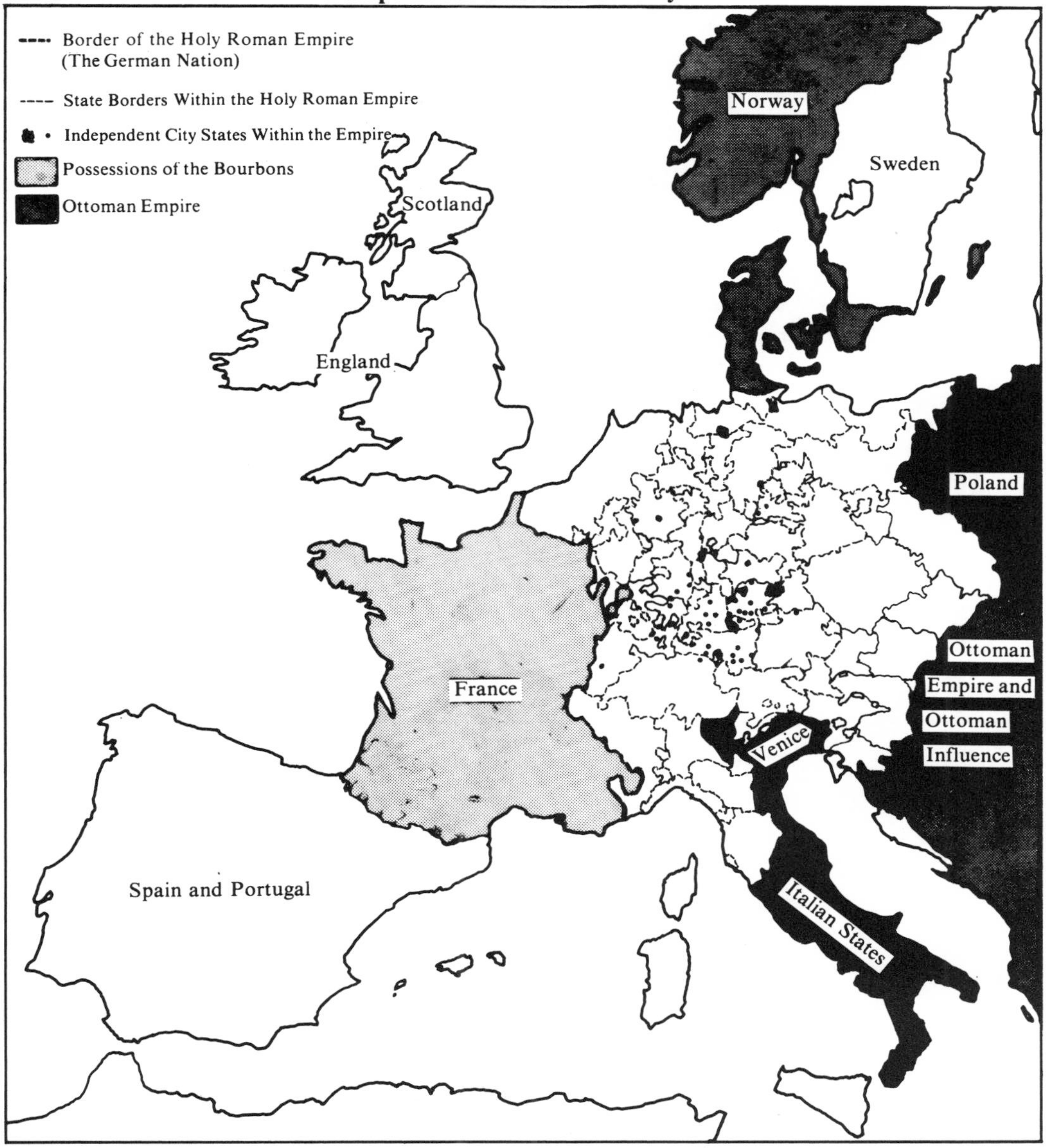

Figure 2-4 Europe in the sixteenth century, a mass of small polities and city-states between the rivers Rhine and Elbe created by the fragmentation of the Carolingian polity. Fragmentation created a barrier to nomadic incursions west of the Rhine and allowed larger, more efficient polities to prosper there. Redrawn from Westermann 1978.

Competent ships, and the infrastructure of harbors, seamen, geographical knowledge, and navigational tools needed to service them, became crucial to Europe, as did wood, the resource most needed in ship construction. The history of Europe until the midnineteenth century is the history of European ships. With occasional exceptions, those ships are either merchant ships or ships of war, a division going back to classical times. When the power of states was weak, as at the height of feudalism, even merchant vessels went well armed. When one polity or a small number of polities could guarantee the security of all ships through naval prowess, merchant ships could go unarmed.

The history of the warship is the history of ship armament. As long as ships grappled and men fought hand to hand on what were really floating combat platforms, the costs of sea warfare made trade unprofitable unless a powerful polity could protect merchant ships with specialized warships. In the classical world the Pax Romana made long-distance trade in bulk goods feasible: wine, olive oil, and wheat moved around the Mediterranean basin under the protection of Roman galleys. In feudal Europe no single polity could provide such security. What made trade profitable was the development of technologies that lowered the labor cost of warfare: the crossbow, the longbow, and the gun (Cipolla, 1965).

Without a powerful central polity, armed merchant ships were a necessity. Although the Hanseatic League had emerged by 1256 with the stated aim of protecting and encouraging the growth of trade (Kemp, 1978, 56), it could not police the seas. The league was an important software innovation, the first multinational corporation, though in a

Figure 2-5 The Hanseatic League. From the 1200s to the 1400s in the absence of strong European polities after the feudal fragmentation, the defensible ships and walled, usually independent cities of the Hansa dominated trade in the Baltic and North seas. Redrawn from East 1950.

period of exceptionally weak polities, it had no strong centralized control. At its height its trade in bulk goods such as iron, copper, timber, fish, wheat, and wool extended from Novgorod in the east through the Baltic and North seas to Bruges and London in the west (fig. 2-5). As Europe emerged from feudalism and polities became stronger once more, the league was challenged and its power diminished. It was moribund by the midsixteenth century (Dollinger, 1970). As Europe, in particular the polities around the Rhine delta that eventually merged into the Dutch republic, switched to an Atlantic focus, the geographic importance of the Baltic and North sea trades to northwestern Europe declined precipitously (East, 1950, 46).

The period from 1500 to 1600 reflects the rise of the modern state system in Europe (Wallerstein, 1974, chap. 5). Although Iberia dominated the first part of the 1500s, Holland and England took control after 1580. Holland and England were islands in the 1500s: Holland in a sea of marsh, England as part of the British Isles. The modern boundary between Holland and Belgium marks the northernmost extent of firm ground in the 1500s, when Dutch reclamation was much less developed than now. The Rhine marshes insulated the emerging Dutch republic quite successfully, although in battle both the Dutch and the Spanish assisted nature. The Dutch perfected the art of guerilla war, luring Spanish troops into marshland and preventing their exit. The Spanish persisted in wearing steel breastplates and helmets, which were useful in their New World conquests perhaps, but of little use against guns, and lethal when blundering around in marshland.

The Dutch also innovated in conventional tactics. Prince Maurice of Nassau, captain-general of Holland and Zeeland between 1585 and 1625, emphasized close-order drill, the division of his army into smaller, more flexible tactical units than had been customary in Europe, and the development of siege engineering (McNeill, 1982, 126–35). In Mumford's terms (1967, chap. 9), Maurice rediscovered the mega-machine, in which centralized direction of a large number of humans allows large-scale work to be performed. Mumford's initial use of the term applied to the means by which the pharaohs erected the pyramids (1967, 196). Maurice's mega-machine was also Hobbes's "Leviathan," the centrally directed state with its awesome power.

England is a genuine island, as Shakespeare poetically reminds us, but the English gave geography a hand. By 1588 English ships were noticeably faster than Iberian ships, as the Spanish noted at the time (Howarth, 1981, 146). The power to outmaneuver and outspeed their opponents allowed the English to prevent the Armada from leaving the Channel after the battle of Gravelines off the Belgian coast. The surviving ships had to return to Iberia by sailing north around the British Isles, a perilous voyage that few survived. The superiority of English warships was well noted by the Dutch, who improved on the English galleon and produced a fighting fleet at least equal to and often better than the English fleet in the three hard-fought Anglo-Dutch wars of the mid-1600s. Dutch merchant ship design likewise benefited.

Improvements in Power: Wind and Water

The primary energy converters for the Eotechnic were the wind-powered ship and wind- and water-powered mills. Wind power was mainly for exchanging economic surplus between unlike regions and exerting geopolitical control. Wind-powered mills were found only in areas where topography mitigated against falling water, such as the Dutch republic and the East Anglian section of England. Elsewhere water power reigned supreme. It was also ubiquitous from an early date. The great Norman record of the conquest of England, the Domesday Book of 1086, lists 5,624 water mills for some 3,000 communities (White, 1962, 84). Until recently historians have underestimated this "mechanical revolution in the eleventh, twelfth and thirteenth centuries" (Braudel, 1981, 353). By about 1750 there were perhaps 500,000 to 600,000 mills in Europe, with a constant ratio across Europe of 1 mill to 29 people (Braudel, 1981, 358). The vast majority of these mills simply ground grain into flour and put out very little power (fig. 2-6). Wooden construction limited the diameter of wheels, and power was usually calculated at 1 horsepower per meter of diameter. Although by the seventh century 10-meter wheels were feasible, the

A Typical Eotechnic Rotary Mill

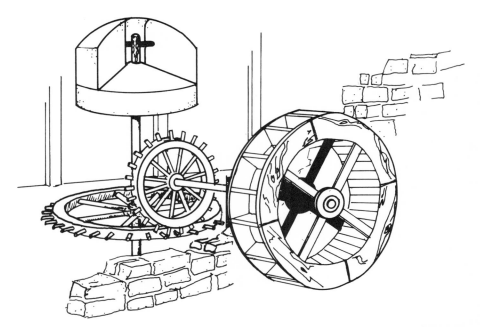

Figure 2-6 A typical Eotechnic rotary water mill for grain. Such mills substituted inanimate for animate labor in one of the most labor-intensive chores in a small-grain economy. By the time of the Domesday Book (1086) there were one or more such mills in every hamlet in England. Redrawn from Burke 1978.

majority of mill owners "preferred to increase the number of wheels rather than deal with all the complicated technological problems involved in the concentration of energy on one single wheel" (Cipolla, 1980, 172). As long as energy had to be transmitted mechanically from the waterwheel to the point of use, friction ensured that it could not be transmitted far. Several small wheels could minimize loss of output in transmission, and wheel diameter usually did not exceed 5 meters.

The water mill provided a good answer to Europe's search for efficiency. Small grains require grinding into flour, then baking into bread. Grinding was a labor-intensive operation well suited to small-scale mechanization because it could be done at any time of the year: it was not, like spring planting or autumn harvest, a temporally situated component of the agricultural economy. In the absence of mills, either human or animal labor was needed. Human labor was relatively scarce in northwestern Europe, which was an undersettled frontier at the onset of feudalism. Animal labor competed with human labor for food and was needed for plowing and warfare. Human labor freed from the task of grinding grain could be profitably employed elsewhere in the economy. The water mill thus played a central part in developing a prosperous agrarian economy in medieval Europe, one that came to support increasingly elaborate towns and cathedrals as well as a comparatively decent standard of living for its agrarian population. Braudel suggests that whereas water milling began as a major component of the agrarian manorial economy in the early feudal period, its real importance lay in supporting trade (1981, 356). The mill freed people to work at non-agricultural pursuits as towns grew.

The mill also helped in manufacturing. Wool textiles require "fulling": hammering water-laden cloth to close gaps between the fibers and give a smooth "hand" to the cloth and a fuller appearance. References to fulling mills are common in the Domesday Book (Usher, 1954, 184–85, 268). Whereas the grain mill used rotary motion, the fulling mill converted rotary to up-and-down motion by the important technical innovation of the camshaft. This classical Greek invention (Usher, 1954, 140) does not seem to have been put to any serious use until its eleventh-century application in the mill. As early as 1010 "water-driven trip-hammers were at work in the forges of Germany" (White, 1962, 84) to reduce the human labor otherwise required to hammer carbon into molten iron and turn it into mild steel. The water-driven bellows forcing air into the blacksmith's furnace to achieve temperatures sufficient to melt iron seems to have come somewhat later: the first is recorded in 1323 (Gimpel, 1976, 67). In combination the water-driven camshaft and bellows revolutionized the labor productivity of the blacksmith, and thereby the production of mild steel for agricultural implements and warfare (fig. 2-7).

Despite such use of inanimate power, the Eotechnic world system was almost totally agrarian. In the equation Land plus Labor plus Capital equals Product, Land loomed very large indeed, Labor less large, and Capital scarcely at all. In such a system the balance among the three Fs was critical, and it was largely controlled by population.

**Cams Produce Up and Down Movement from
Rotary Motion, the Blacksmith Shop**

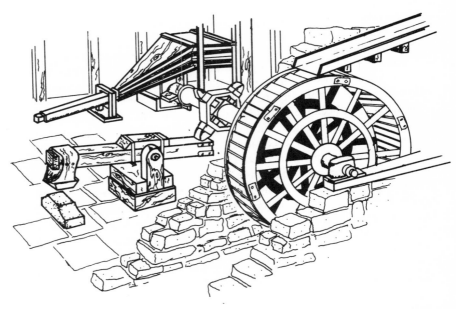

Figure 2-7 The Eotechnic blacksmith's shop. Cams turned rotary motion into up-and-down motion to work bellows to get a fire hot enough to smelt iron. They also lifted the trip hammers that did the heavy work of beating carbon into molten iron to make mild steel. Similar hammers did the equally heavy work of fulling woollen textiles. Redrawn from Burke 1978.

EMERGENCE OF THE MODERN MAN-LAND SYSTEM IN EUROPE

The Impact of the Black Death

If feudalism set the stage for capitalism, it was the catastrophic demographic event known as the Black Death that raised the curtain on the first act. After 1300 the climate of northwestern Europe became cooler and wetter, and the Black Death fell on a population that, in some regions, had been weakened by fifty years of increasingly poor harvests. Southern England and the Paris basin seemed unaffected (Ladurie, 1971, 13), but northern England, Scandinavia, and the northern German plain suffered horrendously. The Black Death transformed the face and future of Europe by altering the food-fiber-fuel ratio against food. Nor was the Black Death a singular event. The first of four outbreaks hit Marseilles in late 1347. By the summer of 1348 the plague had arrived in England, and by 1350 it had run its course through Europe, leaving one-third to one-half of the population dead (Bridbury, 1973, 589). The second pestilence arrived in England in 1361, the third in 1368, and the fourth in 1380. In spite of some demographic recovery from the first epidemic, later plagues again took off large numbers, usually previously unexposed children (Russell, 1972, 56). How many died can never be accurately known, but a conservative estimate would suggest that by the mid-1380s Europe's population was scarcely half what it had been in 1347.

As demand for food waned, landlords sought alternative crops in the form of fiber from livestock. Their only real alternative, as Wallerstein points out, was to convert feudal dues into money rent, but renting resulted in reduced control over the land, sometimes even its sale to newly wealthy independent proprietors (1974, 108). Sheep or cattle allowed retention of the land and total control over it. Tenants paying money rent were not easily alienated from the land, but sheep were. Livestock pasturing required little of the scarce, expensive labor that survived the plagues. It also provided in wool a crop that the climatic downturn of the early 1300s had brought into demand. Before the plagues, wool was an expensive luxury (Baker, 1973, 207). After them, fiber production accelerated rapidly and real wages rose. A much better-off population could buy ever cheaper cloth in a climatic period that encouraged fiber production. Without, however, the opening up of large areas of land by demographic catastrophe, such a wholesale switch from food to fiber would have been unlikely.

Once this mixed, modern man-land system was developed there was no going back to feudalism in northwestern Europe. Landlords' profits were higher with fiber, and they had no tenants to cope with. Thus, when demographic recovery began after 1390, Europe approached its next major transition point. Demand for food rose, yet landlords refused to displace more profitable fiber. Four solutions were possible, and all were tried. These were increasing productivity on existing lands, internal colonization, demographic control, and external colonization.

Sources of Increased Wealth

The switch from feudal agriculture to yeoman farming, with farmers paying money rent or owning their own land, increased productivity. Feudalism was never an efficient labor system, having far too many feast days and saints' days, quite apart from the temptation for those legally bound to the land, the serfs, to gather their own crops before those of their lord at the crucial harvest period. Yeoman farming, with crops gathered by the landowner or tenant paying money rent, was more efficient because labor effort was not split. Feudal labor dues were highly defined but varied markedly from place to place and time to time. Duties amounted to one-third to one-half of a peasant's labor power. Although a day's work usually only lasted until noon, at harvest times it was usually from sunrise to sunset (Bennett, 1937, 108–9). Holy days could theoretically remove fifty or more days from the working year, although there is little evidence that more than fifteen or twenty were taken (Bennett, 1937, 117). The lord had first call on the peasant's labor power in the harvest season. Despite the peasants' resistance the manorial crops were usually harvested before those of the peasants (Bennett, 1937, 110–13). Pounds calculates rapidly increasing grain productivity in England and the Low Countries after 1500 (fig. 2-8). For France, Spain, Italy, central Europe, and Scandinavia, increases were only modest. For eastern Europe they were substantial, but from modest beginnings, and they turned into steady declines after 1600 (Pounds, 1979, 183). Although eastern Europe was a major grain producer, its still feudal labor system denied it the productivity increases of England and the Low Countries.

A second reason for an increase in European crop production and thus wealth after 1500 was internal colonization. The remaining woodland was cleared and the remaining marshes drained, but little new land was taken in (Baker, 1973, 213). Easier increases in production could be obtained elsewhere.

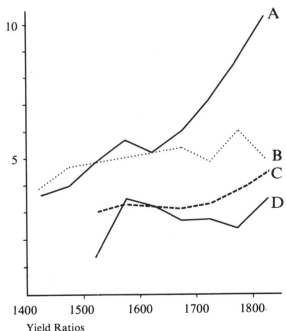

**Relationship of Crop
Harvested to Seed Used, Europe 1400-1800**

Yield Ratios

A = Britain and the low countries

B = France, Spain, and Italy

C = Central Europe and Scandinavia

D = Eastern Europe

Figure 2-8   Relationship of crop harvested to seed used. Europe 1400–1800. *A,* Britain and the low countries; *B,* France, Spain, and Italy; *C,* central Europe and Scandinavia; *D,* eastern Europe. Redrawn from Pounds 1990.

A third force for increased wealth was reduction of demand by demographic control. The extent to which this was conscious is highly debatable, but the population of Europe after the plagues did not breed back with the rapidity it might have. Fertility control through delayed marriage was a conspicuous part of at least English demographic behavior as early as 1541 (Wrigley & Schofield, 1981).

Geographic Expansion: The Baltic Economy

The fourth and most important way in which Europe's wealth increased was external colonization. Productivity gains in the Baltic lands were nonexistent, but production gains were substantial. Wallerstein describes this process as the refeudalization of Poland. The nascent middle and mercantile classes of Poland disappeared, and Poland returned to an essentially feudal condition with only two main social classes. Peasants became again chained to the land by the feudal obligations dropped in the rest of Europe after the Black Death. By the sixteenth century refeudalization was in full flood (Wallerstein, 1974, 122). It was, of course, based on cheap, reliable sea transport out of such Baltic ports as Gdansk to the consuming countries on the Atlantic coast. The only real constraint on the trade between the Baltic and the Atlantic was the sound dues exacted by the Danes for ships passing through the waters between Denmark and Scania, then a Danish province, now part of Sweden—a passage so narrow at Helsingor that it could be chained. After the Dutch and English took control of the Baltic grain trade in the seventeenth century, the Danes threatened closure of the sound, and the Dutch and English kept it open by force. Between 1658 and 1815 English fleets were sent to keep the sound open on nearly twenty occasions (Albion, 1926, 165).

Wheat was not the only staple of the accelerating Baltic trade of the sixteenth and seventeenth centuries. The Dutch and English paid sound dues as much for their timber ships and their naval construction programs as for their grain ships and merchants. The Dutch came to depend almost entirely on the Baltic for ship lumber. England's needs were less imperative because Britain had been less deforested than Holland. By the late seventeenth century England was able to draw heavily on her American colonies, but the wood was inferior to that of the Baltic and transatlantic freight rates were about three times higher (Albion, 1926, 152).

As England and Holland challenged each other for hegemony in the 1600s, the Baltic became a crucial field of conflict. In 1651 the Dutch bought the right to "farm" the sound dues for 35,000 pounds per year, with the right to close the sound to the English in times of war (Albion, 1926, 166). The first of three Anglo-Dutch Wars broke out in 1652, and even neutral ships bound for England were seized in the sound. Unprepared to send an English against a Dutch fleet, Cromwell sent diplomats who, in 1658 and after war between a Denmark backed by Holland and a Sweden backed by England, negotiated a settlement that somewhat favored England. In 1658 the Swedes attacked Denmark and the Dutch fleet intervened, producing a treaty favorable to Holland. In 1659 the second Anglo-Dutch War nearly began five years ahead of schedule as the English fleet made a show of force in the sound. With Cromwell recently dead, however, England's position was far from secure. As one contemporary put it, "When our forces are gone to the Sound, an army may be landed here, and Charles Stuart to head them" (Albion, 1926, 174). But whether the Stuarts returned to the throne of England or not, Cromwell's policy of keeping the sound open was now irrevocably that of England. Without such boldness the Dutch could easily have seized the "turnstile of trade" between the Atlantic and the Baltic (Albion, 1926, 175).

A major issue between England and Holland in the three Dutch Wars concerned the economic theory of trade. England, in 1651 and again in 1660, passed a series of Navigation Acts. These restricted trade between English ports to English vessels except in carefully specified circumstances, a policy known as mercantilism (Albion, 1926, 156). Holland's policy more closely approximated that of free trade, with no restrictions on whose ships might ply what cargoes where. Because Dutch ships were built more cheaply than English, England was at a competitive disadvantage in free trade. The Navigation Acts restored to England what was actually the most profitable part of the timber trade, cartage. As a contemporary commented in the debate over mercantilism, "freight is the most important raw material we possess" (quoted in Albion, 1926, 158). By 1700 the Navigation Acts had successfully excluded Dutch ships from carrying Baltic lumber and wheat to England; such goods could be carried to English ports only by English ships or ships of the Baltic nation originating the shipment.

The combination of the Navigation Acts and a policy of armed intervention to hold the sound open guaranteed that the Baltic trade continued and that its adverse effects upon England's balance of trade were minimized. The latter was a continual problem for mercantilist theory. The Baltic lumber trade was particularly expensive for England, and after 1700, the mercantilists encouraged the substitution of lumber from the American colonies. In 1631 the Massachusetts Bay colonists launched their first seagoing vessel, *Blessing of the Bay* (Whitehurst, 1986, 11). By 1650 more than "a score of ship and boatyards" at the north end of Boston alone were turning out everything from pinnaces to full-rigged ships (Rutman, 1965, 189). By 1750 some 125 colonial shipyards were turning out between 300 and 400 ships a year (Whitehurst, 1986, 14). By the Revolution, 2,343 American-built ships accounted for "a third of the total British registry" (Albion, 1926, 246). Vessels constructed in America were 30 to 50 percent cheaper than ships built in British yards.

Geographic Expansion: The Atlantic Economy

Two products preceded lumber as forces pushing European external colonization westward into the Atlantic rather than east into the Baltic;

they were sugar and fish. As long as good nutrition merely meant a full belly, expansion of the carbohydrate frontier into the Atlantic to produce sugar made as much sense as expansion into refeudalized Poland to grow wheat. Eating fish is usually regarded as of religious rather than geographic significance. Yet an obsession with eating no red meat every Friday in honor of Christ's crucifixion has a huge impact on the three Fs. Even in the unlikely situation that all people ate red meat every day of the week, the reduction of demand for red meat by one-seventh would have been significant. Roughly speaking, meat needs ten times as much area per calorie as does grain (Cipolla, 1978, 41). Fishing needed little more than ships and the labor to sail them, although ship construction obviously needed lumber.

Sugar had long been an important element in the diet of wealthy Europeans. It was introduced into northern Europe from the eastern Mediterranean around 1100, and it was prized as a spice (Mintz, 1985, 79–80). This eastern Mediterranean industry began to lose ground in the early 1400s, first to Sicily, Spain, and Morocco, then to the Atlantic islands (Galloway, 1977, 181). By the late 1400s the Portuguese Azores and the Spanish Canaries had become the largest suppliers of sugar to Europe. As both Portugal and Spain continued their expansion west into the Atlantic upon the discovery of the New World, they naturally took sugar with them into the American tropics, although the Portuguese lagged half a century behind their Spanish neighbors in development of the New World.

Spain's great success in expansion after 1492 was gained by the combination of preconditions for expansion and the geographic nature of its acquisitions in the western Atlantic. Meinig points out the crucial importance of the internal expansion of the Castilian state in the 1400s (1986, 8–9). The Reconquista that drove the Moors from Spain refocused on the Caribbean after 1500. The Spanish "had been gaining experience in combining seafaring, conquering, and planting for some years before Columbus in their subjugation of the Canary Islands" (Meinig, 1986, 9). The Caribbean archipelago was easily exploited by oceangoing ships (Meinig, 1986, 11). They also encountered little or no effective resistance from the indigenous inhabitants, who proved appallingly susceptible to European diseases and died off in large numbers. After 1519 contact with the Aztecs in the Great Valley of Mexico changed the thrust of Spanish expansion from the Caribbean archipelago and the production of sugar to the mainland and the forcible extraction of gold and silver. Even so, the pattern of production agriculture in the tropical New World was firmly set through the early 1800s. The Spanish faced only one serious problem in their drive to produce sugar in the Caribbean, that of availability of labor. As in the Canaries, their initial response was to enslave a local population, but again that population quickly died out from European diseases. They were forced to turn to the importation of slaves, as the Portuguese had done when they needed labor to grow sugar on uninhabited Madeira.

In the steady expansion of the Portuguese down the Atlantic coast of Africa after Gil Eanes rounded Cape Bojador in 1434, Portuguese traders had pioneered the profitable extraction of slaves from sub-Saharan western Africa. Before 1434 the principal market for the warring African tribes that enslaved each other's populations was the Islamic world to the east. Since the major slave production region was the populous west coast, a long crossing of the continent was entailed. The Portuguese short-circuited this by trading directly with the major producing region and transporting the slaves by ship, thus avoiding the cost of moving slaves across a continent and through the hands of many middlemen, each of whom added his own profit to the cost of the slave. The importance of the Portuguese in the Atlantic world has long been underestimated. As Meinig says, "They created the systems and societies that made possible and set the pattern for the creation of an Afro-American world" (1986, 17). After 1430 the Portuguese pioneered new sailing technologies to allow them to begin progress on the goal of Prince Henry the Navigator: circumnavigating Africa to provide a direct link between Atlantic Europe and the Indies. In the process of sailing west into the Atlantic out of sight of land, a process designed to allow the rounding of dangerous Cape Bojador on the African mainland, they encountered the Azores, Madeira, and the Cape Verde Islands, all of which they colonized. They also discovered that the natural circulation of the winds in the eastern Atlantic, which blew them direct to the Canaries, would, if they continued south and west into the Atlantic, then blow them in a great loop first north and then east back to the Iberian peninsula via the Azores (Crosby, 1986, 113). By 1452 Madeira, its lands worked by African slaves, was on its way to becoming Europe's single largest source of sugar (Meinig, 1986, 21). The Cape Verde Islands, some three hundred miles off the African mainland, served well as a forward base for Portuguese slavers, avoiding the need to establish bases on the mainland itself, where tropical diseases endemic to the African environment took a toll on Europeans that reached 80 percent and more (Meinig, 1986, 22).

The first New World market for African slaves was Spanish Hispaniola, but the slaves were provided by the Portuguese. By 1517, only two years before Cortés left Cuba for the Great Valley of Mexico, "Genoese entrepreneurs and Canarian technicians had gotten sugar production under way" on Hispaniola, and the Spanish issued to the Portuguese the first licenses for the direct importation of slaves from Africa (Meinig, 1986, 21). Genoese capital was as crucial to the Spanish Atlantic expansion as Venetian capital was to the Portuguese and Dutch and English capital was to the Baltic trades (Wallerstein, 1974, 49). Venetian and Genoese merchants both saw an Atlantic route to the Indies, whether around Africa or by circumnavigating the globe, as critical to their long-term economic success. Either route would bypass the Islamic world, which soaked up much of their potential profit. That the circumnavigation of the globe turned out to be impractical must have encouraged Genoese entrepreneurs to cut their losses and invest in sugar production in the Caribbean.

Slavery also had ecological implications for Europe. Slaves were im-

ported as adults and were worked to death. Curtin calculates an excess
of deaths over births on English Caribbean sugar islands of between ten
and forty per thousand per year (1969, 28). From a purely economic
point of view, it must have been cheaper to import a replacement than to
breed one. The ecological calculus, however inhumane, suggests that
Europe's agricultural efficiency was greatly enhanced when plantation
owners bought slaves "grown" in Africa on African food, applied their
labor to the production of carbohydrates for export to Europe, and
displayed little concern for their survival past the time when they could
perform useful work.

After a slow start, Brazil became the great consumer of African slaves
and producer of sugar. Portugal gained possession of South America
east of 60 degrees longitude with the Treaty of Tordesillas of 1494, but
the African and Asian trades were more profitable (Meinig, 1986, 10).
Brazil's coastal Indians remained difficult opponents until European
diseases had taken their toll, so that Portuguese settlement only began in
the 1530s. Portugal was a small country with few people prepared to
leave for the New World. It was also poor in timber resources. Sugar was
exported not in Portuguese but in Dutch ships built with Baltic lumber.
In the late 1500s first Antwerp, then Amsterdam became the main
center for financing the Brazilian sugar trade and refining Brazilian
sugar (Wolf, 1982, 150). Dutch ships also carried many of the Portuguese slaves in the characteristic triangle trade that grew up in the
Atlantic. Over the 110 years between 1701 and 1810 close to 40 percent
of all the slaves taken out of Africa went to Brazil and the Dutch Caribbean (Curtin, 1969, 216). Roughly 22 percent went to the British Caribbean, 22 percent to the French Caribbean, 10 percent to the Spanish
Caribbean, and the remaining 6 percent to British North America, the
only area of tropical Caribbean America where sugar was not grown.

To begin with Iberia consumed much more sugar than northern
Europe, though Holland was clearly an exception because of its central
role in the Brazilian trade. As late as 1700 British consumption was only
1.8 kilograms per person per year (Mintz, 1985, 67). British consumption rose steadily thereafter, reaching nearly 8.2 kilograms per person
by 1800. Refined sugar is about as caloric as wheat flour. In a diet of
3,000 calories a day 8.2 kilograms of sugar a year would provide ten
days' calories, or close to 3 percent of total food requirements per year.
Grain crop area in Britain was reduced accordingly. Only in the late
1800s did sugar consumption rise to levels close to those of today, when
it may account for 25 percent of total calories.

Before 1800, however, the Atlantic fishery had a greater impact than
sugar on European production of the three Fs. Dutch and English fishery
expanded rapidly in the North Sea and on into the Atlantic in the 1400s.
John Cabot of Bristol reported the discovery of Newfoundland's cod
resources in 1508. The Grand Banks fishery required no permanent
settlement on the American shore, nor did the Newfoundland coast
encourage it. Nevertheless, Bristol fishermen slowly developed the habit
of wintering over as they cut, salted, and dried their catch (Meinig,

1986, 87). Some of their wintering-over sites eventually developed into permanent settlements, but as late as 1677 the English government could still press to remove all permanent residents from Newfoundland to preserve the interests of the influential Bristol merchants who financed the Grand Banks fishery. This was a low-cost "colonization," costing little more than the ships and providing, in addition to fish, a fertile training ground for seamen (Meinig, 1986, 87).

In geographical terms the calculation is simple. If we assume that every person in the British Isles consumed dried, salted cod once a week, and that portions averaged a generous 100 grams, cod would have had a major impact on the British ecology. Although 100 grams of dried, salted cod represents only 130 calories, it replaced mutton, beef, or pork, all at around 300 calories per 100 grams. Cipolla (1978b, 41) notes that approximately 10 calories of vegetable energy produce about 1 calorie of animal energy. If Europe had not had cod, it would have had to substitute animal protein fed on cereal grown within Europe. All other things being equal, it would have taken 3,000 calories of cereal energy to produce 100 grams of mutton, beef, or pork. Cod allowed a substantial amount of land to be freed for production of items other than food. Although animals were often grazed on lands unsuited to cereal grains, the post–Black Death enclosures had converted much good cropland to pasture. The use of fish means that a substantial area must have been freed up for other uses all over Europe. If about half the cultivated area was in pasture after the Black Death, and if we assume that all of the pasture produced meat (even wool-producing sheep were eventually killed and eaten), we may crudely calculate that fish reduced the pressure on Europe's cultivated area by one day's protein intake per week. One-seventh of one-half is one-fourteenth. Assuming a substantial margin for error I suggest that fish replaced about 5 percent of the calories that would otherwise have to have been grown in Europe.

If England had cod, the Dutch, Germans, and Iberians had herring in various forms. Only France lacked good access to the Atlantic fishery. Combining fish and sugar, we may suggest that by 1800 some 8 percent of Europe's food requirements were coming from the Atlantic world. Given that English sugar consumption was low, the figure for other countries might have exceeded 10 percent, although for France it might have been lower than 8 percent. A substantial amount of European land became available for other uses as a result of Atlantic expansion.

Results of the Geographic Expansions

The reduction in European cropland achieved by the Baltic and Atlantic economies was all the more remarkable given the extremely limited transportation and geography of the Eotechnic world system. Without figures for the Baltic grain imports we can only make informed guesses, but by the mid-1500s Baltic grain and lumber combined with Atlantic sugar and fish were supporting 10 to 20 percent of the population of northwestern Europe. Land thus freed could be turned to fiber. Add to this the impact of the substitution of coke for charcoal in iron smelting in the 1700s, and we may project nearer to 20 than 10 percent by 1800.

In some parts of Europe the impact was sharply higher. Local surpluses might generously be supposed to have freed between 20 and 30 percent of a population from the need to produce their own food in the early 1600s, yet about 50 percent of Holland's population was engaged in nonagricultural activity (Cipolla, 1980, 76). In China only 2 percent of the population was freed from agriculture as late as the 1880s (Stover & Stover, 1976, 110). What was original to northwestern Europe's long-run success was the addition of the "ghost acreage" on which food and industrial raw materials were produced outside its geographic confines at little expense other than the cost of the ships to carry it. Until the mid-1800s the "ghost acreage" was limited to archipelagos, littorals, and the ocean itself because of northwestern Europe's dependence on small-capacity sailing ships. The British Empire on the eve of the American Revolution indicates the insular nature of the Eotechnic world system.

Technology was by no means the only force shaping Europe's destiny in the Eotechnic. Christianity provided a common ideology before the Reformation and, at least for Europe's educated elite, a common language in the form of Latin (Dawson, 1932). It helped unify Europe against Islam and the steppe nomads. As in China many of the nomads settled down and became "civilized." Europe expanded eastward as much by the cross as by the sword in a process that was well understood by European rulers from Charlemagne on. Against the nomads Christianity was a powerful tool, its aggressive egalitarianism well calculated to appeal to a warrior folk among whom aggression and (male) egalitarianism were the norm. Against the equally powerful ideology of Islam, Christianity was almost useless, and the sword had to be used instead, for the most part with little success, as the variable history of the Crusades firmly indicates.

Weber brilliantly described the material impact upon post-Reformation Europe of Calvin's radical Protestantism (1904–5/1958). By reducing the ability of the church to interfere with inquiries into the nature of the physical world, radical Protestantism split Europe into two camps. To the south, in particular in Italy and the Iberian peninsula, inquiring into the physical world was proscribed. The trial of Galileo was an object lesson, and although the arts flourished, science did not. In Holland, England, and, to a lesser extent, France, science and art flourished side by side (fig. 2-9). The reduction of religious interference with scientific inquiry allowed the emergence of practical science in northern Europe. The demonstration in the early 1600s by Otto von Guericke of the Protestant electorate of Magdeburg that air had weight and the creation of a vacuum by the condensation of steam in a closed space by the French Huguenot Denis Papin together led to the creation of a working steam engine and thus to the Paleotechnic.

In a crucial sense, however, technology shaped Europe's destiny in the Eotechnic. Without the stirrup, the feudal state could not have

Concentration of Major Scientists and Artists
born between 1500-1600 and born between 1600-1660

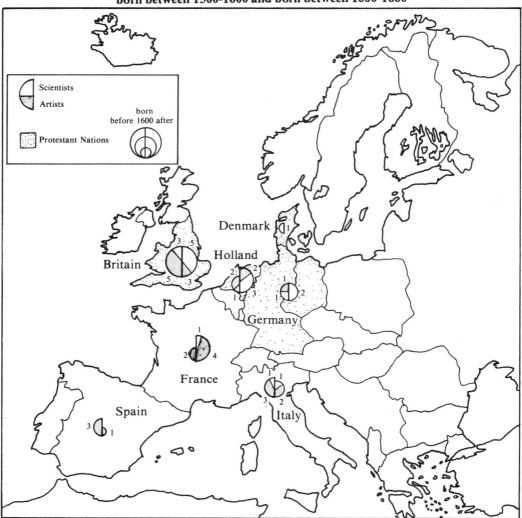

Figure 2-9 Concentration of major artists and scientists in post-Reformation Europe. Catholicism suppressed inquiry into the physical nature of the universe, creative people focused on the arts in Catholic countries. Protestant countries started to produce great scientists as well as great artists.

emerged. Without feudal warfare, the nomads could not have been resisted. Without the surplus of horses generated by feudal warfare, the steppe frontier could not have been pushed east. Without feudalism, Europe could have been dominated by a single polity akin to the Carolingian state. Without ships and trade, the small polities of feudalism could not have survived and the advantages of the "ghost acreage" could never have been enjoyed. Without bows and guns to protect merchant ships, the trade in bulk goods could not have been profitable.

Without mills powered by water or wind, animals could have consumed too much of the available food, crucially lowering the ability of Europe to generate adequate surplus for human activity, in particular to support the growing classes of specialists in warfare, carpentry, and trade.

THE PALEOTECHNIC

The Resource Crisis of Seventeenth-Century England

The origins of the Paleotechnic lie in the technical changes brought on in Europe, and in particular in England, by the increasing imbalance in the food-fiber-fuel ratio of the early 1600s. By then wood for fuel, domestic construction, and shipping was in short supply. Particularly large amounts of fuel were needed to produce charcoal to smelt iron for ships' guns. More merchant ships were needed as trade expanded, more food was needed to feed more mouths, and new and better housing was needed to meet the demands of a wealthier as well as larger population. In such conditions landlords frequently preferred quick returns from food and fiber to the slower return from fuel: woodland clearance accelerated in the early 1600s.

Wrigley and Schofield (1981, 575) plot a rapid, sustained increase in the population of England from 3 million in 1560 to 5.2 million by 1650. By the time that peak was reached Admiral Sir William Monson could write that

> all kinds of wood that belong to the building of ships or other works that have relation to timber, we do and shall find, in a little time, a great want of. For wood is now utterly decayed in England, and begins to be no less in Ireland . . . If money, or wealth, decay in a kingdom there may be means of trade to recover it again; if seamen die so long as there are ships and navigation they will soon increase and make their deaths forgotten; but if our timber be consumed and spent it will require the age of three or four generations before it can grow again for use. (quoted in Albion, 1926, 95)

The very best English oak was reserved for ship lumber and came mainly from the southeastern counties—Sussex, Surrey, Hampshire, and Kent—since rarely would anyone haul ship lumber more than 30 kilometers to the sea (fig. 2-10) (Albion, 1926, 103). Only these four counties had the proper location, climate, and rich loams or clayey soils. As Albion succinctly put it, "Wheat and oak came into rivalry for those same lands. A tree so discriminating in the matter of soil and climate could not be planted profitably in waste places" (Albion, 1926, 18).

The source regions for fuel differed little. In the Domesday Book they were Hampshire, Berkshire, Oxfordshire, Dorset, and Wiltshire. In later medieval times the counties of Essex, Gloucester, Kent, Sussex, Nottingham, and Northampton were added. Only the last two lay outside the southeastern oak belt. Forests more than 30 kilometers from the sea were more important for charcoal, as well as for fuel for lime burning, smithing, salt boiling, and glass making. Even more significant was their

The Oak Producing Region of England

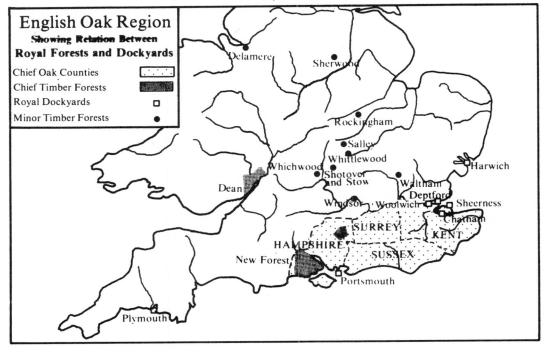

Figure 2-10 The oak-producing region of England during the heyday of wooden ship construction. Land good enough for great oaks for ships was at a premium because it was fine land for growing wheat as well. Redrawn from Albion 1926.

role in chemical engineering: tanning used the bark; and glass making, soap boiling, and cloth making used the alkali for wood ash (Armytage, 1976, 48).

The Substitution of Coke for Charcoal

The key to the Paleotechnic was Abraham Darby's transformation of coal into coke in a process akin to that used by charcoal makers in their conversion of wood. Just as wood could be heated in an oven to gasify and drive off its organic components, leaving only the carbon behind, so could coal. The combined combustion temperature of the carbon and the organic components of both wood and coal was too low to smelt iron. But reducing the wood to charcoal or the coal to coke created material that burned at a high enough temperature. In 1709, at Coalbrookdale in Derbyshire, Abraham Darby began to smelt iron with coke rather than charcoal. He started a crucial process in the development of the modern world: freeing production from the ecological limits imposed by dependence on the organic world for almost all basic raw materials and energy. As with so many of the seventeenth-century persons who promoted science and its practical application, Darby was a radical Protestant, in his case a Quaker.

There was, however, a price for liberation from the organic world. It was exacted both immediately and over the long term. Indeed, it is still being exacted. Dependence on materials that are renewable, if they are renewable at all, only over the very long run of geological time has merely postponed reaching the ecological limits. Over the short term, three serious costs were imposed: first, coal production had to be rapidly increased; second, a competent system of internal transportation had to be developed because many coalfields were far from the sea; third, serious environmental and social degradation had to be faced.

All three short-term costs really hinged on the need for rapid increase in English coal production in the 1700s. In the short run a new technology to pump water out of increasingly deep mines was needed. This technology was available before Darby successfully produced coke; it was used to pump water out of tin, lead, and zinc mines in Cornwall. The demands of the iron industry led to the rapid adoption of steam power to pump out coal mines.

The Paleotechnic world thus quickly came to revolve around the steam engine. The first really successful pumping engine was the atmospheric one designed by Newcomen in 1712 near Dudley Castle, Staffordshire, in the English Midlands. In this engine steam was condensed beneath a piston to make a partial vacuum. The atmosphere then pushed down on the piston to perform the power stroke. But the fuel consumption of the Newcomen engine was so high that it was, at least in its early form, almost unusable away from the pithead.

The second cost of the Paleotechnic lay in creating an internal transport network. The old Newcastle field lay on the coast, and its coal had been carried around England by ship since the late 1100s. But most of England's best coal was inland and could be carted only a few miles before it became too expensive. The English canal system grew piecemeal but effectively, largely at the behest of mine owners (Vance, 1990, 70–71, 88). The canal, of course, had been used for centuries as a tool of political and economic integration. The Chinese had united their spatially extensive polity by linking the Huang-Ho and Yangtze river valleys by means of the Grand Canal, completed in A.D. 610 by the second Sui emperor, Yang Ti (Herrmann, 1966, 27; Needham, 1954, 122). The Italians had integrated the agricultural economy of the Po valley, and thus made Milan a great city, by the Naviglia Grande, begun around 1200 and completed in 1458 (Vance, 1990, 41). England's first serious canals were made not to promote the movement of agricultural surplus, a task easily accomplished by coastal navigation, but to extract coal from inland fields. The Bridgewater Canal, completed in 1761, halved the price of coal delivered in Manchester from the duke of Bridgewater's mines at Worsley (Vance, 1990, 89).

As consumption of coal increased, air and water pollution and their attendant illnesses became common. As people crowded into new industrial cities where they were confined spatially by the necessity to walk everywhere, diseases engendered by crowding and an inadequate technology for human waste removal spread rapidly. Cholera became a scourge in all great cities. No one has painted the bleak picture of the

Newcomen steam engine, 1750. This atmospheric engine operated at 14 strokes per minute and generated 11 horsepower. It was best suited to pumping coal mines. Ford Museum, by the author.

Paleotechnic better than Lewis Mumford (1934, chap. 4). Marxist scholars have been equally critical: "A new barbarism emerged in England as grasping men sought to exploit the labor of the landless proletariat created in England from the sixteenth century on" (Wolf, 1982, 121).

England was also unique in Europe in alienating its peasantry from the land, walking a tightrope between the Scylla of insufficient low-cost labor for industrialization and the Charybdis of violent revolution. In the early 1600s revolution lurked just beneath the facade of Puritanism (Bridenbaugh, 1968, chap. 10). Many landless peasants succumbed to mainstream Puritanism, attracted by its radical egalitarianism and success in the two civil wars of 1642–46 and 1646–48 (Briggs, 1984, chap. 6). The more radical "Diggers" and "Levellers" railed at landlessness: Levellers complained of Cromwell's tying the franchise to landowner-ship; Diggers opposed the private ownership of land. The Diggers were too small and disorganized to occasion much concern, but Cromwell had the Levellers bloodily put down in 1649 (Briggs, 1984, 142). A large pool of landless laborers also required a complex system of social welfare to avoid starvation. Despite the costs, the benefits were great: labor costs were forced down for mine owners, canal builders and, eventually, the owners of the factories that developed after the 1780s.

A third problem was that in organic societies men, women, and children usually worked side by side in the fields with little division of labor by age or sex. This labor system was carried into the Paleotechnic with appalling consequences, particularly in mines, where children and women were preferred because of their small size and ability to traverse narrower, cheaper shafts. The heat in mines was substantial, and most of the labor force worked naked or seminaked. The mines also blurred night and day, resulting in a loss of the organic rhythm of life on the surface. As Marx later summarized it, "all bounds of morals and nature, age and sex, day and night, were broken down" (1954, 264).

The social degradation of labor was completed by Watt's steam engine, which freed production from the spatial and temporal limits of the Eotechnic. Watt, an instrument mechanic at the University of Glasgow, recognized as early as 1764 the shortcomings of the model Newcomen engines he demonstrated to students in practical physics classes; their overall thermal efficiency was less than 1 percent (Armytage, 1976, 88). By 1782 Watt had greatly improved his engine and converted the rocking motion of the Newcomen pumping engine into rotary motion that could be used to drive more complex machinery (Armytage, 1976, 88–91). By the late 1790s Watt engines were better-ing 4 percent fuel efficiency with ease (Thirring, 1958, 54), making it much easier to move the steam engine away from the pithead. Von Tunzelmann conservatively estimates that the social savings to Britain from the switch from Newcomen to Watt engines, which had already occurred by 1800, amounted to 0.11 percent of national income (1978, 149). Production could now be expanded to twenty-four hours a day as long as affordable artificial light was available. Assuming the average length of daylight to approximate twelve hours over the year, and ignor-ing time lost by climatic problems in the Eotechnic, doubling the work-ing day potentially doubled production. It certainly amortized capital far better. Human labor, however, became redefined in terms of the machine: if steam power could produce twenty-four hours per day, humans had to follow suit (Mumford, 1934, 176).

The appalling consequences of this machine domination soon became apparent. In every area of human existence, Mumford reminds us, the quality of life deteriorated sharply. Even sex, once used as much for pleasure as for procreation, came to be defined almost solely in terms of the latter. In an important sense the Paleotechnic became ruled by three drives: procreation, money, and death. The owners of industry maximized all three.

A high birth rate meant plentiful, thus cheap, labor. A high birth rate may have come about because, under the factory system, young children contributed usefully to household income, a fact observed at the time (Marshall, 1929, 267). But entrepreneurs strove to maintain a high birth rate, in particular by legally restricting the availability of contraceptive information. Certainly the English population increased more rapidly after 1756 than before. Wrigley and Schofield calculate a gross reproduction rate rising from a low of 1.81 in 1661 to 2.32 in 1756, followed by a sharp acceleration to 3.06 in 1816 and an abrupt fall to 2.37 in 1846 (1981, 229–30). The remarkable peak in fertility in 1816 was accompanied by a decline in infant mortality (Wrigley and Schofield, 1981, 249) to produce a substantial net population gain. In 1661 the English population was 5.14 million, by 1756 it was 5.99 million, by 1816 it had risen sharply to 10.65 million, and by 1846 it was 15.93

Watt steam engine for mills, 1788. This engine generated 10 horsepower at 25 revolutions per minute and was the first that was efficient enough to use removed from the inexpensive coal supply at the top of the mineshaft. Ford Museum, by the author.

million despite the drop in fertility (Wrigley & Schofield, 1981, 208–9). But the increase was not evenly distributed among social groups. We lack figures for the late 1700s, but in the late 1800s the working classes routinely had more children per family: for marriages contracted in the decade 1890–99 there were 5.11 live births to laborers, 4.85 to manual wage earners, 3.04 to salaried employees, and 2.80 to professionals (Wrigley, 1969, 186–87).

Variations in both reproduction and income were commented upon at the time. The duke of Bridgewater, one of the largest mine owners in England, was once told by one of his workers that he was late to work because his wife had given birth to twins that night. The duke philosophically commented, "Aye, well, we have to have what the good Lord sends us." The worker replied, "Ah notice He sends all t'babies to our house and all t'brass to yores" (Mullineux, 1959, 29).

Death was as central to Paleotechnic society as procreation and wealth. Eotechnics was obsessed with the organic world, thus with life. Paleotechnics was obsessed with the mine and the machine. The mine was the grave both in the popular imagination and, all too often, in the harsh reality of pit accidents. Life expectancy dropped in the early Paleotechnic cities compared to the countryside. Humans became expendable components of machine production, valuable only because they could reproduce themselves. The technology of death improved as rapidly as the technology of production. Breech-loading guns began to fire over the horizon as cartographers produced maps that told gunners what was out of their sight. On a face-to-face basis, the machine gun emerged in the late 1800s as the ideal tool of colonial warfare, capable of massacres undreamed of even with repeating rifles and at extremely low cost in terms of the labor needed to work it (Ellis, 1986). All these chickens came home to roost in World War I, none so spectacularly as the machine gun. Millions of men died in trench warfare, cut down by a machine that mass-produced death as efficiently as the Paleotechnic factory mass-produced goods.

For all its increasing power over the inorganic world, the core still needed organic raw materials and food. Better global transport and communications allowed the production of these to be moved to the peripheries. The organic world did not disappear in the Paleotechnic; it merely moved from the core to the colonies and former colonies of the periphery and semiperiphery.

Removing the Ecological Limits

Although the loss of the natural world began in the mines, it accelerated sharply with the harnessing of steam to drive factories in the early 1800s. Few streams were easily dammed, and they were often isolated. Would-be entrepreneurs often had to build whole villages to use water power, such as Richard Arkwright's settlement at Cromford in Derbyshire. In a winter freeze, or in seasons of low rainfall, mills could not work as long as when water was plentiful. High water levels also shut down production. Von Tunzelmann argues from the well-studied Quarry Bank mill for at least a 5 percent loss, and Quarry Bank was well managed (1978, 171). In the United States, where water power was better developed than in England, the waters that flowed freely off the Appalachian plateau to run mills in upstate New York were also needed for the Erie Canal. A series of running battles ensued between local mill owners, who wanted the level of their upland lakes kept up to see them through the dry summer months, and the Erie Canal Commission, which wanted the lakes drawn down to allow traffic to move in the canal.

A second major problem of the Eotechnic was that it was impossible to concentrate large amounts of power at one site to achieve economies of scale and complementarity. Steam allowed large numbers of factories close to each other, whereas in Eotechnic spatial organization they were dispersed along streams, often in hilly districts with high transport costs. In the Bolton district by 1837, an area topographically suited to water power, sixty wheels generated 1,170 horsepower. It would have been impossible to concentrate this much power in many other small areas using water power alone. Economy of scale in interactions between specialized producers could only reasonably be produced with steam power.

A third ecological limit to Eotechnic production was the working day. Until a cheap method of lighting interior spaces was developed, the working day was defined by daylight. In cloudy England, winter work was seriously curtailed because daylight began after 8 A.M. and ended before 5 P.M. Melbin points out in his essay, "The Colonization of Time," how much this limited production (1978). In 1798 William Murdoch first used the gas that was a by-product of the coking ovens to illuminate the Soho factory in Birmingham where Boulton and Watt built steam engines (Armytage, 1976, 108).

Because Eotechnic production derived almost totally from land, the upper limits on the surplus that could be generated were fixed. Smith and Ricardo both suggested ways in which production could be maximized within these limits, and Malthus pointed out that restrictions on population growth made some societies wealthier than others because their consumption per capita could be higher. Yet the upper limit of production was still fixed by nature. Paleotechnics introduced production based on a veritable storehouse of resources from previous ecologies. Wealth thus increased markedly, although it became much more concentrated than in the Eotechnic. Land ownership is, of necessity, spatially diffuse. Those who could tap into the new energy flows, either directly in the case of the mine owners or indirectly in the case of the factory owners, were able to quickly produce wealth beyond anyone's wildest dreams.

Paleotechnic Software Innovations: The Factory System

Eotechnic industry ran mostly on the "putting-out" system, whereby workers in isolated farmsteads contracted to produce goods on a piece-work basis. Putting out was particularly prevalent in the textile trade, in which much human labor went into spinning. Most thread was spun in cottages on hand-powered spinning wheels by spinsters, then taken elsewhere to be woven on hand looms. Unmarried women were the main source of this labor, and the term *spinster* became synonymous with "unmarried woman." "Weavers' cottages" became a common feature of the late Eotechnic landscape in northern England, their upper stories marked by a considerable area of leaded glass to allow adequate light for the fine task of weaving.

The factory system destroyed the putting-out system by concentrating spinning and weaving in one location. It began in the 1730s with improvements in the efficiency of weaving, in particular Kay's flying shuttle. The increased demand of power weavers for spun yarn heavily stressed the putting-out system. Technical improvements in spinning followed in the 1760s, with Hargreave's "spinning jenny" of 1764 and Arkwright's roller frame of 1769 (Rose, 1986, 8–9). To Arkwright, however, must go the credit for the factory system. Both the spinning jenny and the roller frame could as well have been operated by hand. Arkwright concentrated many such units under one roof and used water power and the factory system to ensure better output per unit of labor. In a factory, labor could be "overseen." In a weaver's or spinner's cottage, work proceeded at an organic pace. People worked also at tending their produce garden, their house, and their family. In a factory the foreman, still called an overseer in many northern English mills, ensured that work was constant. As Mumford put it, "Arkwright's great contribution . . . was the elaboration of a code of factory discipline: three hundred years after Prince Maurice had transformed the military arts, Arkwright perfected the industrial army" (1934, 174).

The evils of the factory system became quickly apparent, and it lasted in England in its original form only some sixty years, until the Factory Act of 1833. This radical document specified the ordinary factory working day as from 5:30 A.M. to 8:30 P.M., with one and a half hours off for meals, six days per week; prohibited the employment of children under the age of nine, with minor exceptions; restricted children aged nine to thirteen to eight hours of work per day, and those aged thirteen to eighteen to twelve hours, again with minor exceptions (Marx, 1954, 265). That these restrictions were heavily resisted by entrepreneurs shows the extent of previous abuse. As the report of the inspector of factories in 1860 said, "The fact is, that prior to the Act of 1833, young persons and children were worked all night, all day, or both *ad libitum*" (quoted in Marx, 1954, 264).

Paleotechnics concentrated industrial production spatially and temporally. Steam power allowed the spatial concentration of inanimate power in new industrial cities, which made for considerable improvements in economy of scale as well as productivity. The spatial concentration of labor in factories allowed a considerable increase in output per worker. The addition of steam power to the factory allowed twenty-four-hour production, at least until minimal legal curbs were applied.

The British Maritime World System, 1800–1900

The Paleotechnic world system was economically, socially, and politically dominated by Britain although British hegemony was not firmly established until 1815 and was under challenge by 1883. British industrial production grew rapidly at the turn of the nineteenth century, far faster than agricultural production. The software innovation of the factory system had little impact on the farm because of the seasonal rhythm of agrarian life. Two crops dominated Paleotechnic capitalist agriculture. Wheat fed the growing industrial working class, and cotton provided the raw material for the factories in which they worked. There was a crude feedback loop between the two crops. More cotton meant more industrial workers, who consumed more wheat.

The demand for both crops was increased by steam power, but so was the supply. Steam-powered machinery spun and wove far more cotton than water-powered machinery could. Steam-powered boats and railroads made the great continental interior of North America accessible to capitalist agriculture. Steamships moved cotton and wheat across the Atlantic faster than was previously possible, and they moved cotton textiles to market around the world.

In the Paleotechnic geographic expansion was a constant necessity to keep pace with the demand for food and raw materials. Cotton was a subtropical crop that could not be grown at home. Despite the best efforts of such British agricultural reformers as Coke and Townsend in the late 1700s, British wheat agriculture simply could not keep pace with a rapidly growing population, and Britain steadily moved toward dependence on imports. In 1805 Britain was still virtually self-sufficient in carbohydrate production. In 1880 it was producing 1,585 calories per head per day and importing 1,227; by 1910 the figures were 897 and 1,617. Wheat could be grown much more cheaply in North America, Argentina, Australia, or Russia, and import dependence was not as strategically unsound as it might seem. As Offer has noted (1989) the ready market for North American and Australian wheat increased the willingness of Britain's major trading partners to fight imperial Germany in World War I. Even so, as geopoliticians such as Mackinder recognized at the time (1900), the continuation of import dependency depended upon the continuation of imports. When the British merchant shipping fleet was threatened by imperial German U-boats, the whole system trembled on the brink of disaster.

Europe as a whole did not move toward dependency on imported food in the Paleotechnic. Countries such as Germany that were without powerful fleets to ensure the free flow of commerce increased the productivity of their domestic agriculture and were willing to substitute other, coarser grains for wheat, or even other crops, such as the potato.

Between 1815 and 1913 cereal production in Europe tripled while population doubled (Pounds, 1985, 188). Even so there was a 12 percent shortfall of production relative to demand by 1913, in part because more grains were being consumed by animals. Productivity per hectare increased, although not so spectacularly as production, which spurted because new technologies enabled lands hitherto unsuitable for cereals to be brought into production. The gains in wheat productivity in Europe are in marked contrast with their absence in North America, where production increases came from the opening up of new lands by the rapid expansion of the railroad net (table 2-1).

The British Isles did not become a major food importer until the 1840s, in part because of productivity gains at home and in part because Ireland could be relied upon for wheat production. Only with the failure of the Irish potato crop and the consequent starvation and death or migration of much of the Irish labor force did Britain turn outside the British Isles for its cereal grains; it repealed in 1846 the tariff on imported wheat known as the Corn Laws. In the late 1800s lowered food costs for the industrial work force became an important component of British struggles to retain hegemony as the productivity of British industry declined relative to that of Germany and North America. Cheap wheat from Russia and North America and meat from Australia and New Zealand allowed money wages to fall in late nineteenth-century England while real wages rose. North America could produce wheat and meat cheaply in the internal peripheries of the plains and the ranching West. German labor had to accept rye or potatoes rather than wheat bread, although German domestic agricultural productivity per hectare rose markedly throughout the 1800s. Germany stressed self-sufficiency rather than low cost and produced far more of the food needed by its workers than did Britain (table 2-2).

The Wheat Economy of the Republic of the North

After 1846 wheat from any part of the United States was free to enter Britain with no tariff. The wheat frontier exploded westwards as new railroads were thrust out into productive virgin soils. Whereas in 1840 Buffalo, New York, was receiving a little more than 27,000 tons of grain a year, by 1850 that figure had risen to more than 95,000 tons, and

Table 2-1 Productivity in small-grain cultivation, 1850–1970 (metric tons per hectare)

Year	Germany[a] (wheat)	France (wheat)	United States (wheat)	Canada (wheat)	Japan (rice)
1850	1.05	1.11	—	0.68	—
1870	1.23	1.07	0.82	0.89	—
1900	1.82	1.29	0.82	0.89	1.57
1910	1.90	1.05	0.92	1.00	1.71
1930	2.27	1.16	0.95	1.14	2.53
1950	2.58	1.78	1.11	1.15	3.27
1960	3.56	2.53	1.70	1.42	4.01
1970	3.79	3.45	2.09	1.79	4.28

Sources: Allen, 1981; Mitchell, 1980; Urquhart, 1965; U.S. Bureau of the Census, 1975.

[a] West Germany only after 1945.

by 1855 it had reached almost 220,000 tons (Meinig, 1966, 167). (Throughout this book "tons" are metric tons.)

As the 1800s wore on imperial Germany and the American republic, though mainly that section of it which made up the republic of the North, challenged British hegemony. Wallerstein points out that nations first achieve hegemony in agroindustrial production, and the successful agricultural development of the Great Lakes region is critical to understanding the U.S. challenge. Although the challenge was based on a wheat economy, it was one with very specific regional characteristics: the ready availability of development capital, transportation, and labor. Critical arguments toward this conclusion have been advanced by three geographers: Michael Conzen, James Vance, and Carville Earle.

The Ready Availability of Capital

Regional supplies of capital in the American republic varied. The South was a net importer, mostly from Britain. New England was able to generate its own capital as early as the 1730s. Shipbuilding and Atlantic trading paid handsome dividends. Maritime New England became the source of capital for the entire republic of the North. Conzen has demonstrated that Wisconsin banks were almost entirely financed from Connecticut and Massachusetts on the basis of kinship by 1850 (1975). As the children of well-off New Englanders moved west, they borrowed development capital from back east (Hugill, 1977, 117, 292–93). Over time this pattern formalized into family banking systems. Kinship links for capital transfer made a great deal of sense in a society in which capitalist institutions were not yet fully developed. Americans resisted a national police force until the formation of the FBI under the shock of Prohibition-induced gangsterism in the 1920s. Meanwhile bankrupts and debtors had plenty of frontier to hide on, and such private police

Table 2-2 Food crops produced per capita per year, 1880–1960, selected polities (millions of kilocalories)

Year	United States	Britain	Canada	Germany	Japan
1880	4.5	0.6	2.3	1.4	0.5
1890	3.7	0.5	2.4	1.6	0.5
1900	4.8	0.4	3.3	1.8	0.6
1910	4.8	0.3	4.0	1.9	0.6
1920	4.5	0.4	6.8	1.0	0.7
1930	3.2	0.3	6.4	1.6	0.6
1939	3.4	0.4	6.2	1.7	0.6
1950	4.1	0.6	5.4	1.0	0.5
1960	3.7	0.6	4.5	1.2	0.6

Sources: Bank of Japan, 1966; Mitchell, 1980; Urquhart, 1965; U.S. Bureau of the Census, 1975.
Amounts reported are corrected for losses in processing. Approximately 1 million kilocalories of food feeds an adult comfortably for one year. Food crops include barley, corn, millet, oats, Irish potatoes, sweet potatoes, rice, rye, soy beans, sugar beet, and wheat.

forces as Wells Fargo and Pinkerton could extract them only with difficulty: "gone to Texas" was a common cry of those in financial trouble in the 1800s.

A second element in this regional availability of development capital was broad ideological agreement on how capital should be treated. New England's Puritan settlers assured the implantation there of one of capitalism's driving forces. In Puritan society money worked for the glory of God, just as did humans. Less radical forms of Protestantism crept into the region in the 1820s (Ellis et al., 1967, 196–98) but were countered by a burst of communitarianism in the 1830s and 1840s that refocused and continued Puritan fervor (Cross, 1950). Communitarian Protestantism deplored the atomistic individualism of early nineteenth-century capitalism (MacPherson, 1962) but never abandoned the attitude that money should work. The only difference between the financial behavior of the Latter-Day Saints and the Calvinist saints is that the Mormons believed in the concentration of wealth and capital in the church, not the individual. As Walzer (1971) has emphasized, the Calvinist saints were revolutionaries in the full sense of the word. They radically reinterpreted royal rights when they beheaded England's Charles I in 1649. The American Revolution was a logical extension of the English civil war (Bendix, 1978, chap. 9). The Calvinist saints believed that virtue was an automatic part of the godly life, but the framers of the Constitution recognized the need to impose virtue on the vast majority (Diggins, 1986, 101). Tocqueville made an exception for the frontier, where opportunity and abundance made restraints on political or economic behavior less necessary (Diggins, 1986, 100–101). This frontier image dominated U.S. society throughout the 1800s and is regionally significant even today. The shift of the idea of Calvinist virtue to the idea of self-interest was crucial to the spread of capitalism outside the narrow religious group of the Puritans. As long as the frontier offered abundance, "nature would absorb politics and allow economic activity to have free rein" (Diggins, 1986, 121). In the late 1800s the closure of the frontier brought on the latent crisis of U.S. democracy identified in Turner's famous essay (1893). Such a crisis was implicit in the imposition of a Constitution demanding strict Calvinist virtue on a frontier where, as Tocqueville so cogently put it, it was "not that virtue is great, but that temptation is small."

Toward the close of the century, as Turner delivered his address, the farmer came to see himself as a Jeffersonian yeoman menaced by international money power. The drama of folkish virtue standing valiantly against commercial vice has classical allusions, and thus one might see in the "adversary" writing of Henry George, Edward Bellamy, and Henry Demarest Lloyd traditional republican fears of imperialism and standing armies and prophetic forebodings of the corruption of politics by business and commerce. But such writers were also steeped in Christian and Emersonian idealism, and their preoccupation with the abuses of wealth, capital, property, mortgages, rent, and "unearned income" followed the spirit of Lincoln and Paine and the labor theory of value which was central to Lockean liberalism (Diggins, 1986, 123).

The expansion of New England after 1793 was, in fact, the first westward expansion of a politically sophisticated culture in the United States. Even with the Constitution they had written to back them up, owners of land and capital stepped warily. Detailed examination of the records of one major landowning family in upstate New York, whose land sales evolved into a family bank, part of which later evolved into the New York Security and Trust Company, makes plain the fears of capitalists. When times were hard payment was taken in kind, usually just enough cattle to cover interest. Letters attest to a refusal to make loans on properties that immediate family members could not visit on a regular basis. Yet there was a constant exception. Loans were routinely made to kin in Illinois and Wisconsin, often of substantial amounts (Hugill, 1977, 292–93). In a society with populist tendencies, where the frontier beckoned the bankrupt, and where there was no national police force, such policies were no more than sound business.

Transportation and Cities

The extension of New England also relied on the existence of cities to provide a wholesaling function, aggregating the surplus product of a myriad family farms into large shipments for the Atlantic trade. Vance has argued cogently for the central importance of this wholesale function, extending it to include an important role for merchants in the capital supply system (1970). Merchants needed to put up a great deal of money at harvest season, but they sold the stocks from their warehouses steadily over the whole year. This practice had several advantages: it focused a second level of profit in the region beyond that accruing to producers, and it generated development capital. Merchants frequently ended up as bankers in the early American republic, providing capital for other ventures, notably manufacture of agricultural implements and improvements in the infrastructure. These ventures further accelerated regional economic growth. The expansion of the northern American wheat frontier and of the republic of the North was a highly urban expansion, at least until it reached Minnesota, after which better communications in the form of railroads and mail-order systems allowed farmers moving further west to look back east for urban services (Borchert, 1987, 61–67). Minneapolis–St. Paul was the last great urban center of the northern plains, although the Canadian border helped create the need for a subsidiary center at Winnipeg to serve the prairie wheat expansion north of the forty-ninth parallel. Functional links between Minneapolis–St. Paul and Winnipeg in any case were strong; the former is still the principal banker for the region (Hugill & Everitt, 1992). Only the relatively late expansion of Canadian prairie wheat farming driven by European demand in World War I, after the Canadian border had become more formalized, encouraged the growth of Winnipeg. In Canada the wheat frontier was not closed until the early

1930s, the last of the virgin prairie northwest of Edmonton being opened in the Peace River valley around Dawson Creek and Fort St. John only in 1931.

Regional wealth was also invested in infrastructure. Vance says that merchants of "misplaced colonial cities" were searching for access to the continental interior (1990, 267–69). This process began haltingly after the Revolution but accelerated sharply once Britain returned to the international marketplace in 1815. Until 1776 eastern seaboard cities prospered in the triangle trade of the Atlantic. After the Revolution, American ships were no longer protected by the Royal Navy and the Navigation Acts kept them, at least legally, out of British ports. U.S. ships were licensed to trade with British ports during the Napoleonic Wars, when Britain had higher priorities than trade, and U.S. merchant shipping filled a crucial need. With Napoleon defeated, however, events caught up with U.S. merchants. Although they flouted the Navigation Acts, in particular in the West Indian and Indian trades, the demand for their shipping was substantially reduced after 1815. U.S. merchants turned their attention inland, seeking to encourage settlement both to generate surplus agricultural product for trade and to create a larger U.S. market now that the British market was more difficult of access. But of the major seaboard cities only New York had viable access to the interior, given the transport technology of the very early 1800s.

New York's Erie Canal, completed between Lake Erie and the Hudson in 1824, caused panic in the mercantile establishments of Baltimore, Boston, and Philadelphia. New York could now annex the trade of the vast Great Lakes region. Philadelphia's merchants pushed for construction of the Pennsylvania Mainline Canal, begun in 1826 and designed to connect Philadelphia to the Great Lakes via Pittsburgh. This Rube Goldberg system had far too many transfer points: from rail to canal at Columbia, 132 kilometers west of Philadelphia; from canal to portage railroad at Hollidaysburg to cross the 708-meter summit of the Allegheny front; and from portage railroad back to canal from Johnstown to Pittsburgh (Vance, 1990, 130–32).

Railroads needed no transfer points as long as route engineers could keep grades reasonable. The first trans-Appalachian line was completed in 1842 with the bridging of the Hudson at Troy, New York, linking Boston to the short, intercity lines that crossed the Lake Ontario shore plain to Buffalo (Ellis et al., 1967, 251–52). Six more trans-Appalachian lines were completed between 1851 and 1853: Montreal to Portland, Maine (the St. Lawrence and Atlantic); Wheeling on the Ohio River to Baltimore (the Baltimore and Ohio); Buffalo to New York City via a consolidation of the Lake Ontario shore plain short lines and a line down the Hudson Valley from Albany (the New York Central); Dunkirk on Lake Erie to New York City (the New York and Erie); Pittsburgh to Philadelphia (the Pennsylvania Railroad); and Chattanooga to Charleston and Savannah (the Western and Atlantic) (Vance, 1990, 286–90). Only the last of these six did not serve a northern route into the Great Lakes or the Ohio River, and only the first and last did not serve one of the four "misplaced cities." Completion of the line along the Hudson from Albany to New York came when New York merchants realized how much upstate trade was being diverted to Boston (Meyer, 1948, 365–66; Ellis et al., 1967, 251).

As westward expansion continued and merchants competed for hinterlands for their cities, the North developed an exceptionally fine-textured network of railroads. The fineness of texture was crucial because animal haulage of farm surplus became prohibitively expensive once distances ranged much above 16 kilometers. A network of railroads 30 to 50 kilometers apart spread through the wheat-producing region, first in the states bordering the Great Lakes to the south, then in those to the west. This net was financed by readily available regional capital, the rising profits from implement manufacture and infrastructural investment, because merchants understood that if cities failed to penetrate an agricultural hinterland they would quickly lose business.

Labor

The agricultural labor system of the republic of the North was defined largely by the ecological demands of the principal crop, wheat, which requires a great deal of attention at the harvest season, a little at planting, and almost none the rest of the year. Because new land was available for wheat until well into this century, there was no need to maximize labor productivity by using labor year round, as in European agriculture. Earle has compared the intensely focused temporal ecological requirements of wheat to the more spread out requirements of tobacco (1978, 57). Both crops needed about twenty-five man-days of labor per year, but tobacco needed it spread over about seventy-five days whereas wheat required only twenty-five days of attendance. The intense temporal focus of wheat argued for family farms as the principal unit of labor, with wage labor hired to cope with the seasonal peaks. This seasonal labor was urban rather than rural, and it earned high wages in the two months of harvest. It could be employed at subsistence rates in cities the rest of the year. This drove the labor cost of industrialization down in the republic of the North. Paradoxically, but logically in economic terms, reduced labor costs allowed U.S. entrepreneurs to devote more of their profits to investment in machinery, which displaced cheap labor and helped keep the price of labor down (Earle & Hoffman, 1980). The labor system of the wheat-producing North had an elegance that does much to explain the remarkable success of the United States in the agroindustrial arena during the 1800s. High-cost harvest labor encouraged farmers to replace humans with machines. Low-cost urban labor encouraged manufacturers to produce agricultural machinery. Agrarian and industrial profits were reinvested in the region and its infrastructure. The kin-based system of capital transfer further retained capital in the region, since few people moved far from their section of birth.

The agroindustrial supremacy of the republic of the North developed because of the intersection of five forces. First, the repeal of Britain's Corn Laws opened the profitable British market to western U.S. wheat. Second, the republic of the North had adequate internal capital for regional development, in part because of its Puritan heritage, in part because of the region's lucrative shipbuilding and involvement in the Atlantic trade. Third, the temporal-ecological labor requirements of wheat provided a cheap labor supply for industrialization. Fourth, new wheat lands were easily available to the west and north, which caused farmers both to resist the use of agricultural labor on a year-round basis and to take exhausted eastern lands out of wheat production. The re-placement of wheat by dairy farming in upstate New York and by corn in Ohio, Indiana, and Illinois encouraged entrepreneurs to produce an ever more complex range of farm machinery. Finally, merchants in different cities competed to expand their agricultural hinterlands by infrastructural investment in the railroad net.

The Cotton Economy of the Southern States

The other great crop in nineteenth-century U.S. agriculture was cotton. Arkwright's factory system increased demand for cotton so much that by 1801 cotton textiles had displaced woollen textiles as the leading sector of the British economy. Whereas most of Europe could keep sheep, none of Europe could grow cotton. Like sugar, cotton was a product of global capitalist agriculture, and it was to the American tropics that European, mostly English, entrepreneurs looked when they sought to develop the cotton trade. Cotton grew well all over the Carib-bean. In the later years of the Napoleonic Wars cotton rapidly displaced sugar on many islands.

Tropical America had been increasingly unstable politically since the American Revolution (Meinig, 1986, 332–38). Political turmoil and soil exhaustion caused by sugar monoculture reduced the region's utility to Europe. Sugar required heavy capital investment in land, slaves, and the sugar factory itself, and the levels of capital needed were not avail-able locally. It was European bankers and merchants rather than Carib-bean residents who profited most from sugar. Cotton needed only land and labor, and even the gin was not much of a focus for slave resentment. Even so the Caribbean was incapable of meeting the rising demands of British factories.

It was on the subtropical margins of the Caribbean rather than on the islands themselves that the answer was found. Between 1783 and 1836 the requisite combination of technology and geopolitical control came into being in what are now the southern United States. Britain was ousted from Florida in 1783. In 1803 the Louisiana Purchase gave the American republic control of the Mississippi. In 1836 Texas was opened for U.S. settlement unhindered by Mexican vetting of would-be settlers, oaths of loyalty, and conversion to Catholicism. Texas independence was a minor theme, subordinated to the American republic's thrust into the subtropics to control cotton production. South of the Appalachians the Creek Indians also resisted the expansion of the American republic. Jackson's military campaigns of 1813 and 1814 opened most of Alabama to cotton production, and his battle-hardened army smashed the last European challenge to the American republic's control over its subtropical continental periphery at the battle of New Orleans in 1815 (Morison, 1965, 393–95).

Four other forces were also at work to control the growth of the cotton economy. Whitney's patented cotton gin of 1794 substantially lowered labor requirements after picking by speeding the separation of seeds from lint. But the gin does not deserve the pride of place still often given it (e.g., Hindle & Lubar, 1986, 81). Cotton was not perishable and needed only to be kept dry after harvest. Once the first frost had killed the cotton flowers, slaves could have spent several of the winter months removing the seed by hand. The task would have been onerous and the capital investment in watertight storage facilities high, but the labor was paid for.

Second, and far more important than the gin, was the emergence of a viable hybrid form of the major tetraploid American cotton, *Gossypium hirsutum*. Between 1800 and 1820 the numerous varieties of American cotton were crossbred to come up with "upland" cotton, a variety that yielded plentiful lint of a medium staple and that could tolerate some frost and relatively saline soils (Fryxell, n.d., 171–73). Upland cotton can be grown as far up the Mississippi as western Tennessee, the limit being reached around the line of two hundred frost-free days. Third, the river systems of the Gulf Coast region offered remarkable access to this region once Shreve in 1815 developed a suitable steamboat for the Gulf Coast drainage.

The fourth control on the spread of the cotton economy was the lack of an adequate labor force. Importation of new slaves was illegal by 1814 (Knight 1970, 50). Needed slaves were bred in the region, a system rife with abuses, but which created by far the healthiest slave population in the Americas. Many were sired by their masters, a habit reflected by the fact that 13 percent of the black population of the American republic was mulatto in 1860 (Morison, 1965, 506). Some large owners specialized in breeding, especially those in the eastern South, where soil exhaustion from continued cotton monoculture was creating severe yield reductions as early as the 1830s. There was a time lag in the breeding cycle because slaves had to be nine or ten years old before their labor was of value. The price of slaves rose constantly before the Civil War. A prime field hand aged between eighteen and twenty-five was worth $500 in 1832, $1300 by 1837, and $1800 by 1860 (Morison, 1965, 504).

Cotton production expanded steadily in the American republic. It reached 100,000 500-pound bales (22,680 tons) in 1801 (table 2-3). British mills consumed 24,000 tons that year, at a time when the republic of the North was just beginning to spin cotton. Because the republic of the North lagged in applying steam power to its mills to gain the advantages of economies of scale and geographic concentration,

U.S. cotton consumption expanded no more rapidly than that of Britain until after the U.S. Civil War (table 2-4). In 1860 Britain consumed more than two and a half times as much cotton as did the republic of the North. U.S. consumption surpassed that of Britain for the first time in 1899.

The American republic dominated the world's cotton markets through the nineteenth century. In 1860 two-thirds of the world's cotton, and more than three-quarters of the cotton entering world trade, was grown in America. Eighty percent of British imports in 1860 came from the United States (Farnie, 1979, 137–38). In a world dominated by the ideal of free trade U.S. ability to grow huge quantities of cheap, high-quality cotton deterred all but the most specialized competitors, such as Egypt with its fine-quality long-staple cottons, from entering the market.

The Labor System in the South

U.S. cotton production was never highly profitable because it emphasized the value of slaves rather than land. Slave owners saw their capital stock increasing in value over time, much as does a houseowner in an inflationary economy (Wright, 1978, 149). The dangers were

Table 2-3   Production and export of cotton by the United States, 1794–1988 (average per year, in thousands of tons)

Year	Production	Export	% Exported
1794–1800	7	4	57
1801–1810	33	19	58
1811–1820	53	31	59
1821–1830	128	92	72
1831–1840	255	196	77
1841–1850	445	317	71
1851–1860	740	535	72
1861–1865	405	31	8[a]
1866–1870	651	336	52
1871–1880	1048	645	62
1881–1890	1519	967	64
1891–1895	1828	1293	71
1896–1900	2276	1466	64
1901–1910	2568	1695	66
1911–1920	2958	1727	58
1921–1930	2988	1757	59
1931–1940	2973	1480	50
1941–1945	2671	651	24
1946–1950	2791	955	34
1951–1960	3170	1105	35
1961–1970	2683	931	35
1971–1980	2583	1138	44
1981–1988	2769	1257	45

Sources: Statistical Abstract of the U.S.; U.S. Bureau of the Census, 1975.

[a]South blockaded by federal navy during U.S. Civil War.

obvious, even had abolition not occurred. Because ecologically the system deemphasized the role of land, soil exhaustion seemed irrelevant.
Slaves were portable wealth that could easily be moved to a new production region further west. Deemphasizing land also meant deemphasizing the public and private infrastructural investments of the wage-labor
wheat economy of the republic of the North (Wright, 1978, 129).

The contrast between cotton South and wheat North is even more
extreme than Wright suggests. From the perspective of the British bankers who financed the cotton South, it made more sense to warehouse
cotton near the Lancashire mills than at the Gulf Coast shipping ports.
The resulting lack of capital accumulation in the region was a serious
problem. Before the onslaught of the boll weevil around 1900 the picking season for "long-season" cotton was ideal for slave labor. Cotton
flowered as many as five times from the end of August to the first frost
and could be picked by a relatively small labor force. The subtropical
climate also produced a magnificent crop of weeds unless the field was
regularly "chopped," that is, hoed and weeded. Labor was needed eight
to nine months of the year (Hugill, 1988d). Earle and Hoffman have
demonstrated convincingly the high price of labor in the South and the
consequent lack of industrialization (1980), but even their argument is
not complete. The complex that emerged around wheat in the republic
of the North included mechanization of agriculture, which failed to

Table 2-4 Consumption of raw cotton, 1800–1988, selected polities
(thousands of tons)

Year	United States	Britain	France	Germany[a]	Japan
1800	—	24	—	—	—
1811	—	56	8	—	—
1820	—	54	19	—	—
1830	—	112	34	—	—
1840	—	208	53	9	—
1850	—	267	59	16	—
1860	192	492	115	67	—
1870	181	489	59	81	1
1880	340	617	89	137	1
1890	571	755	125	227	31
1898	787	799	176	322	153
1899	833	799	175	295	208
1900	836	788	159	279	157
1910	1088	740	158	383	297
1920	1534	783	202	139	470
1930	1567	577	361	346	574
1938	1466	503	288	353	562
1950	2374	461	252	189	356
1960	2375	278	298	419	753
1970	1986	172	248	347	835
1980	1377	69	—	—	795
1987–88	1579	43	—	—	899 ('87)

Sources: Bank of Japan, 1966; CSO *Annual Abstracts;* Mitchell, 1980; *Statistical
Abstract of the U.S.; Statistical Handbooks of Japan;* U.S. Bureau of the Census, 1975.
[a] West and East Germany combined after 1945.

develop in the South. One of the rewards of mechanization was that, as the wheat frontier moved west, agricultural implement manufacturers diversified into new machinery to supplement or even replace their trade in wheat equipment. But mechanization only made sense where a farmer was attempting to reduce dependence on high-priced wage labor needed only for the harvest. Slaves represented capital, not a continuing item on the bill under labor; therefore, they were better worked all year.

Transport in the Cotton Region

The South also faced severe transportation problems once the riverine lands were planted in cotton. West of the Red River, riverboats were unable to penetrate far inland and railroads were needed (Hugill, 1988d). The Texas cotton lands could only be fully exploited by railroads. The lower Brazos Valley was one of the most productive cotton regions in the antebellum period, and the first capital of the fledgling Republic of Texas was located at Washington-on-the-Brazos, the head of steamboat navigation just south of Hidalgo Falls. Railroad construction to open the upper Brazos Valley began in the late 1850s in the form of the Houston & Texas Central, aiming for Waco, then Dallas. Although interrupted by the Civil War, the railroad was completed as soon as the war was over, and the northern Brazos Valley became the major cotton-producing region in Texas, a position it held until after the turn of the century. By the late 1870s Texas was producing more cotton than any other state.

New forms of transport created a new regional geography. Steamboats plying the lower reaches of the rivers of the Texas coast moved easily to Galveston, on the northeastern end of Galveston Island. Galveston was the only site on the Gulf Coast west of the Mississippi delta which offered some shelter from gulf hurricanes combined with deep enough water for oceangoing ships. Railroads did not serve Galveston so well, and the trestle to the mainland opened in 1859 was a major choke point. The logical railroad focus was Houston, "Where Eleven Railroads Meet the Sea." Yet Houston did not displace Galveston for a long time. The hurricane that flattened Galveston in 1900 increased rather than reduced Galveston's importance (Walden, 1990). Dredging the harbor to raise the island allowed the much larger steamships of the early 1900s to dock directly. As ships grew in the late 1800s Galveston had had to lighter cotton out until its harbor was deepened, first by a 7.62-meter ship channel dredged in 1896, then by the 1900 hurricane. Houston achieved direct docking only in 1914, when its channel was deepened to 7.62 meters (Buenger & Pratt, 1986, 22).

Financing the Cotton Economy

Finance had much to do with Houston's rise. Even with modern production techniques, cotton is a "gambler's crop" and requires a large supply of capital. Climatic variation causes wild annual production swings. An early November frost in long-season cotton could prevent one or even two pickings, a loss of 20 to 40 percent of the crop. The replacement of slavery with sharecropping after the Civil War also increased the region's capital requirements. Slaves grew enough food for themselves. Sharecroppers had to grow only the cash crop of cotton. Land and seed were provided only on this basis, and tenants became dependent upon cash advances against their potential crops. At its worst this system was debt peonage, with tenants "enslaved" to the land by the constant cycle of debt. In practice, if any alternative employment became available, lenders could never really collect. In the nineteenth century such alternatives were sadly lacking, and the tenancy rate increased with remarkable speed after the Civil War (Fite, 1984, chap. 1). Only after 1900 and, to begin with, only in certain geologically favored regions of the Gulf Coast did alternative employment appear. The 1901 Spindletop oil boom in eastern Texas and related oil booms in neighboring states helped. World War II saw a government policy of diversification of war production away from the East and West coasts to the Gulf Coast, and the process of southern urban industrialization began in earnest.

Before the Civil War the South got its capital from Britain (Fleming & Tinsley, 1966, 3–4). After 1870 it turned slowly but inexorably to the republic of the North, with the shift becoming very rapid after 1919 (Buenger & Pratt, 1986, 20). The east Texas oil boom provided a source of development capital within the region by the 1920s because the core banks avoided oil finance as too speculative. The cotton South thus moved from the British periphery to the periphery of the republic of the North, then developed semiperipheral status early in this century. For all their strength in the 1800s, Liverpool's cotton banks were devastated by the restructuring of the British tax system in 1919 to help pay for World War I, and most dissolved their businesses (Fleming & Tinsley, 1966, 8). New York bankers stepped into this breach, taking high profits from a charge of $3 to $5 per bale on all cotton leaving the South as if it were physically shipped to New York when it was not. Despite this preponderance of finance from the rising new core, major cotton banks were established in some parts of the region. The most important was Anderson, Clayton and Company, which began in Oklahoma City in 1904 and moved to Houston in 1916. Fueled by oil, Anderson, Clayton had, by 1928, enough capital to break free of the New York bankers (Buenger & Pratt, 1986, 65–67).

In fact, the South had begun to change just before the oil boom because of the arrival of the boll weevil from Mexico. Texas had to come to grips with the boll weevil first and did so at the Texas Agricultural and Mechanical College (now Texas A&M University). Because the boll weevil matured slowly, it could be controlled by switching from long-season cotton to short-season cotton, on which all the buds matured together and the weevils had little time to develop. But the concentrated harvest created a serious labor shortage. After 1900 cotton came to be grown more as a cash crop in small amounts on family farms than as the only crop on sharecropper plantations. The changed labor demand curve also implied mechanized harvesting. International Harvester accelerated efforts to build a mechanical cotton stripper in World War I, when labor became scarce. But while picking by hand separated the boll from the leaves and stalks, machine stripping did not. Leaves contained too much moisture and thus caused trouble in the ginning process. In the early 1940s the Texas Agricultural & Mechanical College research stations, working with implement and chemical manufacturers, came up with the requisite combination of chemical weed control during the growing season, chemical defoliation just before harvest, and mechanical stripping (Hugill, 1988d). Once this was achieved cotton became a capital- rather than labor-intensive crop, just as wheat had in the mid-1800s.

Wallerstein's arguments about the structural primacy in the periphery of fixed-cost labor systems must thus be tempered in two ways. Slave labor represented both fixed and running costs: fixed costs in that slaves were capital; running costs in that they had to be fed, housed, and clothed. Sharecropping reduced capital costs but retained high running costs. Wage labor reduced both. For a crop with a harvest period concentrated in time, such as wheat, wage labor was optimal, sharecropping suboptimal, and slavery useless. The retention by Poland and Russia from the sixteenth to the nineteenth centuries of a suboptimal labor system doesn't really contradict this. Unlike the republic of the North, the Baltic wheat region had little surplus capital and borrowed what it needed from abroad. Without regional capital, regional investment in industry or infrastructure was minimal, so that the advantage for industrialization of paying subsistence wages to the harvest labor force during the nonharvest months was never exploited. Thus the temporal ecology of the crops grown and the structure of the regional economy assume a much more significant role in determining labor systems than do structural relations between the core and the periphery.

THE SLOW MOVE TO THE NEOTECHNIC

The switch to Neotechnics was unmarked by anything like the resource crisis that marked the switch to Paleotechnics. By the late 1800s the Paleotechnic seemed securely established. The Bessemer converter, coupled with the Thomas and Gilchrist process for smelting previously unusable iron ores, had averted a potential resource crisis as the supply of high-grade ores was used up. Railroads carried heavier loads on steel rails, drawn by the more powerful locomotives improved materials technology made possible. Cheaper, more malleable, easily handled steel plate allowed considerable acceleration in ship production. More efficient steam engines powered faster and more reliable ships and locomotives. Coal production easily kept pace, although some of the older producers, such as Britain, began to suffer competition from low-cost producers such as Germany and the United States (table 2-5). The factory system settled down, in particular in Britain, where its worst excesses were curbed by the sequence of Factory Acts begun in 1833. Labor unions had come into being, but they still lacked political muscle. From a British perspective the late 1800s was unalloyed progress, in particular in the material aspects of life. Real wages increased substantially in both industry and agriculture, as did wealth; regional inequities were reduced; pauperism declined substantially (Langton & Morris, 1986, chaps. 6, 20, 3, 21).

Yet there were problems of two main types: spatiotemporal and geopolitical. In spatial terms, long-distance trade and transportation were relatively easy, but most of the population still had to walk to work, and industrial development depended on access to the rail net. Even in small, rich Britain, railroads were by no means ubiquitous: in poorer polities lack of capital meant huge gaps, and in sparsely settled ones the market could not support ubiquity. Steam power suffers severe diseconomies at smaller scales. The second law of thermodynamics ensures that small steam engines can never convert fuel to work as efficiently as can large engines. The efficiency of the steam engine depends upon the size of the heat reservoir (the boiler) and the size and temperature of the

Table 2-5 Inanimate energy consumption per capita per year, 1820–1980, selected polities (millions of kilocalories)

	United States				Britain			
			Natural				Natural	
Year	Coal	Oil	gas	Total	Coal	Oil	gas	Total
1820	0.23	—	—	0.23	9.05	—	—	9.05
1830	0.45	—	—	0.45	10.00	—	—	10.00
1840	0.94	—	—	0.94	12.81	—	—	12.81
1850	2.35	—	—	2.35	16.22	—	—	16.22
1860	4.15	0.02	—	4.17	23.22	—	—	23.22
1870	6.65	0.19	—	6.84	28.17	0.04	—	28.21
1880	10.30	0.73	—	11.03	32.18	0.05	—	32.23
1890	16.23	0.99	—	17.22	34.20	0.13	—	35.33
1900	22.63	1.14	0.46	24.23	36.19	0.29	—	36.48
1910	34.52	3.18	1.50	39.20	36.60	0.35	—	36.95
1920	38.43	7.26	2.07	47.76	35.16	0.80	—	35.96
1930	27.54	10.88	4.37	42.79	31.01	2.00	—	33.01
1938	19.24	12.81	4.93	36.98	30.26	2.86	—	33.12
1950	23.22	20.62	11.26	55.10	30.51	3.99	—	34.50
1960	31.50	63.71	39.62	134.84	27.01	10.43	—	37.44
1970	22.86	54.86	40.52	118.23	18.93	21.02	1.98	41.93
1980–82 (1980)	22.51	49.90	29.83	102.24	12.15	14.36	7.66	34.17
1986–88 (1988)	24.09	43.57	23.89	91.55	12.87	14.01	9.09	35.97

Sources: Bank of Japan, 1966; *Eurostat;* Mitchell, 1980; *Statistical Abstract of the U.S.; Statistical Handbooks of Japan;* Thirring, 1958; U.S. Bureau of the Census, 1975.

Consumption = domestic production of coal, oil, and natural gas, less exports, plus imports.

exhaust (which represents heat lost). Large engines are more efficient simply because a lower percentage of heat is lost upon each extraction of energy from the reservoir to do work. The steam engine also needs a substantial weight of auxiliaries to increase its efficiency. A railroad locomotive is not as efficient as a ship because it cannot usually afford to carry around the weight of a condenser, which recycles not only water but also heat that is otherwise lost in the exhaust (Thirring, 1958, 20–21, 53). The most efficient form of steam engine, the turbine, was not developed until the 1890s and better represents Neo- than Paleotechnic technology. Turbines are even more subject to scale effects than steam reciprocating engines (Thirring, 1958, 67).

Temporal problems were never as well recognized as spatial ones, yet control of time as well as space has long been a hallmark of capitalism. Mumford described the monastic development of clocks to time the hours for prayer as a far more important cultural precondition for the modern industrial age than the steam engine (1934, 14). The clock denied the natural rhythms of the organic world—days, seasons, and human feelings—through its arbitrary imposition of an ordered incre-

France				Germany[a]				Japan			
Coal	Oil	Natural gas	Total	Coal	Oil	Natural gas	Total	Coal	Oil	Natural gas	Total
0.26	—	—	0.26	0.38	—	—	0.38	—	—	—	—
0.56	—	—	0.56	0.46	—	—	0.46	—	—	—	—
0.90	—	—	0.90	0.79	—	—	0.79	—	—	—	—
1.46	—	—	1.46	1.21	—	—	1.21	—	—	—	—
2.77	—	—	2.77	2.72	—	—	2.72	—	—	—	—
3.54	0.01	—	3.55	4.94	0.03	—	4.97	—	—	—	—
5.33	0.02	—	5.35	7.54	0.05	—	7.59	0.18	0.01	—	0.19
6.95	0.05	—	7.00	11.01	0.14	—	11.15	0.47	0.04	—	0.51
9.12	0.06	—	9.18	15.15	0.19	—	15.34	1.24	0.11	—	1.35
10.47	0.03	—	10.50	18.87	0.19	—	19.06	2.32	0.11	—	2.43
10.74	0.11	—	10.85	15.74	0.01	—	15.75	3.90	0.22	0.07	4.19
15.54	1.07	—	16.61	18.68	0.51	—	19.19	3.84	0.47	0.01	4.32
12.15	2.00	—	14.15	22.67	0.85	—	23.52	21.76	0.56[b]	0.01	22.33
11.32	2.86	0.06	14.24	18.15	0.44	0.01	18.60	8.09	0.16	0.01	8.26
				(17.15)	(0.52)	(0.01)	(17.68)				
11.47	6.66	0.59	18.72	24.24	5.36	0.06	29.66	19.72	3.00	0.08	22.80
				(20.53)	(6.55)	(0.08)	(27.16)				
7.54	20.6	1.31	29.45	21.10	20.01	1.71	42.82	6.15	21.00	0.38	27.53
				(16.12)	(23.83)	(2.00)	(41.95)				
5.51	17.9	4.13	27.51	NA	NA	NA	NA	5.74	22.20	2.14	30.07
				(13.98)	(18.75)	(6.59)	(39.31)				
3.62	16.5	4.80	24.95	NA	NA	NA	NA	6.28	18.13	3.32	27.73
				(13.02)	(19.29)	(7.90)	(40.20)				

[a] West and East Germany combined after 1945 (West Germany only).
[b] Japanese oil consumption is for 1935.

ment of seconds, minutes, and hours. Later writers have considerably elaborated on Mumford (Cipolla, 1978a; Mayr, 1986). Cistercian monks extended the use of the clock into agriculture in the early 1100s, using church bells to control work in distant fields. By the mid-1300s control of time meant control of work, thus of production, and medieval cities invested heavily in clocks and bells (Landes, 1983, 68–75). Control of time languished from then until "the invasion of the night" of the late 1700s. Machines worked faster than humans, but still in real time. Audio recording enabled us to recapture times past more accurately by the 1930s (Mumford, 1934, 242–45); video has expanded such abilities since the late 1970s. Multiplexing of messages and time compression in computerized transmission of data have begun to free us from real-time constraints.

Hägerstrand and his students have shown considerable concern with the intersection of space and time. Giddens refers to Hägerstrand in his development of "time-space distanciation" as a way of empirically grounding the world system model (1981, 196–98). Wallerstein in his sixty-day model suggests that the world system has a temporal structure and that "the size of a world-economy is a function of the state of technology, and in particular of the possibilities of transport and communication within its bounds" (Wallerstein, 1974, 17, 349). The theoretical aspects of spatiotemporal systems have been best treated by Hägerstrand and by one of his students, Tommy Carlstein (1982). Carlstein's concern has been with the use of time in preindustrial agrarian societies. Hägerstrand has provided two powerful theoretical overviews: one ties time to the principal social and productive unit of preindustrial societies, the family (1978); the other invokes Goldenweiser's "Principle of Limited Possibilities" (1913) to suggest that a major key to temporality is that decisions taken in the past radically reduce the range of options available in the present (Hägerstrand, 1988). The latter is an extension of Carlstein's basic argument that time used to perform one agrarian function is lost to other possible functions (1982). Transport technology has offered radical spatial extension of the sixty-day model. At the height of the Eotechnic the sixty-day model was subcontinental. At the height of the Paleotechnic such a model was still subglobal (as Verne's *Round the World in Eighty Days* reminds us). Today a sixty-day model covers the globe with ease. The Neotechnic also offers the prospect that, in communications technology at least, time will no longer be lost permanently; nor will we be restricted to performing a limited number of functions at any one time or to working in real time.

Geopolitical problems were also important in the switch to Neotechnics. Marx suggested that capitalism was doomed to overproduction caused by the entry of new firms into the marketplace. His principal solution, the emergence of monopolies, was resisted by polities that subscribed to the marketplace ideology of Adam Smith, though not by others. Marx saw six more counteracting influences: increasing exploitation of labor; deliberate depression of real wages; cheapening of fixed capital; deliberate increase in the number of unemployed laborers; increase in stock capital; and expansion of foreign trade (1959, chap. 14). Mackinder, Hobson, and later Lenin elaborated on the last by elevating it to the status of an "ism." Imperialism, with monopolistic control of prices in colonial markets, would end the crisis of overproduction and the falling rate of profit. Schumpeter later suggested a seventh solution, technical innovation. Innovations cluster not only in time but also in space; hence the importance of the work of Hägerstrand, Carlstein, and Giddens. In the 1880s and again in the 1930s technical innovations were clustered heavily in Germany and the United States (fig. 2-11). These polities used new technology to move ahead of Britain in their drive for hegemony. Britain's economy and society had become too rigid to absorb new technologies. It was simpler to live with a declining rate of return in Paleotechnic industries, where huge amounts of capital were already invested, than to risk huge new investments that only *potentially* offered a high return. Germany and the United States, later Japan, had so little invested in the Paleotechnic that it was easy to invest in the Neotechnic.

The Neotechnic focused around innovations in power generation and distribution and their application in production, transportation,

Technology and Geography in the Elaboration of Capitalism

Concentration of Major Scientists and Technologists
born before 1840 and between 1840-1914

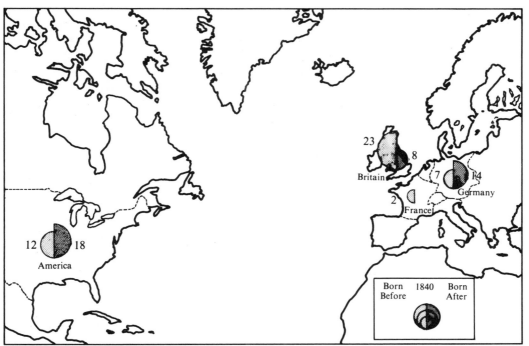

Figure 2-11 Concentration of major scientists and engineers born before 1840 and between 1840 and 1914. Before 1840 Britain dominated the scene, but as Paleotechnics gave way to Neotechnics American and German scientists and engineers came to the forefront.

and communication; around innovations in the organization of production and, later, marketing; and around changes in the nature of raw materials. Each of these had serious consequences for both the internal socioeconomic structure and the external political relations of the principal innovating countries. Compared with earlier technical systems, innovations in the Neotechnic also diffused very rapidly, aided by far better scientific and general communications. So rapid did diffusion become that, at the height of the struggle for hegemony between Germany and the United States, in the second half of the long, interrupted world war between 1914 and 1945, the speed and thoroughness of implementation became more critical than the innovations themselves. For simplicity, Neotechnics are best split into those involving electricity and those involving the internal combustion engine.

Electricity

Electricity is merely a way of distributing mechanically generated power. In the Eo- or Paleotechnic the range of mechanically generated power was a few hundred feet. A major impetus for the factory system was the need to concentrate labor spatially to maximize the advantage of the new water- or steam-powered machines. Machinery was crammed together in the tightest possible space to minimize losses to friction. The result was a dangerous work environment in which loss of fingers, limbs, and eyes was depressingly common. It also reduced productivity because the flow of work through the factory was controlled by the flow of mechanical energy, not the logic of the tasks at hand. In the textile industry raw fibers were examined and sorted on the top floor, where the strongest light was available; bulky, heavy raw materials had to be hauled there before work began. Although gravity then helped the flow of work proceed down the floors, the system was not optimally efficient. Electrical distribution of power freed entrepreneurs to locate machinery

Table 2-6 Electricity consumption per capita per year, 1900–1988, selected polities (millions of kilocalories)

Year	United States	Britain	France	Germany[a]	Japan
1900	0.03	—	—	0.02	—
1910	0.19	0.04	0.02	0.07	—
1920	0.46	0.17	0.13	0.21	0.06
1930	0.80	0.34	0.35	0.39	0.21
1938	0.94	0.63	0.44	0.69	0.40
1950	2.20	1.12	0.69	0.82 (0.78)	0.48
1960	3.53	2.18	1.36	1.88 (1.84)	1.06
1970	6.88	3.69	2.40	3.43 (3.44)	2.95
1980	8.65	3.79	3.95	NA (4.85)	4.24
1987–8	9.43 ('88)	4.25	5.55	NA (5.53)	5.88

Sources: Bank of Japan, 1966; CSO Annual Abstracts; Eurostat; Mitchell, 1980; Statistical Abstracts of the U.S.; Statistical Handbooks of Japan; Thirring, 1958; U.S. Bureau of the Census, 1975.
[a] West and East Germany combined after 1945 (West Germany only).

to maximize work flow and worker safety. Factories could move out of cramped cities to much cheaper land on the urban fringe, significantly reducing the fixed costs of production. Cheaper land allowed single-story factories, in which the flow of work never had to fight the force of gravity and natural light could be provided throughout. Factories no longer had to be on a fall of water or next to a canal or a railroad to bring coal for their engines. Finally, electricity drove streetcars, which allowed workers good access to factories beyond walking distance (Hughes, 1983).

The adoption of electricity clearly differed among countries (table 2-6). Before 1914 the United States produced between three and seven times as much electricity per capita as Germany and Britain. That margin narrowed during and after World War I. By 1921 America outproduced Italy four to one, France three to one, but Britain and Germany respectively just over and just under two to one. Japan surpassed Italian production per head between 1921 and 1925. By 1939 the U.S. lead in per capita production had been narrowed to only one and a half to one compared to Germany, Britain, or Japan. Soviet production, badly hurt by the Revolution, increased only slowly in the 1920s and 1930s. After World War II the United States recovered somewhat of a lead but did not exceed a two-to-one margin over Germany once German war damage was cleared away in the early 1950s. Compared to Germany, Japan's production per head fell sharply between 1939 and 1950, and recovery was slower. The Soviet Union gained rapidly only after 1950, almost tripling production per head in each of the subsequent decades.

Improving Industrial Production

Electricity needed a new breed of entrepreneur to exploit flexible location of machinery within the factory. Some entrepreneurs understood the advantages empirically, Ford foremost among them. The moving assembly line allowed Ford to slash selling prices, raise wages substantially while reducing working hours, and jump profits. The selling price of a Model T roadster was $680 in 1910, $525 in 1912, and $440 in 1914. In early 1914 Ford doubled wages to $5 per day, while reducing the work day to eight hours. Ford paid $11.2 million in dividends in 1913 and $12.2 million in 1914, despite its higher wage bill. The moving assembly line allowed a chassis that had previously taken fourteen hours to be produced in less than two hours. With no increase in the labor force, production doubled from 78,440 in 1911–12 to 168,304 in 1912–13, then nearly doubled again to 248,307 in 1913–14 (Wilkins & Hill, 1964, 52–53). Other manufacturers merely struggled to catch up with Ford until the mid-1920s. Ford made no secret of his technology but others needed time to close the gap (Wilkins & Hill, 1964, 52). Would-be mass manufacturers in Europe after World War I, such as Citroën, Austin, and Opel, simply bought assembly lines in America (Hugill, 1988a, 126). As late as 1938 Nazi Germany bought from America much of the production machinery for what, after the war, would become the Volkswagen Beetle. The engineers and production men to staff the Volkswagen plant were also hired in the United States because "there were not enough German engineers and production men familiar with [American] . . . methods" (Nelson, 1965, 55).

Americans perfected not only the hardware of mass production using electrified factories but also its software (Hounshell, 1984). In 1911, the year Ford was putting such principles into practice in reorganizing his factories, Taylor was publishing *Principles of Scientific Management* and Gilbreth was publishing *Motion Study*. The underlying technology that allowed scientific management was electrical power, which gave freedom to locate machines where they were needed within the factory. A further contributor to the relative success of the United States early in this century was better social relations of production than elsewhere. Caste did not intrude much, and individuals were able to achieve remarkable upward social mobility. This was particularly important in the rise of a much enlarged middle class, many of whom were the middle managers needed for scientific management. The U.S. educational system, although less sophisticated than Germany's, was quite capable of coopting the brighter farm and immigrant children, giving them a rudimentary education in the basics, and allowing industry to complete their education on the job.

Improving Domestic Production

Electricity also greatly improved domestic production. In recent years, it has produced the almost completely technified household. Domestic labor beyond that of the nuclear family, in particular "live-in" labor, is much less required. This development has been an important precondition for the success of feminism in polities where the social relations of production have evolved as fast as the technologies. The bicycle and the automobile freed women as well as men from the spatial tyranny of the home; reliable means of conception control freed women from frequent pregnancy; and the switch from work requiring muscle power to work requiring brain power opened more positions to women.

Domestic production took longer than industry to technify because early electric motors were large. As late as the 1930s middle-class British households still expected to have live-in help despite primitive refrigerators and vacuum cleaners. Domestic labor was much rarer in the United States, and the technification of domestic labor was much faster (Bent, 1938). A more democratic society and more job opportunities relegated domestic labor to immigrant women, who were never in adequate supply. Early in this century feminists proposed commercial kitchens, laundries, and childcare centers, as well as "homes without kitchens and . . . without housework" (Hayden, 1981, chap. 11). Most feminists came from a socialist background and confused socialist ideals with centralized production and the factory system. This was a bizarre conjunction, given the origins of the factory system. Other notions,

Kitchen, c. 1890. Mechanization came very late to the home, and especially in food preparation. Servants were still needed in most kitchens. Ford Museum, by the author.

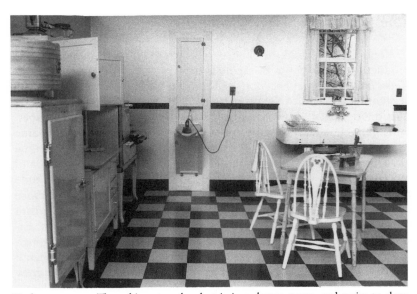

Kitchen, c. 1930. The refrigerator, the electric iron, better stoves, and easier to clean surfaces reduced the labor needed in the home to that manageable by the woman of the house. Such improvements freed the servant population to enter the mainstream labor force. Ford Museum, by the author.

particularly the idea that the family was an archaic form, display the hold that the ideal of factory production had on even intelligent persons. Factory production was clearly more efficient than cottage industry, but feminists in their otherwise understandable drive to free women from domestic production forgot to ask whether it was more humane.

Kitchen, c. 1940. The adoption of smaller and lighter electric motors, begun before World War II but spurred on heavily by the war in the air, completed the technification of domestic production and freed women to enter the labor force on a virtually full-time basis. Ford Museum, by the author.

Neotechnics offered the chance to return to humane, individualized production in both the household and the factory. Although Bent's *Slaves by the Billion: The Story of Mechanical Progress in the Home* is a fascinating account of the progress made in technifying domestic production in the United States in the 1930s, the smaller, lighter, cheaper, more reliable electric motors needed to complete the process came about through the vast U.S. effort in World War II to win the war in the air. Motors were needed to drive gun turrets, retract undercarriages, and move wing flaps and control surfaces with reduced pilot effort. A crude measure of the impact of such electrification is the fact that otherwise comparable U.S., German, and Japanese twin-engined bombers from the late 1930s had, respectively, aircrews of three, four, and seven. Fatigue was deadly and reducing it critical: over Germany bombing missions lasted ten or more hours; over Japan, fifteen or more. Because every pound of excess weight meant less ammunition to defend the crew or fewer bombs to drop, smallness and lightness were crucial. Unreliability was life threatening in an airplane.

The vast productive capacity paid for by social rather than private investment could be turned over to peacetime purposes once the war was over, and the household claimed its share. Washing machines became ubiquitous in the United States and almost so in Europe. Clothes driers followed. Huge amounts of power machinery for the household have tumbled after. A typical U.S. middle-class household before World War II might have had half a dozen electric motors in the form of a vacuum cleaner, a refrigerator, a record player, a fan or two, and an electric clock. A modern household may contain upward of half a hundred: refrigerator, freezer, washer, drier, dishwasher, microwave oven, trash compactor, garbage disposal, blender, hand mixer, bowl mixer, can opener, coffee grinder, food processor, half a dozen or so ceiling or floor fans, fans to force air through the heating and cooling system, humidifier, stereo turntable, compact disc player, cassette tape player, two or three motors in a typical videocassette player, various personal cassette players and toys, power drills, circular saw, saber saw, leaf blower, and string trimmer. All these increase productivity, for the most part of women, on whom most domestic chores still devolve. Such productivity increases are critical to the ability of women to enter the work force in increasing numbers. Some of the increase in household productivity has been achieved by mechanizing tasks, such as blending, mixing, or grinding food. Some has been achieved by technologies that are simply faster, as a microwave oven is faster than a conventional one. Some has been achieved by automating tasks that previously needed much human labor, such as heating the house or washing and drying clothes and dishes. Some has been achieved by technologies that reduce time spent running errands: refrigerators and freezers mean fewer trips to the store.

Technological improvements outside the home have also reduced the time needed for housework and cooking. Supermarkets minimize the number of stores needed to provision a household. Materials have improved, particularly in clothing and furnishings, so that less care is needed. Disposable technologies such as the sanitary napkin and tampon, the facial tissue, and the disposable diaper have improved hygiene and reduced work loads. Factory-prepared foods now dominate our diets and grow more complex by the year.

Improving Communications

Electricity has brought further control of time and space in the form of improved communication technology. Even with hardware and software worked out, the telegraph still required a vast organizational and infrastructural investment. Although the first, overland telegraphs carried railroad scheduling information, it was quickly realized that they could also carry news and commercial information. Farmers, in particular, could get information on the value of their crops at different markets and increase profits by selling to the highest bidder in a large information pool. As submarine cables developed, they became an important element for managing the movement of ships and goods in the British maritime world system.

The principal modern impact of telecommunications has been virtually to remove the time costs of distance from verbal communications and the communication of large quantities of digitized data. This is in line with the Neotechnic trend toward ever greater freedom of location. Even at the beginning of the telecommunications revolution, as the first artificial communication satellites were being launched, people like

Marshall McLuhan were predicting a "global village" (1964). Although reality has yet to fully catch up with McLuhan's ideal, it is clearly moving in that direction.

Part of the ideal that has already been mostly achieved is the homogenization of our vision of what the world should be like. This process began with Hollywood's domination of the movies in the late teens and has continued through U.S. domination of television. Britain has been able to resist U.S. cultural hegemony in this area somewhat, but only by specializing in dramatic productions and documentaries of high quality. Of the world's major television markets in 1983 the United States imported only 2 percent of its programs; western Europe imported 30 percent, nearly half of that from the United States (Tracey, 1985, 19–22). In stylistic terms almost none of the world's movie producers have been able to stand against U.S. technique, and television producers, for the most part, merely follow styles generated in the movie industry.

The Internal Combustion Engine

At the present stage of the Neotechnic, the internal combustion engine has its most visible effect on transportation. Automobiles and aircraft have wrought a revolution in the location of economic activities. The flexibility and speed provided by Neotechnic transportation have also radically affected shipping and railroads. Containerization has altered the geography of port location: unloading of containers and customs functions can now occur well away from a coast. New types of ships have brought economies of scale and specialization to what previously had been a "cottage industry." In some densely settled countries high-speed diesel trains attain average speeds of 160 or more kilometers per hour, fast enough to compete with aircraft on stage lengths up to 750 kilometers when airports are out of town and railroad stations are centrally located. Yet railroads are still inflexible in terms of route; they also have high fixed costs and suffer from diseconomies at small scale. Because they are most efficient when carrying large loads, they can provide only infrequent service except where population densities are high. Oceangoing ships may be remarkably efficient and cheap bulk carriers, but they are far too slow to transport impatient human beings over long distances.

Automobiles, Trucks, and Other Devices

The automobile and, later, the truck resolved the scale problems of railroad transport at a stroke. Internal combustion engines, in particular the Otto-cycle, spark-ignition, reciprocating engine typical in most small trucks and automobiles, are efficient even when small. They are profitable even when used to move small loads. In large trucks, the compression-ignition reciprocating engine developed by Rudolf Diesel offers advantages. Once reasonable systems had been worked out for the management of such engines in road vehicles, success came rapidly. Key technical developments were torque multiplication by the transmission, ease of starting with the electric self-starter, the development of all-weather bodies, and the use of brakes on all four wheels. By the late 1920s all these technologies were in place (Hugill, 1982).

Yet there were marked variations in adoption not explained by national wealth, the process of invention, or the availability of petroleum resources (table 2-5). Automobiles were European in origin, and European countries were wealthy enough to afford them and to import the oil on which they ran. Yet by 1905 the United States, then a somewhat poorer nation, was the world's major producer. In part, U.S. success was entrepreneurial: Ford's adoption of the moving assembly line and Detroit's overall embracing of scientific management and motion study were critical. But it was also infrastructural: the United States had not invested heavily in steam railroads, distances were much greater both between cities and within them, and land ownership patterns were not complex enough to inhibit highway development. In Europe only the Fascist dictators of the 1930s had the power to cut through the web of landowners and impose a highway system. In democratic nations such as Britain, a small number of landowners could easily block highways. Early automobiles were also too heavy, long, or inefficient for the tight Eo- and Paleotechnic infrastructure of European cities or, later, those of Japan.

Mass adoption of the automobile outside the United States and Canada had to wait until lighter, shorter, more efficient automobiles were developed in the late 1950s (Hugill, 1988a). Until the 1950s in Europe, the 1960s in Japan, and to this day in poorer parts of the world, bicycles were far more important than cars for individual transportation. They fit tight infrastructures yet allow considerable freedom of residential and work location. The chief advantage of the automobile and the truck is their much greater range, greatly expanding the geographic scope of the city while maintaining or decreasing time distances (Forer, 1978). The importance of the internal combustion engine is the freedom of location it affords, the compactness of its power source, and its ability to be attached to a vast range of specialized machinery. Modified slightly to burn a "clean" fuel such as propane, it can even be used in closed environments, where the carbon monoxide from a gasoline engine's exhaust would be lethal to humans.

In agriculture tractors have greatly increased productivity, as has later, more sophisticated and specialized machinery. Forklift trucks have made handling bulky, heavy goods in the vertical dimension inside factories and warehouses much easier. Internal combustion engines also changed military productivity. The French used Paris taxicabs to rush troops to the front in 1914 and prevented a German breakthrough akin to the one that engulfed Paris after the capture of Napoleon III and a hundred thousand French troops at Sedan in 1870. British troops moved to and from the front in World War I in converted London busses. Caterpillar tractors and four-wheel-drive trucks towed field guns through the appalling mud of Flanders. Toward 1918 the caterpillar tractor and the gun were married to produce the tank. In "Plan 1919" the British proposed an entirely Neotechnic strategy: ground and air advance coordinated by wireless telephony. Ironically Plan 1919 was, with the much improved radios available by the 1930s, perfected by Germany as "blitzkrieg" (Macksey & Batchelor, 1970, 48, 78). Such improved geographic mobility made a mockery of fixed lines of defense, as the French found out all too late when the supposedly impregnable Maginot Line fell to German flanking advances using tanks and dive bombers in 1940 (Macksey & Batchelor, 1970, 86–87).

Other specialized petroleum-fueled machinery has greatly increased our ability to extract resources from below the earth's surface, in particular to strip-mine coal and iron ore. We can now reshape the earth's surface by dyking, dredging, and draining as well as by raising levees and even entire cities. Serious earth moving began when Galveston was raised to clear subsequent storm surges after the near destruction of the city in the 1900 hurricane. Machinery developed in Galveston helped cut the Panama Canal, the Houston ship channel, and the Intracoastal Waterway. By the end of World War II this last was the longest artificial inland navigation in the world, reaching from the Mexican border to New York City.

Small gasoline engines have also become ubiquitous in the domestic landscape. Outdoor tasks such as lawn mowing, edging, and trimming and tree trimming and removal are now usually performed, at least in the United States, with gas-engined mowers, edgers, weeders, and chain saws. The technification of the household described earlier, which was begun in the areas of cleaning, laundry, and food preparation by the application of electrical energy, has increasingly been carried into the household workshop by electrical tools such as the power drill and saw and into yard maintenance by electric and gasoline-engined tools.

Aviation

Aviation's major contribution to Neotechnic economic development has been in reducing the time costs of distance. Many of the major improvements in aircraft and aircraft systems come from the military and are bound up with geopolitics and geostrategy. Offensive aviation, in the form of long-range strategic bombing, is intimately linked to civil aviation. Strategic bombers have frequently been converted to mail carriers and airliners, and vice versa. Defensive aviation may be linked to strategic geopolitics if it fails to defend a polity against a warring neighbor. By the early 1920s airmail service had begun a radical improvement in the level of political and economic integration of large continental states. It was particularly important in the United States and Germany, and later in the Soviet Union. As longer-range, multiengined, more reliable flying boats developed in the 1930s, airmail service expanded over sea as well as land networks. Britain sought to service a far-flung empire, the United States to expand U.S. business interests in Latin America and the Pacific.

After World War II four-engined, long-range, propeller-driven land planes proved far more efficient than flying boats. They extended the process begun by the DC-3 to a continental and transatlantic level, allowing multinational corporations with highly centralized management systems to emerge. The United States benefited most of all. After World War II the United States was clearly not only militarily but also commercially hegemonous. The high-priced dollar made U.S. purchase of foreign companies attractive, and foreign companies were themselves desirous of U.S. managerial and technical competence. The European economies quickly assumed a markedly American character, and American managers spent more of their careers abroad.

Aviation has had two major impacts on capitalist development. It has allowed truly national, centrally directed corporations to emerge even in the largest polities, or powerful national governments to emerge where the principle of the command economy predominates, or both. It has allowed transatlantic corporations to develop with headquarters either in the Boston-Washington or the London-Paris-Bonn megalopolises. It promises the development of truly global corporations, best centered in the European megalopolis if current technological limitations on supersonic flight speeds remain, but anywhere if speeds can be substantially raised. The reappearance of the command economy in the late 1800s (McNeill, 1982, 269) and of the megamachine in the fascist and communist dictatorships of the 1920s on (Mumford, 1970, 243) has thus been a function of improved transportation and communication. On the other hand, the increasingly complex web of air transport linkages has given corporations much more freedom of location. This process has been facilitated by the tremendous growth and cost reduction of long-distance telephone service and data transmission. In this sense Neotechnics is a two-edged sword. At the same time that it has radically increased the flexibility of societal operations in space and time, it has threatened reduced flexibility through centralized control of those operations.

New Technics, New Resources

A further major force in the Neotechnic has been the ability to reshape the physical world. This development has occurred most significantly since 1940. Before that date the organic world still dominated most horizons. Although industrialization was well under way in Europe, America, and Japan, the rest of the world was still a vital agricultural periphery, providing food and agricultural raw materials for industry elsewhere. Neotechnic ability to synthesize substitutes for organic raw materials radically changed the nature of industrialization, as well as the political and economic order of the world and its social structure. More than any other form of Neotechnics this change has been underremarked by academic observers, even though its ramifications now occupy a huge proportion of the academic work force in science and in such applied scientific areas as geoscience, engineering, and biotechnol-

ogy. The most significant changes have been in substituting hydrocarbons for the products of subtropical and tropical agriculture. Cotton
and rubber are now far less important in production agriculture than
they were only forty years ago. Nylon, polyester, other forms of synthetic fibers, and numerous synthetic rubbers outperform natural materials
and are often cheaper, depending on the world price of oil.

Production agriculture is now more concerned with food than with
raw materials, and biotechnology is poised to make radical changes in
the food chain. Genetic research has improved the performance of existing food crops with respect to yield, sensitivity to pesticides and fertilizers, and has even reduced the growing season somewhat. Artificial
insemination has improved the quality of livestock in a more reliable
way than natural reproduction. Yet there are still barriers, not least in
human taste perceptions. At current technical levels much of the world
animal protein shortfall could be made up by fish, yet high oil levels in
the flesh of many fish make them unpalatable. Oily fish can be dried out,
ground up, and converted to high-protein flour, but the "fishy" taste
cannot be cost-effectively processed out. Genetic engineering offers the
possibility of engineering plant matter to produce high levels of natural
protein, and thus of markedly increasing the productivity of world agriculture. At present, much plant material is fed to animals for conversion
to protein.

CONCLUSION

The general trends of technical development over the past millennium
have been toward decentralization and spatial and temporal extension.
The two have often gone hand in hand. In Europe the process began
with the onset of feudalism, which represented a form of managerial
decentralization, initially for geopolitical reasons, at the state level.
Small polities proved more geopolitically sound, but required new
forms of economic management, spatial extension in trade, and spatial
and temporal extension in production agriculture. Since the late Middle
Ages the European polities, and after 1900, the United States and Japan,
have experienced a dialectical tension created by communication forces
that have allowed increased centralization and that are opposed by the
production and marketing advantages of regional specialization, decision making, and trade. Dependence on the organic world for food,
fuel, and fiber increased the strains on the European economy in the
1500s and 1600s. The slow rate of economic return on lumber for fuel
caused many landowners to switch to food; some switched to fiber
because it required less labor and did not result in many tenants, whom
a landlord might wish to be rid of when the market changed. Although
spatial extension of lumber production was used to resolve the problem
partially, in particular east into Baltic Russia and west to New England,
the technical shift to consuming coke rather than charcoal in iron smelting removed the immediate geographical limits to production.

Increased demand for coal had technical ramifications in encouraging the development of steam power to pump water out of ever deeper mines. Technical improvements made it possible to use steam power away from the pithead and, eventually, in mobile applications. Improved managerial systems, in particular Arkwright's factory system, concentrated labor power spatially and extended its use temporally. This spatial concentration encouraged the application of inanimate power, which permitted large numbers of machines to be grouped close together, thus mitigating the losses to friction which characterized mechanical transmission systems. Temporal extension of the workday, in particular with artificial light, also increased the productivity of both labor and the capital represented by the water wheels or steam engines that drove the new machines. In both a spatial and a temporal sense, steam was preferable to water power because it concentrated power in a limited geographic area and removed seasonal ecological limitations on production.

Steamboats and railroads radically altered production agriculture in the 1800s by opening the continental interiors to world trade. Before steam the head of navigation on rivers was defined by geography as the head of tidewater penetration or the site of rapids. Steam railroads could be built almost anywhere, although grades remained a problem. Steamships could haul large quantities of food and raw materials around the world reliably and with little loss from accidents. Although productivity per acre was low in Paleotechnic continental agriculture, huge new acreages were available, at least until about 1900.

Neotechnic telecommunication systems have allowed a spatial concentration of authority and decision making, whereas technologies that allow relatively frictionless transmission of power have allowed remarkable increases in freedom of production location. Neotechnics, in particular the application of physics and chemistry to production agriculture, first radically increased productivity per acre, then began to find substitutes for agricultural raw materials from the inorganic realm. Careful manipulation of available inorganics produced new industrial materials. Management systems improved radically at the factory level, with more efficient use of labor and capital both in space and time. Far better individual transportation, especially the automobile, allowed much greater locational freedom for all activities within and around cities. The high speeds attained by aviation have allowed the slow emergence of an almost-global economy. The price paid has been the reemergence of huge centralized polities administered with telecommunications and the emergence of their economic equivalent in the form of multinational corporations. The reappearance of the megamachine or the command economy, be it political or economic in nature, is against the general trend of decentralization and flexibility that has served the West very well. Industry suffered heavily in social terms from the recentralization of production that occurred in the Paleotechnic, as Marx was quick to understand. Polities have also suffered from recentralization, in particular those with limited controls on centralized authority. Mussolini's Italy, Stalin's Russia, and Hitler's Germany are all examples of the sophisticated use of communications systems to centralize political authority in polities with inadequate institutional, that is managerial, controls intended to prevent such centralization. Roosevelt employed similar communications technology to rebuild the United States after the Depression, but the system of checks and balances in the U.S. Constitution prevented excessive concentration of authority.

Chapter Three

The Triumph of the Ship

What goods could bear the expence of land-carriage between London and Calcutta? Or if there were any so precious as to be able to support this expence, with what safety could they be transported through the territories of so many barbarous nations? Those two cities, however, at present carry on a very considerable commerce with each other, and by mutually affording a market, give a good deal of encouragement to each other's industry . . . The inland parts of the country can for a long time have no other market for the greater part of their goods, but the country which lies round about them, and separates them from the sea coast and the great navigable rivers.

Adam Smith
The Wealth of Nations

Human beings have long been mobile, even on a global scale. In the early Eotechnic, seas and navigable waterways afforded mobility, but access to the geographical interiors of the great continents remained restricted. Commerce was therefore uncommon except in items that were either self-mobile, such as humans and animals, or of high value relative to their weight, such as precious metals or silk. In arid places, in particular around the Islamic world, camels provided land transportation for somewhat less valuable goods (McNeill, 1988). The overland carriage of bulk goods only became possible with the construction of canals, but the expense of these waterways, particularly in rough topography, greatly reduced their utility. Only with the development of steam-powered riverboats and railroads in the Paleotechnic did commerce truly invade the continental interiors.

The development of ships has been crucial to the development of trade in any goods other than preciosities and at any distance beyond the few miles a person can walk. Since ships' cargoes represented the accumulation of considerable agricultural surplus, essentially in the form of concentrated, commodified labor power, they were targets for thieves. Because trading ships had to be defended, the history of ships is also the history of fighting ships.

Applying power at a distance has been a crucial interest in world history. Practical experimentation led to theorizing, and theorizing led to action. In this chapter I suggest a direct line of theorizing from Sir Walter Raleigh in the sixteenth century to Admiral Alfred Thayer Mahan in the nineteenth. Mahan's naval geopolitics also gave rise to the land-based geopolitical theories of Mackinder in the early twentieth century, views Mahan himself later accepted (Blouet, 1988a).

The ship and its capabilities at different times determined the character and extent of all Eotechnic world economies, profoundly influenced those of the Paleotechnic, and is still geopolitically at the heart of the Neotechnic system. Without an adequate understanding of ship technologies, no account of the rise and fall of hegemonous powers can be complete.

EOTECHNIC WIND-POWERED WOODEN SHIPS

I cannot do justice here to the long prehistory of the ship before the expansion of Europe. The impressively early importance of sea trade in the Mediterranean has been well examined by scholars such as George Bass and his colleagues at the Institute of Nautical Archaeology at Texas A&M University (Bass, 1987). Naval historians such as Morison have noted the essentially maritime nature of the Pax Romana in the Mediterranean basin (1980, 47). The development shortly after the time of Christ of extensive trade networks linking Mediterranean trade to that of the Indian Ocean and ultimately to the Southeast Asian archipelago has been remarked upon by a variety of authors (Cary & Warmington, 1963, 95–96; Wheatley, 1961, 138; Coedés, 1968, 19–21; Majumdar, 1963, 12–16). The aggressive expansion of Islam took this trade all the way to China by the fourteenth century (Dunn, 1986).

In northern Europe Danish and Viking sea nomads ran riot in the Baltic, the North Sea, and the eastern Atlantic from the late 700s through about 1100, raiding, trading, and settling (Thørdarson, 1930; Ekwall, 1936; Darby, 1973b). The Scandinavian habit of burying merchants complete with their ships has left us an unusually good record (Greenhill, 1976, chap. 14). Hansa traders from the Baltic towns, many of which had been Viking settlements, took up in the mid-1200s where the Vikings left off. They were the first to stress the bulk trades, particularly in fiber. Hansa merchants made more profit from woollen textiles than from any other trade item (Dollinger, 1970, 217). The Hansa cities developed the first defensible ships, erecting shipboard "castles" from which defenders could rain arrows on would-be pirates. Bremen seems to have been the origin point of the cog, the most important ship design before the three-masted ship of the mid-1400s. Cogs were beamy, deep hulled, stern ruddered, decked, and plank built although limited in size because they lacked a skeleton frame (Scammell, 1981, 79; McGowan, 1981, 6).

Rigging, hull construction, and navigational technology all improved markedly during the early 1400s, probably at the remarkable technical institution built by Prince Henry the Navigator at Sagres on the Portuguese coast. The three-masted, skeleton-built ship used by the Sagres navigators was part of the evolving maritime tradition of northern Europe (fig. 3-2). Skeleton construction allowed stronger and, eventually, larger ships. On the rear mast was mounted a lateen sail of the type found on Islamic ships. The lateen sail allowed the three-masted ship to beat more easily to windward, as Islamic ships had long been able to do. Coupled with square sails on the front two masts, it also allowed rapid movement before the wind. With the addition of guns to the fore and aft castles of the three-masted ship, an almost wholly new ship type emerged. Skeleton construction, decking, and the stern rudder gave survivability in the bad weather typical of the Atlantic and the North Sea. The new rigging allowed the ship to be maneuvered more readily. Gunpowder weapons allowed a relatively small crew to protect itself against threats or to act aggressively. Such ships, mostly referred to as caravels despite our archaeological ignorance of their actual look and construction, were what made possible the opening of the geographic frontier into the Atlantic in the mid-1400s and, eventually, its extension around Africa by the end of the century. Europe was then able to reach out directly to the Indies, cutting out the Islamic "middlemen" in the long-distance trade between Europe and China via the Mediterranean and the Indian Ocean.

This explosion of European energies had modest beginnings. Although the Franks and the feudal system had pushed the Moors out of Europe proper, the Pyrenees were as powerful a barrier to Christian aggression westward as they were to Islamic attempts to push east. From 800 to nearly 1500, Iberian Christendom engaged in a long and often bloody Reconquista. The Portuguese completed this process by 1415, nearly eighty years before Castilian Spain finally evicted the Moors from Granada. The Portuguese then turned their attention west into the Atlantic (Penrose, 1952, 46). The key to the destruction of Islam was to cut off its flow of wealth, hence its trade, and the Portuguese were aided in this by the Genoese, whose carrying trade in the Mediterranean was, like that of their competitor, Venice, held hostage by Islam. By the early fourteenth century Lisbon had become the great center of Genoese trade (Wallerstein, 1974, 50). Because the Italian city-states of the thirteenth and fourteenth centuries had more experience in colonizing endeavors than any other part of Europe, the link was logical (Verlinden, 1970, 6–7). They had also sent out such explorers as Marco Polo in the late thirteenth century to map the routes to the East, to India, and to China (Polo, 1926). The Genoese alliance with Portugal paid off, not only in access to the Atlantic for trade and colonization, but also through access to Islamic records, then the best in the world with regard to both geography and classical history.

The final link in the chain of Portuguese expansion was the character of Prince Henry the Navigator. Rather than murder his way to a throne inherited by an elder brother, as was all too customary in fifteenth-century Europe, Henry chose a scholarly life devoted to improving

Selected Warships

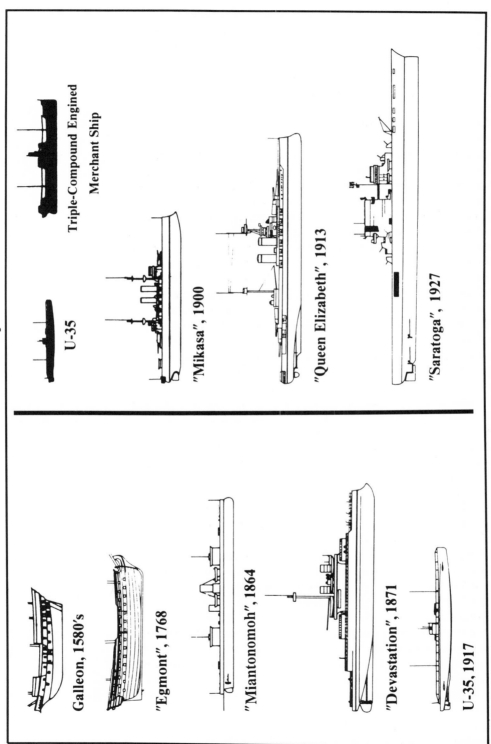

Figure 3-1 Selected warships. The primary role of sea power has been to secure the seas for the profitable commerce that makes sea-based weapon systems affordable, or to interdict the commerce of an opponent. Sail-powered, broadside cannon–armed, wooden warships changed comparatively little over some four hundred years from 1431. Steam-powered iron and steel ships quickly moved their guns to turrets, added missiles (in the form of torpedoes), then switched to aircraft as their weapon systems. Internal combustion and nuclear engines allowed undersea delivery of ever more sophisticated missiles.

geographic knowledge and its practical application in exploration and colonization. At Sagres on the coast of Portugal he established the world's first technical "university." He attracted Christian, Islamic, and Jewish scholars and artisans to translate classical and Islamic accounts of the size, shape, and geography of the world and to refine charts, navigational aids, and, in all probability, the ships themselves (Penrose, 1952, 43–61).

The beginnings of the first phase of Portuguese Atlantic expansion were marked by discovery in 1431 of the wind pattern in the eastern Atlantic that allowed ships to sail south and west to the Canaries, then in a sweeping circle first north and then back east to the coast of Portugal via Madeira and the Azores (Crosby, 1986). Such voyages gave the Portuguese confidence in open-ocean navigation. Thus armed they sailed south past the Canaries to round Cape Bojador, the dangerous meeting place on the African coast of powerful ocean currents that could tear a coasting ship apart. South of the latitude of Cape Bojador they sailed due east back to the African coast, which they reached for the first time in 1434. A further series of voyages begun in 1441 started the second phase of the Atlantic expansion, as the Portuguese discovered that they could tap into the profitable sub-Saharan slave trade. The death of Prince Henry in 1460 and internal disruption in Portugal stayed the process of expansion, although the great western bulge of Africa into the Atlantic had been almost rounded. After 1470 the push south began again, although it was the slave trade that took pride of place for the Portuguese (Penrose, 1952, 55). The Guinea coast was heavily exploited for its "black gold," despite Spanish attempts to cut in on Portugal's lucrative trade. In the early 1480s Cao continued the push south into modern Angola, and in 1487 Dias finally rounded the southern tip of Africa and sailed into the Indian Ocean. Dias's return to Portugal with news that the Indian Ocean could be entered thus marked a return to the European obsession with a route to the Indies bypassing Islam.

The success of the Portuguese expansion east was spectacular, although it was bought at a high cost in human life on the long voyage around Africa and a high cost in ships. Yet it finished the Arab world as a major actor in world affairs until the discovery of oil there in the twentieth century. Until 1500 the Arabs had acted as geographic and economic "middlemen," profitably moving goods between Europe to their west and India and China to their east. Portuguese expansion radically cheapened the transportation costs of such trade by replacing a mixed sea-and-land route with an all-sea route. Transfer costs, where merchants customarily doubled the price of goods at each handling, were reduced to nothing. Before 1500 eight or more transfers were the norm between Europe and China, so that only the most precious of goods could stand the multiple doublings of price. By breaking the Arab "blockade" of trade across the Eurasian region, the Portuguese brought to much of the Eurasian landmass the same Ricardian efficiencies that Greeks and Romans had brought to the classical Mediterranean and the Vikings and Hansa had brought to medieval Europe.

Before the Portuguese expansion, goods from northern Europe were carried by Hansa merchants to Bruges or Ghent, later as far as Lisbon. A Portuguese ship would carry goods to Genoa or Venice, whereupon a ship of one of those cities would take goods to the coast of modern Israel or Egypt. Islamic traders then carried goods across the Isthmus of Suez or into the Tigris-Euphrates river basin. Islamic ships took over, sailing down the Red Sea or out of the Persian Gulf to a port on the western coast of India. From there a Chinese ship moved goods through the Strait of Malacca to China proper. This route became possible in the early Christian era, as the Chinese began to send ambassadors to Malaysia. But it was the expansion of Islam with its powerful links to a mercantile urban way of life that really brought the route to its full glory.

The next phase of European expansion came from Spanish expansion into the Atlantic, first to the eastern islands, then to the Caribbean. The Spanish did not complete their conquest of the Canaries until 1496 (Sauer, 1966, 10). Columbus used the islands as the staging point for his epic, if ill-conceived, attempt to reach the Indies by circumnavigating the globe. His correspondence with the Florentine geographer, Paolo Toscanelli, led him to conclude that China was a mere 3,000 nautical miles across the Atlantic instead of the real, great-circle distance of some 10,600 nautical miles (Parry, 1966, 44). Toscanelli accepted Marco Polo's belief that Asia extended much further east than it in fact does and compounded the error by believing the estimate of the Roman geographer, Ptolemy, that the circumference of the globe was only around 18,000 miles (Penrose, 1952, 99). Columbus compounded Toscanelli's errors by assuming that the length of an equatorial degree of longitude was 10 percent shorter than the figure given by Ptolemy, which is 25 percent shorter than the true figure (Parry, 1966, 44).

In 1492 the crown of Castile completed the Spanish Reconquista. With the distraction of the Moors removed, Queen Isabela was willing to fund Columbus's search for a westward route to the Indies. Columbus had long importuned the Castilian crown to this end. He returned having found land that to this day bears the misleading title, the West Indies, as if it were an archipelago off the western coast of a China viewed from Europe. Columbus himself certainly believed it the western islands of China and outfitted another expedition on the strength that it was. He returned empty-handed, though with fabulous stories of potential wealth and profit. As Sauer so aptly says of him, "As he continued to hold to an illusory geography undisturbed by every evidence that he was wrong, so he invented riches which did not exist. It was well apparent that he was a chronic, compulsive romancer who lived in a world of wishful thinking" (Sauer, 1966, 104).

Yet Columbus's achievements were manifold. He focused the attention of others on the lands out in the Atlantic, though it remained to the Italian navigator Amerigo di Vespucci, who explored from the mouth of the Hudson River in the north to the Golfo San Jorge in the south, to proclaim as a "New World" the landmass that now bears his name (Palmer, 1965, 62–63).

The Americas had no powerfully entrenched polities and societies to raise the cost of extracting surplus. Under the onslaught of European diseases, the indigenous inhabitants died like flies. Wolf suggests a 94 percent decline in the population of Mesoamerica between 1492 and 1650 (Wolf, 1982, 134). European slaving, begun by Columbus against such West Indian groups as the Caribs, mixed as well as demoralized native populations, spreading diseases efficiently as well as lowering human resistance. The collapse of indigenous populations created severe problems for European would-be exploiters. Simple theft of locally accumulated surplus wealth would have been made easier by demographic collapse, but the native American societies encountered before 1519 and Spanish expansion into the valley of Mexico were poor. The Spanish thus tried initially to reproduce in the New World the sugar plantation, slave labor–based society the Portuguese had forged in Madeira in the late fifteenth century, which they were already copying in the Canaries. The Atlantic islands of Spain and Portugal were the laboratories in which European ideas about exploitation and colonization of the world outside Europe were given practical form (Meinig, 1986, 9).

Sugar production began on the Caribbean island of Española around 1515, when it was introduced from the Canaries by a well-off physician of Santo Domingo. Almost immediately the labor problem had to be faced, and by 1518 "the mass importation of Negroes was already underway" (Sauer, 1966, 207). Yet sugar was not a marked success on Española (Mintz, 1985, 35). The success of Cortez's venture into the valley of Mexico in 1519 refocused Spanish attention on tribute taking rather than production in the New World. Gold and silver rather than sugar and slaves occupied the Spanish thereafter.

The Importance of the Cannon-armed, Defensible Ship

The early Spanish success in the West Indies was clearly wrought by superior technology as well as by organizational ability and the spread of European diseases. The West Indian archipelago was peculiarily suited to the new technology represented by the cannon-armed, three-masted sailing ship. Archaeological evidence from the earliest wreck thus far discovered, from before 1513 at the Turks and Caicos Islands, suggests that ships went heavily armed with antipersonnel weapons, presumably to subdue Indian populations being raided for slaves (Keith, 1987). Swivel guns, or *versos,* that were too heavy to use on land offered tremendous firepower against human targets and in all directions.

The real significance of the cannon-armed ship was, however, first realized in the Indian Ocean. In the Caribbean it was used to subdue native populations already weakened by European disease. In the Indian Ocean, the Portuguese demonstrated that this powerful new technology could clear the seas of Christendom's fiercest rival, Islam. In 1498 the Portuguese navigator, Vasco da Gama, circumnavigated Africa and crossed the Indian Ocean to Calicut in the Indian state of Gujarat. His voyage was neither a commercial nor a diplomatic success because of

Islamic intrigues against him. It very clearly demonstrated that, if Portugal were to shatter the Islamic "blockade" of trade between Europe and the Indies, Portuguese arms would have to wrest "commercial control of the Eastern Seas from the Moslem merchants, who held the monopoly throughout the Indian Ocean . . . Calicut, where the eastern flow of trade met the western flow, was the center of the whole system" (Penrose, 1952, 71).

A second Portuguese fleet, led by Pedro Alvares Cabral, reached Calicut in 1500, bombarding that city after a vicious Islamic attack upon an abortive Portuguese trading post. Cabral's return with a highly profitable cargo of spices successfully challenged the Venetians' lucrative near-monopoly of the European/Indies trade in its eastern Mediterranean section. In 1502 Vasco da Gama shattered both Calicut and an Islamic fleet with fifteen Portuguese ships. This continuing success brought the Venetians to action. They helped fund the sultan of Egypt to equip a large fleet of Mediterranean-style oared galleys (Penrose, 1952, 77), which destroyed six of a fleet of eight Portuguese ships in the mouth of the river Chaul on the northwest coast of India in 1508, but at heavy cost. When the Portuguese returned the next spring, they found the Islamic fleet almost abandoned at Diu. The Portuguese boarded and burned every vessel, almost without a fight (Padfield, 1979, 63). This decisively ended the power of both the Islamic and Gujarati traders in the Indian Ocean and established a clear Portuguese supremacy (Pemsel, 1977, 41).

Thus in less than ten years after their first arrival on the coast of India, the Portuguese, aided by the cannon-armed ship, evicted both Islamic and Gujarati traders from the Indian Ocean and claimed its lucrative trade as their own (Diffie & Winius, 1977, 220). Unlike the Spaniards, who brought superior organization and force against a population lacking political organization and weakened by disease, the Portuguese triumphed in the Indian Ocean on the basis of superior technology. The Islamic and Gujarati populations were as well organized as the Europeans, immune to an onslaught of European disease because they shared the Eurasian disease pool, and certainly equal to the Europeans in ferocity. They fell before the cannon-armed, three-masted sailing ship (Cipolla, 1965, 137).

Protection costs, the single largest expense of the preexisting Islamic and Chinese pattern of trade in the Indian Ocean and the China seas, were also lowered by cannon-armed ships. All goods bound between India and China had to pass the Malay peninsula either overland or through the Strait of Malacca, long referred to by Chinese traders as home to the "barbarians of the sea" (Wheatley, 1961, chap. 5). European cannon-armed ships made short shrift of such pirates. The antipersonnel *versos* of the Portuguese allowed a small crew to cut a wide swath through any opposition. Cannon-armed ships lowered protection costs in trade to India and the East as drastically as the direct sea route from Europe around Africa into the Indian Ocean lowered transfer costs. As a result the Islamic world was progressively weakened and

the focus of European wealth shifted from the eastern Mediterranean to
the Atlantic.

The First World System of European Expansion after 1441

The pattern of European displacement of Islamic, Gujarati, and Chinese
trade activity, which began when Portugal cut into the sub-Saharan
slave trade in 1441, was fully established by the early 1500s. Europeans
used their cannon-armed ships to establish trading posts on the coasts of
the heavily populated and disease-resistant lands of Eurasia and Africa.
Where political organization allowed, as in an India fragmented by
religious division, they came to ally with various local states against
others to increase their penetration and control of the internal trade of
the region. Europeans avoided some areas, such as tropical Africa, literally like the plague: diseases vectored by climate-specific insects took a
terrible toll on Europeans who ventured inland. Not for nothing did
tropical West Africa become known as "the white man's graveyard."

Where demography and geography allowed, Europeans established
colonies connected to the core by wind-driven ships. These tended to
fall into two types: plantations and settler colonies. Both were concentrated in an America depopulated by European diseases. Plantations
were by far the more important at first, in particular in the Caribbean
archipelago and along the Brazilian littoral, areas where sailing ships
could move freely. All the European nations scrambled for Caribbean
sugar islands, but Portugal owned Brazil by virtue of the Treaty of
Tordesillas of 1494 (Palmer, 1965, 61). In the plantation economy, one
crop usually dominated at any given time, and importation of labor
from the Old World rapidly became necessary as disease took its toll of
the indigenous peoples. Here the Portuguese experience in the sub-
Saharan slave trade became vital, and a huge flow of black slaves crossed
the Atlantic to Brazil and the Caribbean (Curtin, 1969).

In tidewater North America, settler colonies became the norm. Tropical products such as sugar would not grow in the cold winters typical of
the area. As settlers struggled to survive, they found their living in
diversified agriculture rather than monoculture. Despite the later importance of tobacco in the colonial economy of North America, as late as
the American Revolution the principal export crop from the southern
tidewater colonies was ships' stores in the form of pitch, hempen rope,
and lumber. At least until well into the eighteenth century, therefore,
labor demands were low enough to be met by the importation of inden-
tured servants from Europe rather than slaves from Africa.

This simple, maritime cohesion was pioneered by the Portuguese and
Spanish and perfected by the Dutch, English, and French, although the
French regularly forgot themselves and spent their effort on land-based
expansion in Europe rather than maritime expansion abroad. Within
Europe a series of vicious wars, partly driven by the religious differences
between north and south codified by the Reformation, but always un-
derlain by the imperative of retaining the profits of long-distance sea

trade, obscured the unanimity with which Catholics and Protestants approached the outer world. As the Iberian star waned in the late sixteenth century, that of the Protestant states of northern Europe rose; it was driven by a mix of poverty, religious fanaticism, and intense scientific and technological development that focused most obviously on the ship.

English Improvements in Shipping in the Late Sixteenth Century

The first obvious improvements on the Iberian three-masted ship were English development of what contemporaries referred to as the "race-built" or "low-charged" ship and the use of more, larger, and cheaper cannon cast from iron rather than bronze or brass. After 1500 the addition of more and larger cannon to ships had resulted in a three-masted ship referred to as the carrack or "high-charged" ship. Carracks reemphasized the fore and aft castles, which mounted the heavier, fixed guns to allow fire in any direction. The high castles acted as fixed sails and made it difficult for ships to move to windward, and the concentration of cannon in the high castles made carracks dangerously top-heavy. Designers began to compensate for the castles by the 1540s. The flagship of Henry VIII of England, *Mary Rose*, was rebuilt with gunports no more than sixteen inches above her waterline, supposedly the first ship to use broadside guns. Yet *Mary Rose* retained her high castles and was lethally top-heavy. Sailing out of Portsmouth in 1545 to meet the French fleet, she was caught in a sudden squall, heeled over, and sank with all hands in a matter of minutes (Kemp, 1978, 90). In 1568 Sir John Hawkins had an equally bad experience. He was unable to sail the *Jesus of Lubeck* out of the way of the Spanish at San Juan de Ulloa; in a lower-charged ship, his young cousin, Francis Drake, was able to escape.

Such accidents sparked the "race-built" ship. Hawkins became treasurer of Queen Elizabeth's navy in 1577 and immediately set about cutting down the forecastle almost to nothing and drastically reducing the aft castle. Further improvements were an increase in the length-to-beam ratio and the first evidence of a serious attempt to streamline the hull underwater by emulating the shape of the fish (Kemp, 1978, 93–96). Both improvements increased speed. The "race-built" ship, named for its razed superstructure rather than its superior speed, became the backbone of the English fleet in the 1570s. Such ships, known as galleons (fig. 3-1), were enthusiastically adopted by the Dutch, and more slowly by the French, Spanish, and Portuguese (Kirsch, 1990).

Twenty years after the loss of the *Jesus of Lubeck*, Hawkins and Drake, both of whom had sworn vengeance against Spain, had it in full measure when the joint Spanish/Portuguese Armada of 1588 was destroyed. The Iberians were horrified by the ease with which the "race-built" English vessels moved against their high-charged fleet (Howarth, 1981, 126). Although the English lacked the firepower to decisively finish the Armada in open battle, their superior seamanship bottled up the Iberians at Gravelines on the Belgian coast, then forced them north around Scotland in a vain attempt to return home. Scottish weather blew many Iberian ships onto rocky coasts, and malnutrition, dehydration, and cholera resulted in severe undermanning. Drake's destruction of the Iberian stores of seasoned barrel staves at Corunna in 1587 helped: during the long voyage, water, wine, and oil all leaked away from barrels built with green staves. Of the 130 ships that set out fewer than half returned home, and "many, perhaps most, of those . . . were too badly damaged to be repaired" (Howarth, 1981, 243). On the ships that did return the loss of life, mostly from dietary problems and disease, was appalling: "of the thirty thousand or so who sailed, twenty thousand died" (Howarth, 1981, 243). It was the beginning of the end for Iberia as a major maritime power.

The second major English improvement was iron cannon. Bronze guns were lighter and safer than iron ones at the beginning of the sixteenth century, though they were between three and five times as expensive (Cipolla, 1965, 42). The Iberians were slow to adopt iron guns, in part because they had the wealth of the Americas at their disposal and believed they could afford the luxury of bronze. When they needed iron cannon the Spanish usually bought them from England, or they imported gun founders from the Spanish Netherlands and sent them home when the emergency was over (Cipolla, 1965, 34). Because England could not afford bronze cannon, it persevered in perfecting a smelting technology that produced safer iron cannon. By the end of the sixteenth century English cast-iron cannon were nearly as light and reliable as bronze guns.

Against the Armada the English mistakenly believed they could use superior seamanship and maneuverability to stay out of the range of the heavier Spanish and Portuguese guns, and could destroy the Armada with lighter, longer-ranging guns. The first part of the equation worked, but the second did not, and the English had to totally rethink their tactics after 1588. Subsequent English warships carried numerous heavy iron cannon, and their captains closed in for the kill.

Holland Achieves Hegemony in the World System

The other great actor in the demise of Iberia was Holland (Wallerstein, 1980). Holland capitalized quickly on English improvements in ship design and cannon founding, adopting the galleon in the last decade of the sixteenth century and buying iron cannon from the English, then licensing English technology. Exploiting their long-standing links to the Baltic, the Dutch set up cannon founding in Sweden, using high-grade local iron and the plentiful timber supplies of the region (Cipolla, 1965, 48). Superior ships to those of Spain, coupled with the reforms in the structure of land warfare wrought by Prince Maurice of Nassau early in the seventeenth century, brought the Dutch decisive victory over the Spanish in the early 1600s. Since Spain and Portugal were by then united under one crown, the Dutch were able to move in on the Portuguese carrying trade in the Indian Ocean and beyond and focus the wealth of the Indies upon Amsterdam rather than Lisbon.

The Dutch also made significant improvements of their own in shipping, most notably the design of the fluyt (fig. 3-2), the most significant bulk carrier since the beamy, deep-hulled cog (Cipolla, 1980, 274–75). The shallow-draft, three-masted caravels of the early fifteenth century were good exploratory vessels, and carracks and galleons were good warships. But none of the three were good merchant vessels, devoting too much space to armament, needing too many crew to raise and lower the sails, or being just too small. In the fluyt the vestigial castles of the galleon were reduced to nothing. Although broad in the beam, fluyts had even higher length-to-beam ratios than galleons: four to one in the first decade after their introduction in 1595; six to one thereafter (McGowan, 1981, 53). This gave them a good turn of speed, even with their relatively short masts and reduced sail area. Masts were short and pulleys were used extensively to control yards and sails in order to reduce the size of the crew needed, as well as the expense of masts (Cipolla, 1980, 274–75). Finally, fluyts were built in large numbers, out of cheap Baltic raw materials, to a standardized design using "organized capital [that] made it easy for would-be owners to borrow funds and enterprising builders to build speculatively" (McGowan, 1981, 53).

Fluyts were cheap to buy and cheap to run, thus highly profitable in the bulk trades. In 1605 "Sir Walter Raleigh reported that the ratio of tons carried per man of crew was 20:1 on Dutch ships by comparison with a ratio for English ships of 7:1" (McGowan, 1981, 53). Fluyts have been described as both slow and too weakly built to carry guns, yet at least one critical group of users regarded the fluyt as the best ship on the oceans for its purpose short of a genuine warship. Pirates found fluyts ideal: even with twenty to thirty guns added, a fluyt could outrun and outfight anything short of a warship or an East Indiaman, a highly defensible ship usually pressed into naval service during wartime. The fluyt's low labor requirements allowed more pirates to man the guns and to board victim ships, and a capacious hull awaited booty. Lightness and shallow draft allowed pirates to come ashore in shallow harbors to careen their vessels, normal harbors being, of course, denied them (Senior, 1976, 26–28).

The Reemergence of the Specialized Warship and the Emergence of Europe's Maritime States

By the midseventeenth century the old Mediterranean tradition of separate merchant and naval vessels began to reexert itself, at least in northern Europe (McGowan, 1981, 50). The Dutch and English navies kept the peace in northern waters so that their merchant ships could move unarmed. Only on the routes to the West and East Indies were defensible ships a necessity, since piracy was endemic in the Caribbean, along the Malabar coast, and in the Strait of Malacca. The protection costs of trade outside Europe were lowered by the use of cannon-armed ships, but cannon lowered the carrying capacity of West and East Indiamen and an understandable preference for a good turn of speed compromised hull design (McGowan, 1980, 24). Defensible ships ran to higher length-to-beam ratios and narrower underwater cross sections

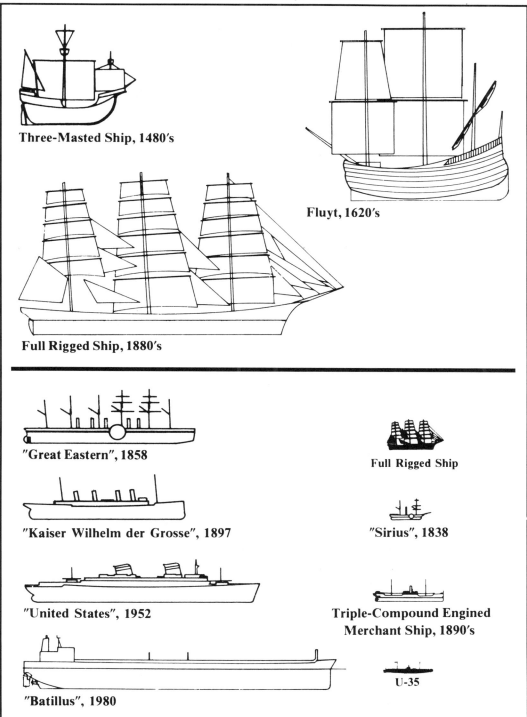

Three-Masted Ship, 1480's

Fluyt, 1620's

Full Rigged Ship, 1880's

"Great Eastern", 1858

Full Rigged Ship

"Kaiser Wilhelm der Grosse", 1897

"Sirius", 1838

"United States", 1952

Triple-Compound Engined Merchant Ship, 1890's

"Batillus", 1980

U-35

Figure 3-2   Selected merchant ships and liners. The power needed to move a ship at a given speed increases much more slowly than its carrying capacity. This guarantees that, if power and materials technologies allow, commercial ships will steadily grow bigger. Metal allowed the limits imposed by planetary ecology on hull size and mast length to be transcended, and steam and diesel engines gave power to drive metal vessels. Size was also the key to installing enough engine power to carry passengers across major oceans at great speed.

than European traders to allow them to outrun pirates. Such ships were fine for a trade in preciosities or even such relatively valuable goods as sugar but were not profitable enough for the bulk trade of northern Europe.

This situation persisted for some two hundred years after it developed in the early 1600s until the British navy, no longer encumbered by war with France, was powerful enough to impose a genuine Pax Britannica on the oceans of the world. After 1815 pirates could be certain that, when they were caught by a British naval captain, justice would be swift. The punishment for piracy was to be hung from the yardarms.

Yet the sea and ships were important not only to the long-term success of the economies of Holland and England, but also to the very emergence and survival of these sovereign states. Once a competent polity had emerged in England after the Saxon and Danish seaborne settlements were completed, England proved difficult to invade. Only William the Conqueror managed it, in 1066, and then only because the Saxon King Harold had let the navy slip into disrepair. Shakespeare summed up England's geographic advantage neatly in John of Gaunt's deathbed speech in *King Richard II*.

> This fortress built by Nature for herself
> Against infection and the hand of war;
> This happy breed of men, this little world;
> This precious stone set in the silver sea,
> Which serves it in the office of a wall,
> Or as a moat defensive to a house,
> Against the envy of less happier lands;
> This blessed plot, this earth, this realm, this England.

If England's nature as an island fortress in a world where ships were the only effective means of transoceanic travel is obvious, Holland's is not. Yet by the technical standards of the late sixteenth and early seventeenth centuries, the seven provinces that would eventually forge the Dutch republic in 1609 were as much an island as England was, but an island in a sea of marsh rather than in an ocean. The Spanish Inquisition, directed most against the Calvinist heretics of the Spanish Netherlands, broke its own back on the intricate network of river and polder along the mouth of the river Rhine. "The armies of Spain, although they reconquered much of the southern Netherlands, had insufficient strength to push beyond the line of the lower Scheldt, since the great rivers involved them in serious transport difficulties and since, moreover, the Dutch held command of the sea" (East, 1950, 259).

Coupled with the Dutch development of guerilla warfare, which was particularly effective in marshlands, and Prince Maurice of Nassau's reform of the Dutch army, the Spanish hopes of regaining the mouth of the Rhine, with its profitable entrepôt trade into the heart of Europe, faded after 1600.

> From successful defense, based on a web of river and marshland and on the wealth of financiers who fled Antwerp, the United Provinces passed to attack,

sending fleets to ravage the Spanish Indies and the Portuguese bases for the slave trade on the African coast, the sugar trade in South America and the spice trade in the Indian Ocean. While they fought they made their real gains by entering direct trade with the Spanish and Portuguese colonies, under-cutting the monopoly slave suppliers, undercutting the Seville traders in manufactured goods needed by the colonists, bringing the products of the plantations and the silver from the mines back to Amsterdam for processing and re-export—meanwhile selling their enemies the Baltic timber, naval stores and cast guns which they needed to counter their own naval assault! During this sustained commercial/naval offensive in the first half of the seventeenth century the Dutch practically beat the two Iberian powers from the seas, raising themselves in the process into the foremost naval, commer-cial and financial power, successor to the sea powers of the Mediterranean, but on a world scale. (Padfield, 1982, 5)

Dutch success brought envy from abroad, however. England's mer-chants wanted the same entrepôt trade for London, and England fought a series of wars with Holland over the issue. Although the three Anglo-Dutch Wars failed fully to resolve the issue militarily, ending in a "draw" rather than a clearcut victory, the Dutch hegemony over world trade established early in the 1600s slipped away to England in the last part of the century.

If the first Anglo-Dutch War of 1652–54 seemed indecisive militarily, not least because the English lacked the commercial muscle to pay for a long fight, it was not commercially. By taking huge numbers of Dutch ships as prizes the English merchant fleet grew almost equal to the Dutch. The second Anglo-Dutch War, of 1664–67, was decisive mili-tarily. The Dutch ran riot in the mouth of the river Thames, entering the principal naval dockyard at Chatham, burning several major ships, and removing the *Royal Charles,* flagship of the English fleet, to Holland (Padfield, 1982, 60–62). So freely did the Dutch move in the Thames that the surveyor to the navy, Sir William Batten, exclaimed, "By God I think the Devil shits Dutchmen" (Hough, 1979, 39). Dutch gains in the East Indies and on the west coast of Africa also loomed large and, for contemporaries, were not offset by the loss of Dutch power in North America when the English successfully took New Amsterdam, renam-ing it New York, to consolidate their hold on the Atlantic coast of that continent (Padfield, 1982, 63). Yet despite these successes, the Dutch merchant fleet did not fully regain control of economic power.

The third Anglo-Dutch War of 1672–78 tipped the balance back in favor of England, mainly because France intervened. England came to terms with Holland in early 1674 after the strategic loss of the battle of the Texel, when the French fleet abandoned their English allies (Pemsel, 1977, 54). The war between France and Holland continued until 1678, when internal dissension forced the Dutch to the treaty table. A result of that internal dissension was the rise to power in Holland of a young prince, William, who later acceded to the throne of England. French duplicity in this war, coupled with increasing English distrust of the later Stuart monarchs, brought about the alliance of the two Protestant powers against France, an alliance sealed when William and Mary came jointly to the throne of England in 1689.

The importance of the Anglo-Dutch Wars lay, however, not in the military sphere but in the economic. In the first war the British took as prizes some 1,700 Dutch merchant ships, most of them fluyts, and lost only 440 ships (Pemsel, 1977, 48). The weight of commercial shipping in the North Sea was thrown to England's favor (McGowan, 1981, 56). More prizes, though nowhere near so many as in the first war, followed in the second and third wars. By the end of the seventeenth century not only were England and Holland allied by royal marriage, by the common bonds of Protestantism and hatred of France, but also by relative equality in commerce. "After 1688 . . . Dutch capital began to flow increasingly to England, where it was invested in the English East India Company, the Bank of England, and the British national debt, as well as in newly developing industries. In this, the Netherlands paid 'the penalty of the commanding lead'. Dominance passed into the hands of her chief rival" (Wolf, 1982, 117).

Dominance in trade required dominance in numbers of merchant ships, and the number of British registered ships increased rapidly in the late seventeenth and early eighteenth centuries. Rapprochement with Holland gave London access to accumulated Dutch capital and British shipbuilders improved access to Baltic lumber. For merchant ships, if not for naval vessels, Baltic oak was acceptable. It was cut according to a careful quality control system developed by the Dutch and was cheap and reasonably plentiful. The Baltic supply, however, was not enough. The second component of the increased British registry was merchant ships built in New England. By 1700 there were about 1,000 of these, and by 1776 there were 2,343, "a third of the total British registry" (Albion, 1926, 246).

New Englanders innovated in ship construction. Because shipwrights were not among the early settlers around Massachusetts Bay, house carpenters took on the essentially similar job. They preferred quick construction with iron nails and often green lumber to the laborious doweling of well-seasoned wood. English ship construction techniques were conditioned by a lumber shortage and a need to make ships last, although the average life was under ten years because of the inevitable accidents. New Englanders conserved labor rather than raw materials, and even ships made of green lumber lasted four or five seasons. New Englanders also used water-powered saws, whereas the sawyers' guild prevented such labor-saving devices in English shipyards. The result of all this was that vessel construction costs were "30 to 50 percent less than in England" (Whitehurst, 1986, 14).

England's Drive for Hegemony in the World System

The late seventeenth and early eighteenth centuries saw a long struggle between England and France to succeed Holland as hegemonous power in the world system. Both cargo vessels and warships underwent only slight technical changes in that period (McGowan, 1980, 5). The larger warships of the last Dutch War of the 1670s could have fought in the line alongside the ships at the battle of Trafalgar in 1805. Also by the last Dutch War, the line-ahead tactics of naval warfare advocated in England's "Instructions for the Better Ordering of the Fleet in Fighting" of 1653 had been fully developed so that the melees of earlier engagements were avoided (Pemsel, 1977, 48).

Eighteenth-century warships were refined and enlarged rather than radically improved. First-rate ships of the line had three gun decks and mounted one hundred or more cannon. These ships were the most powerful weapon of their time and were expensive to build, man, and run. The most common warship of the eighteenth century and the best "all rounder," was the third-rate, two-deck, seventy-four-gun ship, of which the French seventy-fours were the greatest (table 3-1, fig. 3-1). By the time of the Napoleonic Wars the seventy-fours were the backbone of the fleet. At the battle of the Nile in 1798, for example, "8 out of the 12 French and all 13 of the British ships-of-the-line were 74-gun vessels" (Howard, 1979, 177).

The most spectacular warship development of the eighteenth century, however, was the frigate, a fast, handy ship with a single gun deck mounting thirty to sixty guns. Such a ship was not intended to stand in the line of battle. Although the British used many frigates, particularly on blockade work against the French coast, the finest frigates ever built were those of the newborn American republic. The American navy was born not out of the needs of war but out of the need to protect the huge merchant fleet of the new republic once the success of the Revolution removed the protection of the British fleet. American frigates were built for global service and were cheaper and easier to man than the seventy-fours used by Britain and France.

The American republic avoided the need for ships of the line and their attendant expense by avoiding major battles. The American frigate was ideal as a commerce protector or a skirmisher and, in the latter role, performed valiantly against British ships in the War of 1812. Against British and French frigates American frigates were markedly superior, in part because British frigates of the Napoleonic Wars and the War of 1812 were carefully copied from an earlier generation of American frigates that had been used to good effect against Britain in the American Revolutionary War. The American frigate the *Hancock* of 1776 was captured by the British in 1777. Under the British flag, the *Hancock* was responsible for the British capture of a second of the first thirteen American frigates, the *Trumbull*. These two ships were the model for all future British frigates, but American shipbuilders moved on to such big frigates as the *Constellation* and the *Constitution*.

The long struggle for hegemony between France and first England, later Britain, did not end until 1815, nearly 130 years after it had begun with King William's War of 1688–97. The years 1702–13 brought the War of the Spanish Succession, in which England and France each lost some fifteen hundred merchant ships. But whereas "French trade was

Table 3-1 Selected warship statistics

Ship	Country	Year	Ship type[a]	Length (m)	Beam (m)	Displacement (tons)	Horsepower	Engine type[b]	Fuel[c]	Maximum speed (km/h)	Radius of action (km)	No. guns	Broadside weight (kg)
Zeven Provicien	N	1665	3	—	—	1,423	sail	s	w	—	unltd	80	—
Invincible 74	F	1744	2	52.2	15.0	1,822	sail	s	w	24.0	unltd	74	387
Victory	GB	1778	3	69.1	16.0	3,500	sail	s	w	—	unltd	100	536
Constitution	USA	1797	f	62.2	13.3	2,200	sail	s	w	—	unltd	55	341
Warrior	GB	1861	f	128.0	—	9,283	5,267	SE	c	26.0	—	40	714
Miantonomoh	USA	1866	M	76.2	—	3,348	1,426	SE	c	16.7	—	4 × 15 in	816
Devastation	GB	1873	B	86.9	—	9,479	6,640	SE	c	25.6	4,425	4 × 12 in	1,282
Havock	GB	1893	d	56.4	5.6	279	4,600	TE	c	48.2	4,000	—	—
Royal Sovereign	GB	1891	B	125.1	22.9	14,376	11,000	TE	c	30.5	5,143	4 × 13.5 in	2,835
Arab	GB	1900	d	70.7	6.8	478	8,600	TE	c	57.0	6,000	—	—
Viper	GB	1899	d	64.1	6.4	350	—	T	c	62.5	—	—	—
Mikasa	J	1902	B	131.7	—	15,383	15,000	TE	c	33.4	8,530	4 × 12 in	1,542[d]
Dreadnought	GB	1906	DB	160.4	25.0	18,186	26,350	T	c	40.0	12,600	10 × 12 in	3,084[e]
Invincible	GB	1908	DBC	172.8	24.1	17,607	41,000	T	c	51.9	—	8 × 12 in	3,084
Courbet	F	1913	DB	168.0	27.0	23,470	28,000	T	c	38.9	15,940	12 × 12 in	4,627
Texas	USA	1914	DB	174.7	29.0	27,432	28,100	TE	c	38.9	14,800	10 × 14 in	6,350
Admiralty S	GB	1917	d	84.2	8.1	1,092	27,000	T	o	66.7	—	—	—
Queen Elizabeth	GB	1915	DB	196.9	27.6	27,940	75,000	T	o	46.3	13,840	8 × 15 in	6,968
Repulse	GB	1916	DBC	242.0	27.4	28,394	126,300	T	o	60.6	—	6 × 15 in	5,226
Hood	GB	1920	DBC	262.8	31.7	41,859	151,280	T	o	59.5	9,600	8 × 15 in	6,968
Nagato	J	1920	DB	215.8	29.0	33,245	80,000	T	c/o	49.6	10,550	8 × 16 in	7,944
Nelson	GB	1927	DB	216.8	32.3	34,493	45,000	T	o	43.5	26,500	9 × 16 in	10,044
Bismarck	D	1940	DB	251.0	36.0	42,344	150,170	T	o	55.8	16,500	8 × 15 in	6,000
King George V	GB	1940	DB	227.1	31.4	38,608	111,000	T	o	54.2	24,000	10 × 14 in	9,617
Javelin Class	GB	1936	d	106.1	10.7	1,717	40,000	T	o	66.7	—	—	—
Yamato	J	1941	DB	263.0	39.0	69,098	150,000	T	o	50.0	13,320	9 × 18.1 in	13,230
Iowa	USA	1943	DB	270.4	32.9	45,273	212,000	T	o	61.2	33,350	9 × 16 in	11,025

Sources: Brodie, 1943; Gibbons, 1983; Hogg & Batchelor, 1978; Silverstone, 1984.

[a] 2, sailing battleship with two gun decks; 3, sailing battleship with three gun decks; B, battleship; d, destroyer; DB, all big-gun dreadnought battleship; DBC, all big-gun dreadnought battle-cruiser; f, frigate; M, monitor.

[b] s, sail; SE, single expansion steam engine; T, steam turbine; TE, triple-expansion steam engine.

[c] c, coal; o, oil; w, wind.

[d] Muzzle velocity increased greatly after the *Royal Sovereign*, giving greater hitting power with less broadside weight. *Mikasa* was British made and had standard British twelve-inch guns.

[e] Only eight guns on broadside.

ruined and the country financially broken," England "could more than absorb her losses—indeed, the war brought England vast profits" (Pemsel, 1977, 63). This success set the seal on England's emergence as hegemon in the third world leadership cycle just as union with Scotland was securing for England full control over the British Isles.

Nevertheless Anglo-British hegemony was not unrestrained in the third world leadership cycle. The War of Austrian Succession pitted France, Spain, Prussia, and Bavaria against Britain, Holland, and Austria. The trade war resulted "in a loss of over 3,300 ships on each side: France's trade was now in ruins and even Britain had suffered more than usual" (Pemsel, 1977, 66). Colonial tensions caused skirmishing between England and France to break out in 1754 and renewed warfare in the Seven Years War of 1756–63. This, the French and Indian War to Americans, marked the high point of Anglo-British hegemony toward the end of the third world leadership cycle. The French were evicted from the Americas and the Indian subcontinent, French merchant ships were virtually driven from the seas, and the Royal Navy gained control of the seas of the world (Pemsel, 1977, 68–69).

The near collapse of Anglo-British power that followed was therefore all the more surprising for its suddenness. In the Royal Navy much-needed reforms were delayed and incompetence was rife in the officer class. The British lost control of the seas just long enough to allow the French to successfully support the colonists in the American revolutionary war of 1776–83. This success denied to Britain the third of its merchant fleet that sailed under colonial registry. Although Britain retained the rich lumber resources of Quebec and Ontario, it suffered a major setback. Britain won the economic struggle of the 1780s only by switching to a mineral-based economy and developing the factory system. In this crucial arena of economic conflict France lagged very badly, as Wallerstein's third volume makes clear (1989).

The military transition to British hegemony in the fourth world leadership cycle began with the French revolutionary war of 1793–1802 and concluded with the Napoleonic Wars of 1803–15. In the French revolution the British perfected the tactic of the blockade. Shortage of food in France required large imports of grain from America, which the British blockade aimed to intercept. Although the French broke the blockade in 1794 to bring a large merchant fleet into Brest, French losses were severe, and the blockade was not lifted again. As Alfred Thayer Mahan proclaimed of the British blockade, "Those far distant, storm-beaten ships, upon which the Grand Army never looked, stood between it and the dominion of the world" (Mahan, 1892, 118).

The Napoleonic Wars also saw the shattering of the combined French and Spanish fleets at the battle of Trafalgar in 1805, when British ships of the line were outnumbered thirty-three to twenty-seven and British guns three thousand to twenty-five hundred. Sixteen French and Spanish ships, most of them seventy-fours, were taken as prizes, and two French ships were lost, leaving Britain "in unlimited command of the sea" (Pemsel, 1977, 84–85). Control of the sea meant the total success of the blockade and the resumption of British control of merchant shipping worldwide. With Napoleon defeated decisively on land ten years after Trafalgar at the battle of Waterloo, British hegemony was complete and the Pax Britannica began.

The stunning success at Trafalgar was timely. A critical lumber short-age had developed, in particular when large numbers of ships were needed for blockade work. As Nelson commented in 1803, "the *Victory, Bellisle* and *Donegal* are the only ships fit to keep the sea; the rest are unfit for service until docked, altho amongst the finest and certainly best manned ships in the service" (Albion, 1926, 372–73). Admiral Collingwood said in 1804 that his flagship, the *Venerable,* was "so completely rotten as to be unfit for sea. We have been sailing for the last six months with only a copper sheet between us and eternity" (Albion, 1926, 377). The *Venerable* went to the bottom the following year. At Trafalgar Collingwood flew his flag on the *Royal Sovereign,* with which he broke the enemy line. Repairs to *Royal Sovereign,* which amounted "to nearly her original cost, were begun three months before Trafalgar, and were completed in eight weeks" (Albion, 1926, 384). Trafalgar seemed to relieve the lumber problem somewhat, but the blockade had to be maintained, and the War of 1812 renewed the problem, not least

United States frigate, *Constellation,* 1797, thirty-six guns. Among the finest sailing ships of war ever built, the U.S. big frigates could outrun any ship they could not outfight. American live oak construction guaranteed strength and durability. By the author.

when HMS *Guerriere* was captured by the USS *Constitution*. Part of *Constitution*'s success came from its being made of the finest American live oak, a wood so durable and strong that *Constitution* earned the nickname, "Old Ironsides" (Albion, 1926, 391).

After the Napoleonic Wars British shipbuilders recognized for the first time that lumber from other parts of the world, in particular American live oak and tropical hardwoods, was superior to English oak. Yet ship design remained essentially frozen at the levels reached in the 1790s. Military ships were the largest of their time, and their dimensions were fixed by the timber available for such structural members as the knees, which braced the ship's crossbeams to its sides, and the beams themselves. Knee timbers were scarce because their distinctive L shape came only from large, isolated trees, not from those in forests. The beam timbers limited the beam of the ship; no reputable shipbuilder would scarf (join) such a key structural piece. In a heavy beam sea a scarfed crossbeam could snap in an instant, causing the loss of the ship. The size of the beam timbers controlled the length of the ship as well as its beam because of the necessity to keep to a reasonable length-to-beam ratio. Scarfing was also avoided for mast timbers, and great masts were the key to the propulsion of any large, wind-driven ship. Put simply, the maximum size of wooden, wind-driven ships was fixed by the lumber sizes the planetary ecology afforded, and most ships were built well below that maximum.

THE DEVELOPMENT OF ENGINED, METAL-HULLED SHIPS

From almost its first appearance attempts were made to apply the steam engine to the ship. Denis Papin, a French-born Protestant, built a boat with a paddle intended to be moved by steam in 1707: "it was destroyed by Luddite professional oarsmen" (Cipolla & Birdsall, 1979, 148). Newcomen's first working engine appeared in 1712, and in 1736 Jonathan Hulls patented a design for a Newcomen-engined tugboat powered by paddle wheel to move warships in harbors, though it was never built (Armytage, 1976, 86). The more efficient steam engines developed by James Watt in the 1760s and 1770s first offered the possibility of low enough weight-to-power ratios to do useful work. Even so, the first commercially successful steam vessel had to wait until 1807. Such steamboats as Robert Fulton's Hudson River *North Star,* sometimes called "Clermont," were, however, devices intended to penetrate land masses, not to cross oceans; they properly belong with the land transportation that is discussed in chapter 4.

Steam-powered seagoing vessels—steamships as opposed to steamboats—had to wait for significant improvements in six areas: construction materials, water supply, boilers, transmission systems, engines, and fuel. These factors constantly interacted with each other even while developments in each one individually produced spectacular improvements in carrying capacity, reliability, speed, and range.

Construction Materials

All else being equal, the power needed to drive a ship at a given speed increases with the square of the frontal area of the immersed section of the hull (height × beam). Carrying capacity increases with the cube of its total frontal area. Simply doubling the immersed height and beam from two to four units of measurement requires a 1,500 percent increase in power (4 squared versus 16 squared) while it gives a 6,300 percent increase in load (4 cubed versus 16 cubed). As long as ships were built of wood their size was limited by the planet's ecology. They were not large enough to benefit from steam power, which required that they carry a large fuel supply in addition to passengers and cargo.

Large steamships were pioneered by Isambard Kingdom Brunel, the chief engineer of one of England's premier railroad networks, the Great Western Railway. By 1835 the Great Western was substantially complete and Brunel was looking for new ventures (Rowland, 1970, 73). When the railroad directors asked what he intended to do now that their line reached Bristol, then Britain's major port in the Atlantic trade, Brunel replied, "Why not make it longer, and have a steamboat to go from Bristol to New York, and call it the 'Great Western'?" (Brunel, 1870, 233). As a geographical strategy it was logical and obvious. Many of the passengers and much of the mail and commerce flowing west out of London to Bristol on the Great Western's rails were bound ultimately for New York. Brunel planned a large steamship to continue that service and to provide more comfort, regularity, safety, and speed than even the finest sailing packets.

Great Western was longer but narrower in the beam than the Royal Navy's first-rate ships of the line. The scarcity of appropriate lumber made such huge warships expensive to build and controlled their size and number. Brunel innovated in his hull design by using a substantial number of iron bolts for longitudinal strength (Rowland, 1970, 73; Smith, 1938, 41–42). He also adapted techniques developed by Robert Seppings, one of the greatest of the Royal Navy's directors of naval construction, to alleviate the lumber shortage in battleship construction (Lyon, 1980, 15–16); the combination of Seppings's diagonal riders with the use of "four staggered rows of iron bolts, 1.5 in diameter and 24 ft. long, which ranged the entire length of the ship throughout the bottom frames" (Rowland, 1970, 73), gave *Great Western* a hull that was strong, capacious, and affordable.

Great Western was not the first ship to cross the Atlantic under steam. *Sirius* preceded it by a matter of hours (fig. 3-2, table 3-2). *Sirius* was chartered by the British and American Steam Navigation Company when it became apparent that their custom-built ship, *British Queen*, would not be ready in time to win the crucial mail contract. The 700-ton *Sirius* had been built for service between London and Cork in southern Ireland and was hard pressed to carry enough fuel to make the Atlantic crossing, even with every available space crammed with coal (Kemp, 1978, 150). *Great Western* was bigger than *Sirius*, big enough to carry the coal needed to cross the Atlantic. Yet it was also at the maximum practical size for a wooden ship. A one-way crossing on the *Great Western* in the early 1840s cost a passenger $200, a phenomenal sum when a working man made just over a dollar a week. *Great Western* had strong competition from the Cunard Line and the Royal Mail Line, competitors who prospered because they operated at least four identical ships to give a reliable transatlantic service and because they were able to secure lucrative mail contracts (Emmerson, 1977, 25).

Along with *Great Western,* the wooden Cunard and Royal Mail ships revolutionized ocean travel. Average passage times fell by nearly fifty percent. Passengers paid a substantial premium to be at sea only an average of 16.5 days westbound or 13.5 eastbound; the best of the sailing packets averaged 30.5 days westbound and 20.5 eastbound (Smith, 1938, 47). More important was the steamship's amazing improvement in regularity. In the late 1830s a top-class sailing packet could take up to 48 days westbound and up to 36 days eastbound (Smith, 1938, 47). Once steam navigation was over its minor teething troubles, such lengthy voyages were unheard of. The time range was much compressed, and fourteen days became the "normal" voyage.

Brunel innovated even more seriously in his second and greatest ship, *Great Britain,* for which wrought-iron hull construction allowed a ship larger than even the largest first-rate wooden battleship. Besides being relatively small, wooden ships displaced more water than iron ones of the same weight. The greater the displacement, the greater the draft and the greater the amount of power needed to drive the ship through the water. For sailing ships, a deep draft was an advantage except when running before the wind because, with the wind anywhere except directly behind the ship, the hull has to "grip" the water well if progress is to be made. These considerations were irrelevant in a powered ship, and iron became an advantage.

After 1876 Siemen's method of producing mild steel allowed it to be substituted for wrought iron, first in naval ships, later in merchant ships, for reasons of expense. Steel was even lighter than wrought iron for a given strength and thus allowed lighter, stronger ships that rode higher in the water. They thus required less power to drive them, or they could be more heavily loaded (Lyon, 1980, 40).

The final improvement in construction materials came with the large-scale smelting of aluminum, a product mainly of World War II and the demand for lightweight aircraft. Aluminum superstructures on steel hulls gave lowered drafts and a lower ratio of hull deadweight to useful cargo weight. Expensive construction and problems caused by corrosion where aluminum and steel join in a salt water environment have usually limited aluminum superstructures to naval vessels. Two transatlantic passenger ships have enjoyed the luxury of aluminum construction, the *United States* of 1952 (table 3-2, fig. 3-2) and *Queen Elizabeth II* of 1970. Both ships are shallow enough to pass through the Panama canal (Hughes, 1973, 170). *United States* was also the last ship to hold the "blue riband" for the fastest crossing of the Atlantic (Hughes, 1973, 172). It was intended as a high-speed troopship and, hence, was over-engined. High construction and running costs rendered it unprofitable.

Table 3-2 Selected commercial liner statistics

Type of steam engine	Typical ship	Year	Horsepower	Displacement (tons)	Length (m)	Beam (m)	Days to cross Atlantic	Coal consumption (kgs/hp/hr)	Comments
Low-pressure	Great Western	1838	400–750	1,341	72	10.7	15–16	2.95	As large as a first-rate battleship
Expansively worked	Britannia	1840	440–740	1,173	63	10.4	14	1.99	First Cunarder, first mail contract
Expansively worked	Great Britain	1843	2,000	3,676	98	15.4	12		First iron hull, first screw drive
Expansively worked	Great Eastern	1858	8,297	27,824	211	25.2	10	1.95	No regular use on Atlantic
Compound	Britannic	1874	4,971	5,084	139		8	1.13	
Compound	Etruria	1884	14,500	8,250	158.5		6	0.91	
Triple expansion	City of Paris	1889	19,500	10,479	172		6	0.57	British built, American owned
Quadruple expansion	Kaiser Wilhelm der Grosse	1897	30,000	14,579	191		5.5	0.45	First superliner, first non-Anglo-American ship to hold Atlantic record
Turbine	Mauretania	1907	68,000	38,610	239	26.8	4.5	0.45	Held Atlantic record 22 years
Turbine	Queen Mary	1936	158,000	72,953	311	36.3	4	0.27 (oil)	
Turbine	United States	1952	240,000	53,850	302	31.1	3.5		First (and last) American "superliner"; all aluminum superstructure[a]

Sources: Hughes, 1973; Smith, 1938.

[a] To allow passage through the Panama Canal.

Aluminum construction has been largely restricted to warships, where draft is important if a ship is to work close inshore, and where speed and the extra warload that can be carried are of paramount importance. The danger of aluminum—that it burns more easily than steel—was underlined by the loss of the British destroyer *Sheffield* after being hit by an air-launched Exocet missile during the Falklands War with Argentina in 1982.

Water Condensers and Boilers

The second major improvement that made possible long-distance open-ocean travel by steamships was the development of the marine condenser. Until the late 1830s seagoing ships had to use salt water in their boilers and had to stop or switch to sails for twenty-four hours in order to clean out the salt every seventy-two to ninety-six hours (Smith, 1938, 130). Condensers allowed fresh water to be reclaimed after each conversion to steam. Hall's surface condenser was used in the *Great Western* and was central to the conquest of the Atlantic by steam. All Atlantic steamers used condensers of one sort or another from then on.

A third necessary improvement was in the efficiency with which boilers turned water into steam. The higher the pressure at which a boiler operates, the more efficient it is. In the 1830s the "normal" pressure in a marine boiler was about 5 pounds per square inch (psi). Contemporary railroad engineers were already using higher pressures, but they needed to save space and weight in their much smaller machines. Steamship engineers preferred the safety of lower pressure and could afford more space. Marine boiler pressures doubled in the 1840s and doubled again in the 1850s (Smith, 1938, 133). Thereafter progress slowed; 30 psi was normal until the 1880s, when pressures quickly accelerated to 200 psi, with a consequent great increase in efficiency (Smith, 1938, 189).

Improvements in boiler technology then came steadily. The French company, Belleville, began producing exceptional boilers in the 1890s (Smith, 1938, 253). Most other manufactures licensed or copied Belleville designs. The *Mauretania* of 1907 developed 68,000 horsepower using twenty-five boilers operating at 195 psi (Rowland, 1970, 175). The *Queen Mary* used twenty-four boilers operating at 400 psi to develop 180,000 horsepower in 1934. Only four years later, the *Queen Elizabeth* could use only twelve boilers at 450 psi for the same output (Kemp, 1978, 264). By 1962 the *France* could generate 158,000 horsepower using eight boilers at 910 psi, and the *Queen Elizabeth II* of 1970 needed only three boilers operating at 650 psi for 110,000 horsepower (Rowland, 1970, 214, 229). The last four ships had comparable passenger accommodation and speed, but at 65,000 tons, the *Queen Elizabeth II* was much lighter than the earlier ships because of her aluminum superstructure. The "Q.E.II" was not built as an Atlantic record breaker but as a multipurpose ship, working the Atlantic in the summer and the Caribbean and Pacific in the winter; a shallow enough draft to pass through the Panama Canal was needed.

Transmission

Two types of power transmission have been used in steamships: paddle wheels and screws. Paddle-driven ships had two advantages and many disadvantages. Their first advantage was in maneuverability, which is particularly important in vessels used in coastal service. They could be moved easily alongside piers because they could be steered by applying different amounts of power to each wheel (Paterson, 1969, 159), and they could be quickly reversed to avoid collision (McAdam, 1959, 17). In the early days of steam they had a second major advantage: they did not require the hole in the bottom of the hull which bedeviled early screw transmissions. Disadvantages included bulk, drag when under sail, the fact that paddle wheels came clear of the water or dug too far in when the ship rolled, and (for warships) susceptibility to battle damage.

Screw-driven ships were inherently more efficient once a way was found to keep the sea from entering the hull along the drive shaft bearing. The experimental screw-driven *Archimedes* of 1839 so impressed Brunel that he gave the *Great Britain* screw drive, but other builders were more cautious. Drive shaft bearings had to be replaced every few hundred hours, and accidents and near sinkings were common. Only in 1856 was a solution found by using an iron shaft working in a bearing made of a tropical American hardwood, lignum vitae (Smith, 1938, 78). By the early 1860s brass bearings had gone, and screw-driven ships reigned supreme.

The choice of transmission was also affected by the type of steam engine used. Even the largest reciprocating steam engines had major advantages in power transmission: they gave full torque or turning power from rest. The drive shaft from the engine to the wheels or screws could therefore run at engine speed. When steam turbines appeared, the output shaft had to be geared down to the much slower speed needed by the screw. Despite major advantages over reciprocating engines in other ways, turbine ships had to have expensive, unreliable, and bulky reduction gears in their transmission (Rowland, 1970, 181–82). Alternatives to mechanical transmission of turbine power include electric and hydraulic transmission, the latter being used only briefly in Germany for tugs. Turboelectric drive was first used in America in 1908 for fire tenders built for the city of Chicago (Rowland, 1970, 182). As with the German use of hydraulic transmission for tugs, the purpose was to improve maneuverability. Turboelectric transmission, however, was easier to look after, and it required less hull space than mechanical transmission. Its most famous and effective use was in the great French Atlantic liner, the *Normandie* of 1932. Its problems were complexity and requiring water in close proximity to electricity.

Engines

Four major types of steam engine have been used in ships: the simple engine regularly or expansively worked, the compound engine, and the turbine. The simple engine was used to begin with and was markedly inefficient at turning fuel into steam when regularly worked. As boiler pressures rose, however, engines could be worked "expansively." This meant letting a little high-pressure steam into the cylinder, then letting it expand as it cooled and thus work the engine more efficiently. Although expansive working was first used on the Atlantic in the steamer *Britannia* as early as 1840, experiments continued until the early 1860s. Proper expansive working was shown to nearly triple the horsepower available from a given engine and to reduce the consumption of coal per horsepower hour by one-third (Smith, 1938, 152–53).

The third important type of marine steam engine was the compound engine. Although the first seagoing compound engine dates to 1854, it was another twenty years before compounding was generally adopted. It took ten more years for the most efficient form of compound, the triple expansion engine, to be developed (Smith, 1938, 174). The problem was complexity, the advantage reduced fuel consumption. Expansive working cut fuel consumption by a third. Two-cylinder compound engines, which used steam first in a small, high-pressure cylinder, then again in a larger, lower-pressure cylinder, halved it. Triple expansion engines halved fuel consumption once again. Merchant steamships with such engines, built cheaply to standardized designs, could compete with even the finest steel-hulled sailing ships from the late 1800s (fig. 3-2).

The finest reciprocating engined ships ever put into service were the series of crack German liners of 1897 through 1906 that wrested supremacy on the Atlantic from Britain. *Kaiser Wilhelm der Grosse* (table 3-2, fig. 3-2), *Deutschland, Kronprinz Wilhelm,* and *Kaiser Wilhelm II* all held the blue riband in their turn, and all were driven by quadruple expansion reciprocating engines. Unfortunately, such huge reciprocating engines made these ships vibrate uncomfortably and took up a huge amount of hull space. Few such ships are preserved, but a visit below decks to the triple expansion reciprocating engined American battleship *Texas,* moored in San Jacinto Park just outside Houston, emphasizes just how much of a ship's hull was consumed by such engines when speed became of paramount importance.

By the 1890s radically improved steels and machine tools allowed precise engineering and very high-pressure boilers. Charles Parsons demonstrated the steam turbine, blowing very high-pressure steam at a complex "windmill." In 1897 he demonstrated *Turbinia* (illegally) at the naval review for Queen Victoria's diamond jubilee (*The Times,* London, June 28, 1897). With 2,000 horsepower *Turbinia* attained nearly 63 kilometers per hour, and not even the fastest destroyer in the Royal Navy could have caught her.

Initial attempts to install turbines in two Royal Navy destroyers, *Viper* and *Cobra,* met with little success. *Cobra* broke in half because of structural weakness (Smith, 1938, 278). In 1901, three years after these ships were laid down, the Glasgow and South Western Railway began construction of a turbine-powered steamship for service on the Fairlie to Campbeltown run. On the rocky fiord coast of western Scotland railroads were hard to build. Most of the population lived in fishing villages, and the most direct routes between settlements were by sea. With the rapidly increasing industrial wealth of the late nineteenth century, a lively holiday trade was springing up out of Glasgow. Speed was essential, 20 knots or 37 kilometers per hour being called for and achieved. The new steamer, named *King Edward,* "was markedly free of the rhythmic surging motion and vibration detectable even in the best of the [reciprocating-engined] paddle steamers" (Paterson, 1969, 159).

The success of turbines in the *King Edward* encouraged the Cunard Line to install what most contemporaries thought of as an untried if not downright dangerous technology in its new Atlantic liner, *Mauretania.* Together with her sister ship, *Lusitania, Mauretania* was deliberately designed to win back the blue riband for Britain after a decade of German dominance of the Northern Atlantic. To defeat the Germans the Cunard Line "took the unprecedented step" of setting up, in late 1903, a commission of experts to advise them (Hughes, 1973, 108). In March 1904 this commission reported unanimously in favor of turbine propulsion, and construction began later that year. Launched first, *Lusitania* held both the east- and westbound records briefly in 1907. *Mauretania* took the blue riband in 1907 and held it until 1929, the

Parsons steam turbine, 1902, 470 horsepower. Compact, efficient turbines made much faster ships possible as well as mass production of electricity. Ford Museum, by the author.

longest unbroken record in the history of the Atlantic (Hughes, 1973, 182).

Throughout the nineteenth century technical improvements, in particular in such high-technology areas as shipping and ship propulsion systems, came first in the commercial sphere. At the beginning of the twentieth century such innovations shifted to the military sphere. In *The Pursuit of Power* (1982), William H. McNeill emphasized that this was a crucial juncture in the rise of the West. From about 1000 to roughly 1900, Western Europe operated primarily as a market economy: after 1900 it began to return to an old principle in human societies, that of a command economy. As McNeill pointed out, military requirements became paramount after 1900 because the naval establishments, who were then the principal consumers of high technology, began to demand higher and higher levels of performance in the race with their adversaries. This happened first in an England struggling to hold on to hegemony, was quickly copied in Germany, and more slowly in America.

The development of the turbine neatly illustrates McNeill's point. Although Parsons believed the military would be his main customer, commercial interests adopted turbines earlier in order to carry passengers on lines where speed gave a comparative market advantage. The military lagged slightly. After the unsuccessful experiments in *Viper* and *Cobra,* the Royal Navy did not return to the turbine until the Cunard Line had made its decision of 1904. In January 1905 the Admiralty Committee on Designs met to consider proposals for a new battleship, and on October 2 of that year the keel was laid for the revolutionary, turbine-driven *Dreadnought* (Smith, 1938, 282–83). Compared with the Cunarders, *Dreadnought* was small, slow, and not very powerful (tables 3-1, 3-2). Thereafter, however, the power output and speed of military turbines and ships climbed in spectacular fashion compared to commercial ships. Military ship displacements increased only relatively slightly because of limits imposed on British ships by the need to pass through the Suez and Panama canals, and on German ships by the need to pass through the Kiel Canal. Atlantic liners had no such problems.

Fuel, Range, and Displacement

Wood and coal were the first major fuel sources for steamships, and coal was clearly predominant on British ships. The problems with coal were carrying enough to make the next coaling port, and the laborious and filthy task of coaling. Fuel consumption dropped steadily as engines improved, so that more and more powerful ships had greater and greater range.

Coal was never a problem on the Atlantic run. Good coal deposits were easily accessible to ports on either side. For voyaging outside the Atlantic coaling ports were important. Until the Suez Canal was opened in 1869, steamers bound for Australia around Africa had to refuel at coaling depots stocked by sail-powered colliers from England (Emmerson, c. 1980, 11–16). Interest in the Australia run strengthened after gold was discovered there in 1851. The fastest sailing ship took 100 to 120 days, a time that steamers reduced to about 60 days. The first steamers on the Australia run had, however, to refuel at St. Vincent, St. Helena, and Capetown.

Brunel designed his last great steamship, the *Great Eastern,* to have great range without recoaling and enough speed to reduce the journey time from Britain to Australia to thirty to thirty-five days (table 3-2, fig. 3-2). The *Great Eastern* carried just over 12,000 tons of fuel. At a predicted consumption of 185 tons per day (Emmerson, c. 1980, 20), this amount would have allowed the ship to steam around the world nonstop. But at a real consumption of 388 tons per day, *Great Eastern* used far too much fuel for such a trip. Australia's great coal deposits had not been discovered in 1860.

Other problems emerged with the *Great Eastern.* Ports were not set up to handle such large ships, and captains were used to ships no more than a third the length and one-eighth the bulk of *Great Eastern.* The ship's failure encouraged the British government to buy the Egyptian khedive's half share in the Suez Canal Company in 1873, four years after the canal was completed (Armytage, 1976, 165). The khedive's love of gambling and women had led him into financial deep water. When the French began the canal in 1859, while the *Great Eastern* was still being fitted out, British engineers had ridiculed the idea of such a canal, not least because of their confidence in Brunel. Certainly a sea route around the southern tip of Africa was geostrategically sounder than one through the Mediterranean, Suez, and Red seas. The failure of the *Great Eastern* discouraged large ships for some time. No ship exceeded the *Great Eastern*'s tonnage until the *Carmania* of 1905. The *Carmania* was built exclusively for Atlantic service, drawing too much water to pass through Suez.

Oil firing also helped the development of great steamships. Less weight had to be carried as 10 kilograms of coal were replaced by 6 of oil. Oil was cleaner and could be simply pumped on board without large amounts of human labor. Oil also needed much less labor at sea as it required no stoking (Brinnin, 1971, 429). Stokehold crews fell from about 350 to 50 on most major liners. The Russian, French, and American navies all experimented with oil firing of steam engines as early as the 1860s, and by the 1880s the Russian navy had converted most of its ships to it. Russia and America had good native supplies of oil; Britain, France, and Germany did not. Most British ships were converted to oil firing during and after World War I. During Winston Churchill's tenure as first lord of the Admiralty (1911–15) Britain invested heavily in Middle Eastern oil fields to fuel the best battleships of the Royal Navy.

Hull displacement was controlled by a complex set of interactions: structural material, engine type, boiler type, and fuel source. Minimizing displacement was critical if ships were to pass through the geostrategic bottlenecks at Suez and Panama, which greatly shortened the routes between the Atlantic and Indian and the Atlantic and Pacific oceans for northern hemisphere nations. Of the major structural materials, wood displaced the most for a given hull weight and size, then iron, then steel. The replacement of steel with aluminum in the superstructure further lowered displacement. Lightweight water-tube boilers had become normal by the early twentieth century (Smith, 1938, chap. 16), but numerous improvements occurred thereafter. Turbines were also lighter than comparable reciprocating engines: the *Dreadnought* was 5.3 percent lighter than it would have been with reciprocating engines of similar power. Oil was 40 percent lighter than coal for a given caloric value. Some measure of the improvements technology gave to hull displacement can be seen in the ability of three ships to just pass the Panama Canal. The coal-fired, reciprocating engined *Texas* of 1912 displaced 27,432 tons; the oil-fired, turbine-engined *Hood* of 1920 displaced 41,859 tons; and the oil-fired, turbine-engined, aluminum-superstructured liner *United States* of 1952 displaced 53,850 tons (tables 3-1, 3-2).

The Steam-powered Warship

The problems of the warship are not quite those of the commercial ship. Range and speed are important, but so are offensive and defensive armament.

Range

A military ship's range, in the sense of distance between refuelings, must be nearly three times that of a commercial ship. Range is from secure base to secure base, allowing a specified amount of fuel for combat. Sail-powered ships had truly global range, with no need to enter port for fuel. Commercial steamships could refuel wherever fuel could be purchased. As early as 1838 commercial steamships had transatlantic range because fuel was readily available on both sides. The first warship to have transatlantic range was HMS *Devastation* of 1873, which could have steamed across the Atlantic, fought, and steamed back again. Warships thus relied for long-distance cruising on wind power many years longer than commercial ships. The military often pioneered technical improvements in engines to increase range, but foremost in the success of steam warships was the development of secure refueling bases. Two fuel sources have been important with regard to range throughout the Paleo- and Neotechnic periods of warship development: coal and oil. Nuclear power became available in the Neotechnic.

Coal was easily available many places in Europe and eastern North America. In the southern hemisphere and in the Pacific it was more restricted, and coaling depots had to be established to serve military and commercial shipping in those areas. Because Britain was the major trading nation, the major naval power, and the country that most benefited from large-scale world trade in the nineteenth century, most coaling depots were British (Brodie, 1943, 116). Control over these depots was strategically important, as many German ships learned in World War I. Sailing ships carrying coal from England initially served the British coaling depots. The Falklands served ships bound around Cape

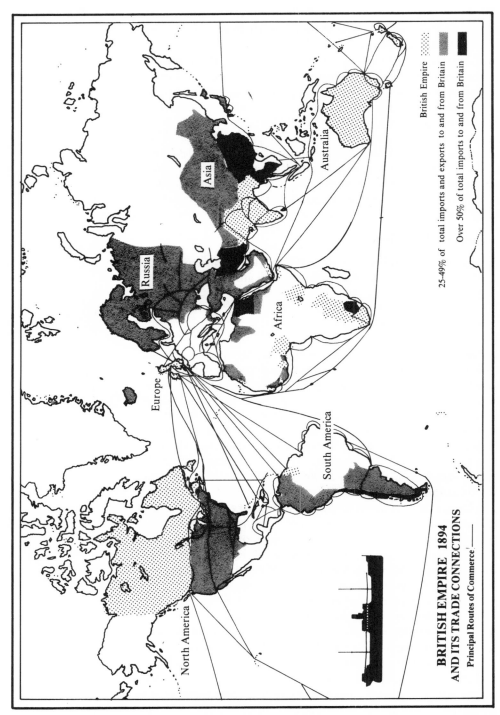

**BRITISH EMPIRE 1894
AND ITS TRADE CONNECTIONS**
Principal Routes of Commerce

Russia

Asia

Europe

Africa

Australia

South America

North America

British Empire

25-49% of total imports and exports to and from Britain

Over 50% of total imports to and from Britain

Figure 3-3   The web of British maritime commerce in the late 1800s. In the British maritime world system, British-built steamships burning British coal collected food for British workers and agricultural raw materials for British industry grown on land developed by British capital and transported out of the continental interiors largely on British-supplied and -financed riverboats and railroads. Redrawn from Parkin 1894.

Horn, South Africa served those bound around the Cape of Good Hope. Local coal eventually became available at Calcutta in India, Newcastle in Australia, Esquimault on Vancouver Island on the west coast of Canada, and Natal in South Africa. Other major coaling ports in the Pacific region included Singapore, Hong Kong, Hawaii, Samoa, and the Philippines. Most were stocked from England.

The bizarre imperial squabbles over seemingly worthless pieces of real estate in the Pacific during the late nineteenth century only make sense in the context of the need for secure coaling stations. Britain had an excellent network by the late nineteenth century (Parkin, 1895) (fig. 3-3). The United States acquired an equally good network, in part by careful dealings with Britain, which resulted in the acquisition of Hawaii, and in part from appropriate maneuvering with regard to the Spanish colonies as the Spanish empire collapsed. Of the Spanish possessions, Cuba provided a major base in the Caribbean, the Philippines one off the Pacific coast of Asia.

On other occasions the United States took a classically imperial stance. As the United States and Germany began to challenge Britain for hegemony, they came into conflict for the first time over the Samoan islands, which, strategically located in the central Pacific south of the Equator, were a critical coaling station. In 1889 the United States and Germany backed different sides in Samoa's civil war. In the best tradition of European jingoism U.S. sabers were rattled over German gains: the *New York World* demanded a U.S. protectorate over Samoa, and the *San Francisco Examiner* lamented that if only Americans would approve expenditures akin to those of Britain, the U.S. Navy could "reduce Bismarck's armada to a pile of iron filings" (Herwig, 1976, 15). In the event power was split three ways, among Germany, Britain, and the United States, a situation that was reinforced in the second Samoan "incident" of 1899. Then Rear Admiral Tirpitz, soon to be the major architect of German naval development, gave to Samoa "an importance that would be radically reinforced upon the opening of the Panama Canal" (Herwig, 1976, 38).

As fuel for military ships, coal seemed to have a major advantage. With no operational experience to contradict the supposition, it was thought that a full bunker would offer good protection against shells. A major disadvantage to coal, however, was the time consumed in coaling (Brodie, 1943, 117). Ships could not be refueled at sea but had to return to a secure base. Up to a quarter of a squadron's ships could be en route to or from their stations, or engaged in coaling (Hough, 1983, 28).

Coaling problems were considered to have been a factor in the appalling losses of the Russian navy in the Russo-Japanese war of 1904–5. The Russians lacked secure coal supplies along the route from the Baltic to Vladivostok, and reserves at Vladivostok were poor. The Russians overloaded their ships at Saigon, a port controlled by their French ally. Overloaded ships could not maneuver well against the Japanese, and losses at the battle of the Tsushima Straits (1905) were horrendous (Brodie, 1943, 133). Most of the Russian ships had coal on their decks and some were so overladen that their armor belts were submerged. Even more significantly, the Japanese predicted that the Russian ships, because of their limited range, would pass through the straits of Tsushima between the Korean peninsula and Japan proper and were able to close the trap on them (Brodie, 1943, 97).

A ship caught coaling could also be in grave danger. At the first battle of the Falkland Islands in 1914 the British battle cruiser squadron only just avoided being caught coaling by the Germans (Hough, 1983, 117). The German squadron destroyed by the British at the Falklands was hampered by the fact that it had to be accompanied by slow and unprotected colliers, whereas the British had the luxury of a relatively secure land base.

Oil fuel freed great ships from many of the problems of coal. Admiral Fisher, the architect of Britain's naval resurgence after 1900, argued for oil firing as early as 1882 on the grounds that it would "immediately increase the fighting capacity of every fleet by at least 50 percent" (Mohr, 1926, 3). A reduction in stokehold crew meant that a ship could carry fewer nonfighting men. Finally, refueling could take place at sea in all but the worst weather, and the total number of ships required to patrol a given area was reduced.

Although Britain had excellent coal reserves, the North Sea oil was undiscovered in the early twentieth century. Winston Churchill's decision during his tenure as first lord of the Admiralty to fuel the *Queen Elizabeth* class battleships of 1915 (table 3-1, fig. 3-1) with oil rather than coal was thus radical, if understandable. It required heavy British investment in the Middle Eastern oil fields, particularly those of Persia (Iran), and a concerted British military involvement to guarantee the flow of oil out of this area through the Red Sea, the Suez Canal, and the Mediterranean, to England (Mohr, 1926, 121–39; Brodie, 1943, 117). The decision to use Middle Eastern oil also depended on the realization that Mexican oil, to which the British had decent access in the first decade of this century, would be denied Britain by U.S. expansionism.

Germany, with no military control over the seas of the world, resolutely avoided oil-fired ships in World War I. The United States, with excellent internal supplies of oil, enthusiastically adopted oil firing a year after Britain. Oil firing was particularly important to the United States's rising naval power in the Pacific. Good steam coal was not available on the West Coast, but the southern California oilfields were.

Range thus increased rapidly. The best ranging class of coal-fired battleships ever built were the French *Courbets* (Gibbons, 1983, 196). The *Courbets* are perhaps best compared to the British oil-fired *Nelsons*. Despite carrying much heavier armament and having a higher cruising speed, the *Nelsons* had a 66 percent better range than the *Courbets* on only 47 percent more displacement (table 3-1). Oil fuel thus produced the possibility of a steam battleship with global range, if extremely high speed was not sought.

The major geostrategic impact of oil fuel in naval warfare was, however, under water. In 1900 the French had started building large numbers of submarines that were powered by electric motors and storage batteries. Although cheap, their range was very short and they were of use only in coastal defense (Brodie, 1943, 285–86). Electricity was the only "fuel" usable under water, cut off from supplies of fresh air. Attempts to combine steam for surface cruising with electricity for underwater propulsion never worked: furnaces rendered the interior of steam submarines intolerably hot (Garrett, 1977, 26). The proper combination turned out to be the internal combustion engine for surface cruising and electricity under water.

This was first used by the American John P. Holland in 1900. Holland's designs were quickly adopted elsewhere, particularly in Britain after 1904. Holland, however, used gasoline-fueled, spark-ignition engines. Highly volatile gasoline and the airtight submarine turned out to be a lethal combination, and many early submarines were lost to explosions (Garrett, 1977, 31). Much less volatile kerosene could be substituted for gasoline in spark-ignition engines, but such engines were hard to start and electric ignition was always a problem in a damp submersible. Compression ignition (diesel) engines also used a less volatile fuel, and they started easily and had no electric ignition to give problems. Germany began production of kerosene-engined submersibles in 1906, but their engines smoked heavily on the surface and they were easily spotted. Given the events of World War I, it is therefore ironic that it was the British who produced the first boats with diesel engines and electric transmissions in 1908 (Compton-Hall, 1984, 169). The first German diesel *Unterseeboots* did not appear until 1912 (fig. 3-2). From then until the development of nuclear power, submarine development was dominated by the use of diesel engines for surface cruising and electric motors for subsurface work.

Speed

Naval theory in the nineteenth century preferred large, relatively slow, heavily armed ships in combination with small, fast, lightly armed ships. Warships of the late nineteenth century were often slower than passenger ships of the same time. Only after 1908 and the HMS *Invincible* did warships begin to outpace passenger liners.

In nineteenth-century naval theory, speed was held to be a prerogative of small boats. French *jeune école* naval theorists challenged British conventional wisdom that battleships were the key to naval hegemony. Other naval powers struggling to catch up to Britain in the nineteenth century—Germany, the United States, and Japan—also adopted *jeune école* reasoning. Such thinking produced first the torpedo boat, then the torpedo-boat "destroyer," later shortened to just "destroyer." Both the republic of the North and the Confederacy used spar torpedo boats to great effect in the Civil War, although spar torpedo boats were essentially *kamikazes*. The self-propelled automotive torpedo invented by Whitehead in the late 1860s seemed a far better solution. The British produced the first boat that was custom-designed to use automotive torpedos in 1875 (Brodie, 1943, 278). Recognizing that hegemony in a maritime world system depended on control of the seas, Britain neglected no form of naval warfare in the nineteenth century. It was Britain that produced the first torpedo-boat "catchers" in 1886 and "destroyers" in 1893 (Smith, 1938, 262–63).

This obsession to remain ahead of all potential challengers in naval technology is well illustrated by the career of Admiral Sir John Fisher. Fisher demanded not only new battleships and battle cruisers, but also destroyers and submersibles. He specified higher and higher speed for Royal Navy ships as well as higher firepower, believing that a ship should be able either to outrun or to outgun potential opponents. The first ships to be specified in terms of speed were the 27-knot class (50 kilometers per hour) destroyers that were essentially all engine and boiler. Demands for more speed resulted in the 30-knotters. These were little more than "hot rods," and accidents were common. Turbines seemed to offer improved reliability, and *Viper* was laid down in 1898. On trial *Viper* showed nearly 68 kilometers per hour, but when her companion ship, *Cobra*, was lost because of poor hull design resulting from excessive weight saving, the navy hesitated to fully adopt the turbine. Thus, passenger ships first proved the potential of the new power plant. Nevertheless turbine-engined destroyers raised the speed of displacement warships to its practical limits at the first try. Thereafter displacement and seaworthiness increased and speed changed little. The main British fleet destroyers of World War I, the M, N, O, and P classes, made only 63 kilometers per hour, and those of World War II, the K class, were no faster, although much more powerful. Destroyers used acceleration and maneuverability as a defense, not guns or armor.

Speed in any type of ship is a function of two variables: engine power and hull size and shape. Some engine designs perform reliably at full speed; others do not. Reciprocating steam engines were theoretically not highly reliable, although they could be persuaded to move a ship as fast as a turbine (Smith, 1938, 283). The turbine was held to be more reliable at full speed, and in battle full speed is the speed a warship always uses. However, at the one great naval battle of World War I, Jutland, no vessel was disabled by mechanical failure (Hough, 1983, 280). German reciprocating engines proved as reliable as British turbines, although significantly slower.

Speed is also a function of hull design. Radical improvements in hull shape began in commercial shipping in the 1850s. In the late 1830s John Scott Russell perceived that if a ship made waves it was losing energy. He thus paid great attention to the way in which his hull designs cut through the water and propounded his "wave-line" principle between 1837 and 1843 (Emmerson, 1977, 22–26). In 1847 he began a serious career as a naval architect and designed several "wave-line" hulls for Brunel, most notably the *Great Eastern* hull (Brunel, 1870, 295). No radical changes in hull design have occurred in more than a century. Moving a hull through water imposes practical limits on speed. One solution, for light ships only, is to lift the hull above water on hydrofoils. Speeds up to 130 kilometers per hour are possible when the water is reasonably smooth and free of floating debris.

Armament

Offensive armament includes the ship itself, which can be used as a ram, guns, torpedoes, aircraft, and missiles. All were used in the Paleotechnic, and all were limited in one way or another.

From the mid-1800s guns and armor alternated in development. Shell-firing guns made wooden warships lethally vulnerable, but iron armor negated shells almost entirely. In the U.S. Civil War no shell pierced an armored ship (Brodie, 1943, 211), but guns fired slowly and accuracy was often poor. At the battle of Lissa in 1866 the Austrian admiral sank the Italian flagship by ramming (Brodie, 1943, 86). Immediately all navies started to build ram prows on their warships and even unarmed but heavily armored ships designed only to ram. Enthusiasm died down later in the century after two particularly nasty accidents. The German *Grosser Kürfurst* was accidentally rammed in 1878 by *König Wilhelm,* and 264 lives were lost (Gibbons, 1983, 87). HMS *Victoria* was rammed by *Camperdown* in 1893 during a normal but ill-conceived fleet maneuver, with 359 lives lost (Gibbons, 1983, 121).

Guns. The rapid improvement in guns from 1860 through about 1900 had three main stages: muzzle-loading smooth bores, muzzle-loading rifles, and breech-loading rifles. After 1900 improvements in fire control greatly increased the percentage of shells hitting their targets and the range at which guns could effectively be used, but guns themselves improved little.

Metallurgical problems in gun construction meant that a gun cast in a single piece was much more reliable than one with a removable breech block. Muzzle loading was thus preferred for safety reasons, but it was hard to adapt the idea of rifling to muzzle loading. Before 1880 and the development of nitrate-based explosives, gunpowder was used to throw the missile. Because gunpowder burns quickly, short gun barrels were used to reduce friction on the missile. Range was limited: 1,000 meters was considered extreme, and the inaccuracy of smooth-bored guns meant that ranges of 100 to 200 meters were preferred. The last huge smooth-bore guns were Dahlgren guns, used in the U.S. Civil War. Hoisting a missile into a muzzle loader was difficult and time consuming: often nearly ten minutes passed between shots (Hogg & Batchelor, 1978, 65).

By the late 1860s smooth bores were being replaced by rifled guns that could throw their missiles farther, although most naval officers were still trained to come in for the kill. The combined weight of guns and turret became a serious problem. Since range was a concern, designers of turret-armed ships were initially reluctant to give up sails. Unfortunately the combination of heavy, turret-mounted rifles and sails made ships lethally top-heavy. After HMS *Captain* went down with 473 men in

1870, all further development along these lines was abandoned (Gibbons, 1983, 73).

HMS *Devastation* of 1873 was thus the first recognizably modern, oceangoing steam warship. She had no sails and carried four muzzle-loading rifles of 305-millimeter bore weighing just over 35 tons each. The guns were mounted in two steam-driven turrets. She had transatlantic fighting range, not least because the U.S. monitor *Miantonomoh* had crossed the Atlantic in 1866 to underscore the United States's coming of age as a military power after the Civil War (table 3-1, fig. 3-1) (Hogg & Batchelor, 1978, 86; Gibbons, 1983, 42; MacBride, 1962, 39).

Breech loading and nitrate explosives developed almost side by side in the 1880s. Nitrate explosives did the same work with less than one-third the weight of the gunpowder they replaced, thus greatly lightening ammunition. Improved techniques for loading at all turret and gun angles developed in the 1890s, and intervals between shots dropped from minutes to twenty seconds or less. Guns, which had shrunk when breech loaders were developed, began to grow again. Improvements in metallurgy and gun construction, particularly by Armstrong in England and Krupp in Germany, allowed muzzle length (caliber) to double in a few years (Brodie, 1943, 223). Long-caliber guns and slow-burning powders gave tremendous muzzle velocity, thus great range.

The first great sea battle fought with such weapons was in the Straits of Tsushima in the Russo-Japanese war of 1904–5. The Russians opened fire at just over 13,000 meters and the Japanese at 8,000 meters, astounding contemporary observers (*The Times*, London, June 10, 1905). Range had increased thirteenfold in thirty-five years. It increased another 50 percent by the battle of Jutland in 1916 but then stabilized at around 20,000 meters. After Tsushima more emphasis was placed on the weight of high explosive thrown in a battleship's broadside, and on the rate at which long-caliber gun barrels wore. Guns almost identical to those at Tsushima were still in use in the naval battles of World War II.

Tsushima confirmed something understood in theory even before: the use of long- and short-range weapons at the same time was a mistake. The smoke and water thrown up by short-range guns obscured range finding for long-range guns. The Italian naval architect Cuniberti predicted this result in 1903. In 1905 HMS *Dreadnought* was laid down as the first ship to dispense with secondary armament; she thus lent her name to the class of future all-big-gun battleships (Jane, 1903, 407).

Broadside weight rose rapidly as four big guns gave way to eight or more. The eight 305-millimeter guns that *Dreadnought* could use on a broadside threw a total of 3,085 kilograms of high explosive at a muzzle velocity of 861 meters per second. The eight 380-millimeter guns on the *Queen Elizabeth* at Jutland in 1916 threw 6,967 kilograms at 747 meters per second, which meant a substantial reduction in wear on the barrel (Hogg & Batchelor, 1978, 117–18).

The range as well as hitting power of a ship's guns became a crucial factor in sea warfare. At the battle of the Falkland Islands in 1914 the British battle cruisers *Invincible* and *Inflexible*, with eight 305-millimeter guns each, opened fire at more than 16,000 meters. Despite their vast superiority in speed and firepower, they needed 1,174 heavy shells to sink the German squadron. The German Vice Admiral von Spee maneuvered with great skill but could not close the range to 13,500 meters, where his smaller guns could come into action (Hough, 1983, 115). The opening shots of the battle of the Dogger bank two months later were fired at more than 19,000 meters by the 343-millimeter guns of HMS *Tiger* (Hough, 1983, 133). The last encounter between dreadnought battleships occurred in 1944 when the USS *Mississippi* opened fire on the imperial Japanese navy's *Yamashiro* at 25,000 meters (Hogg & Batchelor, 1978, 129). Both ships carried 356-millimeter guns, and both were old: *Yamashiro* entered service in 1915 and *Mississippi* in 1918 (Gibbons, 1983, 213, 223). The greatest battleships ever built, the Japanese *Yamato* and *Musahi* of 1941 mounted 457-millimeter guns with a range of nearly 44,000 meters (Gibbons, 1983, 263).

Torpedoes. Torpedoes were used for the first time in the U.S. Civil War. The spar torpedo was almost a kamikaze weapon. It was an explosive device mounted at the end of a long spar attached to a light steam launch, like the lance of a medieval knight only with an explosive tip. Whitehead developed his "automotive torpedo" in the late 1860s; it was driven by compressed air and had a hydrostatic device to hold it submerged at a set depth. Many naval theorists, in particular the French *jeune école,* saw the torpedo as the ideal naval defense; they assumed that large numbers of small, fast, and, above all, inexpensive torpedo boats would be able to destroy large, expensive, big-gun battleships. To this day many European ports have extensive "breakwaters" intended to provide defense against surprise attack by torpedo boats rather than against the sea (Lyon, 1980, 51). Battleships thus had to be fitted with many small guns to be used at the close range at which torpedo boats had to fire their torpedoes. An even better defense was setting a thief to catch a thief, and torpedo boat "destroyers" were developed.

Destroyers also carried torpedoes for offense, but their lack of success against battleships in World War I defined their role as defending merchant shipping convoys against submarines. The major victim of the torpedo-boat destroyer thus turned out to be a rather different form of torpedo carrier, the submersible boat. Submersible boats had been experimented with in the American Revolution, in the Napoleonic Wars, and again in the U.S. Civil War. Like spar torpedo boats, these primitive, human-powered submersibles had a strong kamikaze element. The Confederate man-powered and spar torpedo–armed submersible *H. L. Hunley* succeeded in sinking the Union navy's *Housatonic* in Charleston harbor with the loss of five of the latter's crew: unfortunately *Hunley* and her crew of seven went down as well. In all, *Hunley* sank out of control five times with the loss of thirty-two lives (Garrett, 1977, 18).

The diesel-electric submersible armed with automotive torpedoes seemed to be another matter when it matured early in this century. But it was too slow to be a serious threat to the battleship, despite hopes for the submersible as great as the hope for torpedo boats twenty years earlier. Fisher, for example, believed that the submersible would allow dreadnought spending to die away to nothing. In fact the real victim of the submersible was the merchant ship, although this realization came late. Submersibles did, however, have a fundamental impact on the balance of world power. When imperial Germany finally adopted unrestricted submarine warfare in 1917, it nearly brought German hegemony to pass.

Between 1512 and 1917 a code of conduct existed for the treatment of unarmed enemy merchant ships traveling without escort. Propounded by King Henry VIII of England, and eventually supported by all nations, the "cruiser rules" allowed ships to take enemy crews and passengers as hostages and the vessels themselves and their freight as war prizes. In fact "prize money" was an important supplement to meager naval salaries in the era of sail warfare. If the encounter was far out to sea or the naval vessel could not spare a prize crew, the unmanned enemy merchantman could be sunk. The sinking by German U-boats of such British passenger ships as *Lusitania* and *Arabic* in 1915 and *Sussex* in 1916 caused U.S. public opinion to veer against Germany because of its obvious violation of the cruiser rules, as well as the loss of American lives (Garrett, 1977, 57).

Imperial Germany vacillated on U-boat warfare until her "high seas fleet" of battleships retreated after the battle of Jutland in late May of 1916. The unrestricted U-boat attacks on merchant shipping begun in January 1916 then intensified despite U.S. protests. One of every four British merchant ships to leave port was lost, as was a substantial fraction of neutral (mainly U.S.) shipping (Garrett, 1977, 62). It seemed to

Table 3-3 Submarine campaigns in World Wars I and II

A. World War I	1914–15		1916		1917		1918 (10 mo)
British merchant ships	231		288		1,052		527
British fishing ships	168		134		200		76
Allied merchant ships	76		344		708		314
Neutral (mostly American) merchant ships	93		332		679		186
Total	568		1,098		2,639		1,103
German U-boats	24		25		66		88
Kill ratio	23.7:1		43.9:1		40:1		12.5:1

B. World War II: Atlantic	1939 (3 mo)	1940	1941	1942	1943	1944	1945 (5 mo)
British, Allied, and neutral merchant ships	95	822	1,141	1,570	597	205	97
German U-boats	9	22	35	86	237	241	153
Kill ratio	10.6:1	37.4:1	32.6:1	18.3:1	2.5:1	0.85:1	0.63:1

C. World War II: Pacific	1941 (1 mo)		1942	1943		1944	1945 (8 mo)
Japanese merchant ships	6		134	308		549	155
American submarines	1		6	15		19	8
Kill ratio	6:1		22.3:1	20.5:1		28.9:1	19.4:1

Source: Polmar & Friedman, 1981.

ZEICHNET
KRIEGS-ANLEIHE
FÜR U-BOOTE GEGEN
ENGLAND

German propaganda poster from World War I emphasizing the geostrategic importance of the U-boat after unrestricted U-boat warfare began in 1917. Britain's maritime-based world system was vulnerable to such innovatory warfare. Texas A&M University Archives.

Germany that this was the key to defeating Britain, as it nearly was (table 3-3). Without control of the seas the British maritime world system could not continue. Even the price of U.S. entry into the war in April 1917 seemed to Germany worth denying Britain control of the seas.

The Germans made two miscalculations. They built too few U-boats

(Garrett, 1977, 55–56), and they seriously underestimated U.S. organizational skills and industrial capacity (Herwig, 1976, 122), both products of U.S. software innovations in managerial technology that Europeans were as yet unable to copy or, in some cases, even imagine. Running all merchant ships in convoys was an immense bureaucratic undertaking completely avoided by the British despite a thorough knowledge of the advantages of convoying from a long naval history. Building new merchant ships and protecting them with destroyer escorts built at the rate of one every two weeks was only possible with a large industrial capacity boosted by the work of Taylor, Gilbreth, and Ford.

In World War II Nazi Germany, although wedded to unrestricted U-boat warfare from the outbreak, repeated the two cardinal errors of World War I. Although the battle of the Atlantic (1942) nearly brought the British to their knees again (table 3-3), the Germans still had too few U-boats. The United States again built huge numbers of merchant ships and antisubmarine weapons, destroyers now joined by long-range reconnaissance and bombing aircraft (van der Vat, 1988, 352).

By 1941 the Germans had an advantage they lacked in 1916, U-boat pens on the Atlantic coast of France. World War II U-boats were essentially 1918 designs, but traveled nearly 1,600 kilometers less to their hunting grounds than in World War I, when their bases were on the North Sea coast of Germany. They could range as far afield as the Gulf of Mexico, whereas in World War I they could go no farther than the eastern seaboard of North America.

The United States pursued the same policy with greater success against Japan in World War II (table 3-3). Japan's maritime world system was even more vulnerable than that of Britain, Japan having no raw materials to speak of. The United States was able to strangle Japan's supply lines much more thoroughly than Germany could Britain's. Japanese industry was just not capable of producing enough destroyers and long-range aircraft to resist U.S. undersea attacks. U.S. submersibles were used against warships as well as merchant ships. By 1945 the latter were in such short supply that U.S. submersibles could not achieve a higher kill ratio than German ones.

Submersibles have, in the past generation, been replaced by true submarines—boats designed to spend almost their entire lives under water. Nuclear fuel generates electricity both to propel the submarine and to extract air from water for the crew. Submarines are now capable of traveling submerged as long as their food supply holds out (Garrett, 1977, 115). There are two types of submarines. "Hunter-killer" submarines find and destroy with underwater missiles the second type, missile-carrying submarines, which are designed to attack land targets. The ability of submarines to pass under the North Pole has drastically changed the strategic geography of the planet. By far the quickest route from Europe or the east coast of North America to eastern Asia is the great circle route over (or under) the North Pole (Garrett, 1977, 14). The polar ice also offers security to cruising missile submarines. Under the ice submarines are relatively safe from detection by earth-orbit satellites, which can detect them in the open ocean. Vast sums have been spent to quieten submarines, which, under the polar ice, are vulnerable only to each other. One major scandal has erupted over the sale of such U.S.-developed technology to the Soviet Union by Swedish and Japanese licensees.

Naval Aviation. Despite its success the submersible was an unglamorous, often unwanted stepchild of Neotechnics. By the end of World War I many forecasters believed that the aircraft carrier would become the naval weapon of the future. This hybrid of Paleotechnic steam power and the Neotechnic internal combustion—engined aircraft it carried seemed to prove its worth in World War II.

Some politicians came to believe that one of the root causes of World War I was the naval armament race that began in the late 1890s. At the Washington Naval Treaty of 1922 the United States, Britain, France, Italy, and Japan agreed to reduce their battleship strength and limit the size of new warships in the hope of preventing future sources of conflict. The United States and Britain were to have parity in capital ships, but Japan was to be held to three ships to each of the others' five. This 5:5:3 ratio was the source of much later friction between the United States and Japan. France and Italy were each allowed half the Japanese total (Preston, 1980, 25).

The Washington Treaty was more relaxed about aircraft carriers. Both the United States and Japan were permitted to convert the incomplete hulls of two large capital ships into carriers, whereas Britain had to make do with the hulls of two smaller ones (Preston, 1980, 31). If the Japanese lacked parity in battleships after the Washington Treaty, they made up for it with parity in aircraft carriers.

Although the treaty focused attention on the aircraft carrier, so did the needs of geostrategy. The failure of the United States to assume military as well as economic hegemony in 1919 created a power vacuum. The Germans regrouped and made their second bid for hegemony, developing their high-technology version of the Neotechnic world system. The Japanese saw a power vacuum in the Pacific, with the United States clearly unwilling to play what many Americans saw as the role of a colonial power in the European mold. Japan and the United States therefore moved from reluctant alliance during World War I to increasingly open hostility in the 1920s and 1930s.

Both faced the possibility of war in the Pacific. Ships could easily "lose" themselves in such an open ocean, and no land-based aircraft could provide reconnaissance coverage. The aircraft carrier was a logical fleet scout, but it proved to be more than that. In the 1930s improvements in the quality of aircraft engines, airframes, and weaponry turned airplanes into formidable offensive weapons. Dive bombers and torpedo-carrying attack aircraft were impossible to stop except by carrier-based fighter aircraft.

This development had implications for sea war everywhere. In 1941 the German *Bismarck* was one of the finest battleships in existence, taking only eight minutes to sink the British battle cruiser *Hood,* a ship with more firepower if less armor. Yet *Bismarck* was spotted from the air at a crucial juncture in its attempt to reach the safety of the French coast and German land-based airpower and its rudder was jammed at 15 degrees by a torpedo launched by a plane from the carrier *Ark Royal.* *Bismarck* was reduced to steaming in a circle, allowing two British battleships to move in for the kill (Hough, 1979, 206–15). Naval historians who argue whether *Bismarck* was scuttled or despatched by torpedoes launched by a British destroyer miss the essential point. Airpower created the opportunity for the older technologies of naval warfare to come into play.

Six months after the British demonstrated the possibilities of naval airpower in an open ocean, the Japanese demonstrated its realities. On December 10, 1941, land-based "Nell" bombers and "Betty" torpedo-bombers of the Japanese naval air force attacked and sank one of the most powerful battleships in the British fleet, the brand new *Prince of Wales,* together with the battle cruiser *Repulse.* The British logic for sending these ships into Japanese-controlled waters to try to relieve the attack upon Singapore was implacable. "No major warship had yet been sunk by attack aircraft while at sea and many had survived such attacks" (Middlebrook & Mahoney, 1979, 241). If the *Prince of Wales* and *Repulse* had got in among the Japanese invasion barges at Singapore, they would have wreaked tremendous havoc. As it was, they were the first ships to demonstrate how critical the new technology of naval airpower had become. The Japanese lost three planes and 18 crew from the sixty-six planes that pressed home their attack. The British lost two capital ships, 840 lives, the battle for Singapore, and the eastern component of the British Empire (Middlebrook & Mahoney, 1979, 220–35).

The central role of the aircraft carrier in naval war was first convincingly demonstrated by the British attack on the Italian fleet at Taranto in November 1940. Torpedo-carrying "Swordfish" aircraft slaughtered the anchored Italian ships even more thoroughly than Japanese carrier-based aircraft slaughtered the U.S. ships caught in Pearl Harbor on December 7, 1941 (Pemsel, 1977, 121, 125). Both Taranto and Pearl Harbor were, however, attacks on sitting ducks. The carrier war that developed between Japan and the United States in the Pacific was a mobile one. At the battle of the Coral Sea in May 1942, neither fleet saw the other and all attacks were by aircraft. At the battle of Midway one month later, the Japanese lost so many carriers to a brilliantly led U.S. attack that they were thereafter on the defensive (Pemsel, 1977, 128–31; Keegan, 1989, chap. 3). Without superiority in carriers the Japanese progressively lost control of the Pacific, and Midway marked the turning point of the war.

After World War II carriers seemed to be the critical weapon of war. Problems adapting them to jet aircraft were solved by improved launching and landing systems, and carriers controlled the seas. The nuclear submarine, however, has now joined the carrier as dominant weapon (Modelski & Thompson, 1988, 83–85). Armed with missiles and provided with earth satellite reconnaissance information, the submarine had the additional advantage of surprise. Carriers can no longer hide in the open expanse of the ocean. Although nuclear fuel has given carriers the same free range enjoyed by nuclear submarines, the range of attack aircraft is limited, and the endurance of the carrier is limited by the aircraft fuel it can carry. Carriers today, like battleships after World War I, have been challenged by new technology. They remain useful in circumstances where land-based airpower cannot be deployed, such as off the coast of Vietnam in the Vietnam War of 1964–73 or in the Falklands War of 1982, but they are immensely vulnerable outside such limited wars.

Neotechnic Commercial Shipping

Commercial shipping has benefited from the Neotechnic in many ways. A shift to internal combustion engines has been almost universal. Steam boilers and fresh water supply were long recognized as the weak points of the steamship, and attempts were made in the first decade of the twentieth century to substitute internal-combustion engines powered by gas produced directly from coal (Smith, 1938, 325). Producer-gas engines used electric ignition, however, which was always troublesome in a wet hull. The problem was resolved by Diesel's compression ignition engine.

By 1912 there seem to have been some three hundred diesel-engined merchant ships at work, although it was in that year that the first really successful diesel ship appeared, the Danish-built *Selandia* (Smith, 1938, 330). Diesels are ideal for ships. They need no expensive, leak-prone boilers or water condensers, nor do they rely on electricity for ignition. Like the steam-powered reciprocating engine and unlike the steam turbine, they turn slowly and thus do not need complex and expensive mechanical, hydraulic, or electric transmissions. The screw can be coupled directly to the engine's output shaft. After World War I German manufacturers had a surplus of diesel power plants built for submarines (Smith, 1938, 337). Such shipyards as Blohm und Voss quickly adapted former submarine engines to commercial use (Prager, 1977, 120–21).

The real revolution in merchant shipping began, however, after World War II. Two major events in the 1950s, and one more in 1962, conspired to produce it. After World War II it became clear that a genuinely global economy based on the United States as the critical "hinge" between the long-established Atlantic trading world and the new Pacific world was emerging. The rebuilding of Japan as a major industrial and commercial power was the key to some of this trade, but U.S. and Canadian raw material resources were also critical, Japan having no raw material resources of consequence.

The second event affecting shipping in the 1950s was the closing of the Suez Canal by the Egyptians in 1956. The United States refused to support British and French efforts to reopen the canal militarily, and the immediate consequence was a cutoff of the flow of Middle Eastern oil to Europe. Shipowners began to invest in highly specialized bulk-oil carriers to transfer oil from the Persian Gulf to Europe around Africa. With size limits no longer imposed by the Suez Canal, larger ships proved more profitable than the earlier generation of small ships. Very large crude (oil) carriers emerged in the years that followed Suez. Until 1956 ships up to 35,000 tons had been able to transit the canal. The first specialized bulk-oil carrier of 1950 weighed in at 28,000 tons. The first 100,000 tonner was launched in 1959 and by 1980 the first 500,000 tonners were in service, monster carriers such as *Batillus,* 414 meters long and 63 meters in the beam (fig. 3-2). Operating costs fell drastically. In 1956 the *extra* cost of moving one ton of oil around Africa instead of through Suez was $7.50. By 1970 the *total* cost of moving one ton of oil from the Persian Gulf to Europe around Africa had fallen to $3 (Corlett, 1981, 25–26).

The third event was containerization. The old break-of-bulk ships, where small amounts of goods were slowly and unreliably loaded and unloaded by largely manual labor, were swept away. Ships that spent a day in port for every day at sea were replaced by ships that spent less than one day docked for every ten days at sea (Corlett, 1981, 9, 13). Labor efficiency improved drastically, both in the ports and on ship. A third of the old dock work force handled goods in containers at ten times the old rate, and crew costs on shipboard dropped to 10 percent of the old figure per ton-mile. Finally, breakage and loss fell to 10 percent of the figure for break-of-bulk handling. "This was even more of a revolution than that . . . following the introduction of really efficient steam engines" in the late nineteenth century (Corlett, 1981, 13).

Yet containerization was as much a software as a hardware revolution. What had been a business composed of independent shippers was converted into one component of a totally integrated transport infrastructure. By 1962 the American Standards Association had developed standardized containers, and they were adopted by the International Standards Association in 1965 (Corlett, 1981, 12). As with so many developments of the Neotechnic, containerization has emphasized management skills and has entailed high capital cost but has delivered a standardized product in volume at low unit cost. The origins of containerization came from a U.S. trucking company, MacLeans, not from the shipping industry. In 1956 MacLeans bought a shipping line to improve movement of goods between Houston and New York and fitted two oil tankers with platforms above the tank deck to carry standard thirty-five-foot truck trailers detached from their chassis (Corlett, 1981, 11).

Between the 1950s and the 1980s shipping was revolutionized in terms of cost, speed, efficiency, reliability, and economies of scale as the tonnage of merchant vessels and the number of miles traveled accelerated. As with much else in the Neotechnic, however, this revolution developed on a military base. Ships such as the landing ship tanks

(LSTs) of World War II were the basis for modern drive-on, drive-off ferries as well as for the more important roll-on, roll-off (Ro-Ro) ships that have made the global trade in motor vehicles possible. The large-scale organizational systems begun in World War I to manage an internationalizing military economy were extended in World War II. After both world wars such large-scale efforts became "privatized" as multinational corporations became a supranational form of business organization. Managing the global peacetime economy seemed no different than managing a global war economy: both operated on the principle of command rather than market systems.

From around A.D. 1000 shipping has been central to the emergence of the capitalist world system. Sir Walter Raleigh's dictum has been demonstrated time and again. Each succeeding hegemon has dominated the seas of the world and, for the first four hundred years of global capitalism, has been the major nation trading on the seas as well. Only the advent of the U.S. continental version of the world system and the German technified world system have changed this.

Pax Britannica

The most convincing demonstration of the importance of sea trade and sea power in the world system thus far came with British hegemony after 1815. Britain was also the only polity to have developed a fully mineral-based economy by 1815. As Germany and the United States developed mineral-based economies in the mid-1800s, Britain innovated in maritime trade. Cheap, steel-hulled merchant ships, efficient triple-expansion engines, standardized designs, and a global network of coaling stations and telegraph lines to keep ships fully utilized helped expand British commercial power even as British industrial power declined.

Only industrial Britain had the productive capacity to build the merchant steamers needed to replace sailing ships and the warships to protect them. Steamships represented a radical increase in carrying capacity and reliability and thus greatly lowered transportation costs. They were probably as revolutionary a change as that initiated by the Dutch in the early seventeenth century with the development of the fluyt and the return to the specialized warship. Steamships had relatively short range and so needed coaling stations around the globe. Steamships also needed good transmission of information to ensure that expensive capital did not lie idle. By the late nineteenth century, as Parkin showed at the time (1894), Britain had created an efficient global network of refueling stations and submarine telegraphs to achieve these ends.

Britain's position as hegemon was challenged throughout the nineteenth century by the new American republic, France, and Germany. For various reasons, most notably Britain's productive capacity in both cotton textiles and steamships, but also because the challengers frequently allowed themselves to be diverted, none of the challenges succeeded.

The new American republic came closest to a successful challenge in the Napoleonic Wars and the War of 1812. Britain was too engrossed in war with France to cope with the steady encroachment of New England's shipowners on global trade. But when the Napoleonic Wars were over, British shipping quickly returned to dominance. The Navigation Acts, the first of which in 1651 began the first Dutch War of 1652–54, now denied British ports to American flagships. American merchants were pushed into a search for new markets, which led them to the diversions of natural population increase and the settlement of their own continental interior.

America also mounted a relatively successful military challenge, albeit on a very small scale, with a powerful new class of frigates, such as the USS *Constellation* and *Constitution,* which humiliated the Royal Navy. Ship for ship the American navy was a formidable opponent up to the level of the classic third-rate battleship, the two-deck seventy-four-gun ship. The thirty-eight-gun *Constellation* bested a French fifty-four-gun frigate, and the forty-four-gun *Constitution* proved superior to any English frigate. *Constitution* had a broadside weight only slightly lower than a seventy-four and much finer lines, which gave great speed. But the new American republic had no first-rate battleships, and those that were built after the War of 1812 proved dismal failures.

The American republic posed a second but lesser challenge after the Civil War. The federal navy emerged from that struggle with the most powerful class of battleships afloat, the low-freeboard, screw-propelled, turret-armed ships that owed their class name to Ericsson's *Monitor* of 1862. *Monadknock* class monitors could dictate terms to any ship they met, and the American republic deliberately displayed them to the world after the Civil War. *Monadknock* steamed to San Francisco via Cape Horn in 1866, and *Miantonomoh* was towed to Europe. The English response was *Devastation,* a ship with genuine transatlantic range (fig. 3-1, table 3-1) (*Conway's,* 1979, 23). Ericsson recognized the English threat when he said that *Devastation* could "steam up the Hudson in spite of our batteries and our monitors, and dictate terms off Castle Garden" (Sandler, 1979, 243).

In the real world of the U.S. Civil War, however, raiders such as the CSS (Confederate States Ship) *Alabama* devastated the merchant fleets of the republic, forcing the survivors either from the seas or "to seek shelter under foreign flags" (Sandler, 1979, 159). The British move to large-scale steamship production that followed the U.S. Civil War finished the job the Confederate navy had begun, and the United States lost any chance of commercial dominance. Even though the Navigation Acts were repealed in 1849, the United States did not yet have the industrial capacity to compete with Britain at building steamships, nor was she yet able to put together the global coaling and communications infrastructure such ships needed.

As in 1815, the American republic after the Civil War thus largely retreated to the exploitation of the continental interior. Military expenditures were begrudged, and the fleet of monitors, most of which were ironclads built with green wood hulls, soon rotted away. As Britain's naval attaché in Washington said at the time, Americans were soon lured away from the sea by "high outside wages, and elastic and responsive internal commerce, [and] new fields of industry and labour in the West" (Sandler, 1979, 69).

France, whose merchant fleet was shattered during the Napoleonic Wars, had the same commercial and industrial disadvantages as the American republic, plus the problems posed by Germany, an increasingly hostile neighbor. French industry was never equal to that of Britain in the nineteenth century. Although the French were the first to clad a wooden-hulled warship (the frigate *Gloire* of 1860) with iron, the British responded by building the first iron-hulled frigate, *Warrior*, in 1861. Every French advance was more than matched by the British, who produced it in larger numbers. By the end of the nineteenth century the French were reduced to copying the British. Advances propounded by *jeune école* theorists, such as the torpedo boat and the torpedo-armed submersible, found even greater acceptance in the Royal Navy. Like the Americans, the French therefore stressed the development of an internal market, although unlike their neighbors they held their population levels down (Flandrin, 1979; Parkin, 1894). They had ceased to be a major challenger by the end of the nineteenth century.

It was thus Germany that posed the major threat to Britain in the late nineteenth century. Vast industrial investments, on a scale even the United States did not match at the time and paid for in part by the reparations exacted from the French after the Franco-Prussian War, allowed the Germans to begin to match British industrial production, then the British merchant fleet, and finally the British battle fleet.

U.S. industrialization aided German advancement. After the Civil War the republic of the North launched, behind tariff walls erected by the Republican governments of the time, a period of massive industrial growth fueled by the vast profits of U.S. agriculture. Yet the United States was desperately short of labor, and most of the natural increase of old-stock Americans was soaked up by the westward expansion of the farming economy. Hordes of immigrants were necessary to fuel industrial growth. The states around the Baltic provided the bulk of these immigrants after 1870: first Germany itself, then the Scandinavian countries, then the states caught between a Germany that was expanding aggressively east and a Russia that was expanding aggressively west. German shipbuilders and carriers prospered. The Hamburg-Amerika and Norddeutscher Lloyd shipping lines built a steady business, initially using British-built ships. By 1895, however, most of the Norddeutscher Lloyd fleet was aging, and the line turned to Baltic shipyards at Stettin and Danzig for replacements (Hughes, 1973, 89). The Stettin-built *Kaiser Wilhelm der Grosse* was destined to wrest the blue riband from British hands in 1898. In 1898 Norddeutscher Lloyd landed more first- and second-class passengers at New York than the Cunard Line and carried 24 percent of all the passengers in the northern Atlantic (Hughes, 1973, 92). Between 1881 and 1891 Norddeutscher Lloyd landed 738,668 passengers at New York and the Hamburg-Amerika Line, 525,900. The White Star, Cunard, Inman, and Guion lines out of Liverpool carried 1,291,859 between them: barely 27,000 more than the two Baltic carriers (Hughes, 1973, 89). The significance of this buildup in commercial shipbuilding was that any shipyard capable of turning out crack Atlantic liners was also capable of turning out first-rate battleships.

In the Franco-Prussian War of 1870 the Prussian army won a decisive victory over the French. The North German federation had a small fleet that grew out of the Prussian navy after 1867, but most of the ships were built in England and France (Hansen, 1975, 50–52). At the founding of the German *Reich* in 1871, this fleet became the imperial German navy. The year 1872 saw the last foreign-built first-rate battleships enter German service, as well as the first German-designed, German-built armored ships. It was not, however, a navy with imperial ambitions, but a navy designed for coastal defense. Only after the accession of Kaiser Wilhelm II in 1888 did this focus begin to shift. Wilhelm was particularly impressed by the theories of the U.S. naval historian, Alfred Thayer Mahan, and had a copy of Mahan's work placed at the bedside of every German naval officer. Wilhelm's promotion of Tirpitz from commander of a torpedo boat in 1877, to chief inspector of the torpedo branch in 1886, to secretary of state for the navy in 1897 set the seal on German imperial ambitions (Hansen, 1975, 86). The Kaiser saw the fleet as the key to his ambitions for imperial Germany. Blocked by British and American action from full acquisition of the strategically significant Samoan islands in 1899, he announced, "We have bitter need of a powerful German fleet," and stated that "after twenty years, when [the fleet] is ready, I will adopt a different tone" (Herwig, 1976, 38, 13).

The Samoan incident of 1899 as well as the outbreak of the Boer War in that same year allowed Tirpitz to push a bill through the *Reichstag* in 1900 doubling the required number of ships of the line to thirty-eight (Herwig, 1976, 54). Although this represented a substantial navy, only twenty-four such ships were built to Britain's fifty-one of the same type (Gibbons, 1983, 174). Britain laid down and launched the first all-big-gun battleship, *Dreadnought,* in 366 days, a phenomenal industrial achievement that no other country at the time could match. The first German dreadnought, *Nassau,* took more than two years to launch, and the first U.S. dreadnought, *South Carolina,* took more than three. Both retained reciprocating engines and were significantly slower than the turbine-powered *Dreadnought.* The Japanese planned an all-big-gun ship, *Satsuma,* which was laid down five months before *Dreadnought.* Much time was wasted in a conversion to mixed armament, and *Satsuma* took nearly four years to build (Gibbons, 1983, 170–81).

As long as no one could match British industrial capacity, no one could outbuild the Royal Navy and challenge Britain at sea. At the climactic battle of Jutland in 1916 Britain fielded twenty-eight dreadnought battleships against Germany's sixteen and nine dreadnought battle cruisers against Germany's five. The Germans also incautiously included six lightly gunned predreadnought battleships in their line of battle (Pemsel, 1977, 107). Although the German ships were outnumbered only thirty-seven to twenty-seven, they were hopelessly outgunned. The British ships could fire a combined broadside weight of just under 180,000 kilograms of high-explosive shells to the German ships' approximately 82,000 kilograms. The German guns were also of shorter range. Despite brilliant German gunnery and seamanship, which cost Britain three lightly armored battle cruisers against one battle cruiser and one predreadnought for Germany, the result was a massive strategic victory for the Royal Navy. Only four of the twenty-eight British dreadnoughts were damaged, compared to six of the sixteen German ships. All the damaged British ships were ready for action within two months, whereas the last of the German battle cruisers was out of action for six (Hough, 1983, 296). The British Admiral Jellicoe ensured that the Royal Navy retained control of the seas, thus apparently ensuring the continued operation of the British maritime world system.

It is ironic that the great architect of the German battle fleet, Tirpitz, began his career as an officer of the *jeune école,* fascinated by torpedo boats. Tirpitz was seduced by Kaiser Wilhelm II's obsession with a high seas fleet to match the Royal Navy. Jellicoe's victory pushed the Germans to return to the unrestricted submarine warfare they had first experimented with in 1915 (Botting, 1979, 32–36), and a handful of cheap, torpedo-carrying submersibles almost brought about the victory denied the surface fleet. In the long term, U-boats destroyed Britain's geostrategic security by rendering maritime world systems untenable.

The most spectacular achievement of World War I, however, was the rise of Japan. By the war's end Japan had expanded her merchant fleet dramatically to fill the vacuum left as Europe involved itself in more serious events. The profits from commercial activity paid for a rebirth of the navy, which had two excellent predreadnoughts and a total of ten dreadnought battleships and battle cruisers by 1917 and was embarked on a construction program to add eight of each. In 1917 the U.S. fleet comprised only fourteen dreadnought battleships, though three more were added in 1918 under an ambitious building program begun in 1916 to combat Japanese expansion. Expansion for both nations came to an abrupt halt with the Washington Naval Treaty of 1922, but the Japanese turned their attention to aircraft carriers, in which area the treaty slightly favored them. Japan and the United States "were each permitted to convert two carriers of 33,500 tonnes to make use of existing hulls" (Preston, 1979, 35).

The big, fast carriers that the Americans and Japanese were able to convert from battle cruiser hulls were ideal flying-off platforms that could carry a useful number of aircraft and gave both polities invaluable operational experience in the mid-1920s. The U.S. *Lexington* and *Saratoga* made 64 kilometers per hour and could carry up to ninety aircraft and just over 600,000 liters of aviation gasoline (table 3-1, fig. 3-1) (Gibbons, 1983, 211, 234–35; Preston, 1979, 38, 42). Japanese carriers handled no more than seventy-two aircraft but were just as fast

(Preston, 1979, 40). The British converted smaller, slower battleships with speeds of 42 kilometers per hour. *Eagle* was too slow for efficient flying-off and could carry only thirty-six aircraft and a miserable 36,368 liters of aviation gasoline. All the aircraft carriers converted from World War I capital ships saw service in World War II, and most were lost.

The rise and fall of the carrier illustrates a further important point about world economies. They depend on skilled labor and thus on educational systems to perform certain crucial tasks. It was not the loss of carriers that halted Japanese expansion in World War II; it was the loss of pilots. The Japanese entered the war with a superbly trained, small, elite group of naval pilots and a training program capable of producing about one hundred pilots a year. At Midway they lost nearly a three years' supply of pilots when their carriers were sunk, not from aerial combat. These losses were made up by accelerating pilot training, with the result that, after the middle of 1943, accidental losses of pilots rose spectacularly. By the battle of the Philippine Sea in mid-1944, some pilots could fly off a carrier but were too inexperienced to land safely (Preston, 1979, 132–33). The Americans had no such problem: the basic U.S. educational system was far less elitist and supplied plenty of excellent candidates. By the middle of 1943, nearly forty-five thousand pilots were "in the pipeline" (Preston, 1979, 133).

The British began World War II with problems akin to the Japanese: in 1939 the Royal Navy had only 360 qualified pilots, with 332 in training. Part of the problem, however, was caused by the legacy of a single air force, the old Royal Naval Air Service of World War I having been combined with the army's Royal Flying Corps in 1918 to make the Royal Air Force. Although the fleet air arm was split off again in 1937, it was still dependent upon the RAF for its training until the naval flying school was moved to Canada at the beginning of World War II. Later, a lend-lease agreement was signed that provided for training in the United States as well. By the middle of 1942 there were 1,632 Royal Navy pilots (Preston, 1979, 134).

In terms of industrial production, the shift to the United States was even more noticeable than it was in terms of human capital. In World War I British production remained well ahead of U.S. or German production, whether of trained men or material. In World War II the production advantage moved quickly to the United States and Canada and stayed there. In 1941, British shipyards were at capacity, and they suffered constant interruptions from bombing and had an unreliable steel supply (Preston, 1979, 134). American yards had no such problems and moved to full production immediately after Pearl Harbor.

Pax Americana

The most notable feature of the world system between 1945 and the early 1970s was the final achievement by the United States of military as well as industrial hegemony and the consequent imposition of the Pax Americana, a period similar to the Pax Britannica that followed the

British achievement of hegemony in 1815. Although the United States lacks a great merchant fleet, the U.S. Navy has guaranteed the security of the seas of the world since 1945. Despite the loss of the ill-conceived and badly managed Vietnam War and the threat to U.S. economic hegemony since the oil embargo of 1973, U.S. military hegemony appears relatively secure. Yet militarily the United States is like Britain after 1900, spending a high percentage of a declining economic base on conventional as well as nuclear forces. As one historian has persuasively put it, hegemons tend to overextend themselves "geographically and strategically"; the United States is merely the most recent to do so (Kennedy, 1987, 539).

The U.S. economy resembles that of Great Britain in 1900 in another dangerous way. As the events of the oil embargo showed, the United States is now a net importer of raw materials, in particular, for energy. The country is reasonably self-sufficient in food, with the value of imports about equal to exports. Most imports are substitutes or luxuries: an increase of domestic sugar-beet production would quickly replace cane from tropical client states, and tropical carbohydrates such as manioc for tapioca are scarcely necessary, nor are imported beers, wines, or liquors, however pleasant. Only in the case of U.S. habituation to coffee might loss of a major import create problems.

Raw materials pose significant problems, but outside the energy sphere the principal supplier remains Canada, a nation tied closely to the United States by geography, settlement history, politics, and economics. Some strategic raw materials must come from South American or African suppliers. The major world reserves of copper, a vital necessity in electronics, are in the United States, 17.9 percent, and Chile, 19.3 percent (Espenshade, 1986, 41). Chromite is necessary in high-performance jet engines; just over 30 percent of world production is in the Soviet Union, 27.6 percent in South Africa, 5.3 percent in Zimbabwe, and 4.1 percent in the Philippines (Espenshade, 1986, 43). Nickel, which is equally necessary in high-performance steel alloys, is available in large amounts in Canada and Australia (Espenshade, 1986, 42). Diamonds, which are central to the machine tool industry, are produced primarily in the Soviet Union and South Africa. In all these cases the United States can meet its needs without controlling the seas of the world because air shipping can handle the amounts needed in strategic applications. Only political disruption could cut off these supplies, and the United States therefore places South Africa, Zimbabwe, Chile, and the Philippines in critical positions in the political calculus.

It is in the raw materials of energy that U.S. dependence upon sea trade is most critical. The flow of oil out of the Persian Gulf states is the single most critical component of the current world system because the economies of the major U.S. client states in Western Europe and East Asia almost totally depend on it (Espenshade, 1986, 46–47). Although U.S. production remains substantial at 16.4 percent of the world total, U.S. reserves, at 4.1 percent, are very low. Mexico accounts for a further 7.2 percent of reserves and Venezuela for 3.7 percent. Britain and the other European nations that share the North Sea oilfield account for about 3.5 percent of reserves and about 7.1 percent of production. The United States is still a substantial net importer of oil, and Europe and Japan are huge net importers. As OPEC, the Organization of Petroleum Exporting Countries, demonstrated in 1973, oil can be a potent economic weapon. Although the United States, Europe, and Japan have drastically curtailed the growth of oil consumption since 1973, to the point where there is a current glut and low prices prevail, the closure of the Strait of Hormuz would abruptly turn a glut into a shortage. If Britain was geostrategically vulnerable to the closure of sea trade in the period of British hegemony, the United States, Europe, and Japan are today much more geostrategically vulnerable to the closure of the trade in oil out of the Persian Gulf. On the other hand, the continued flow of oil can be and is being assured by a heavy concentration of U.S. and European naval vessels in the Persian Gulf, whereas British forces in the nineteenth and early twentieth centuries had to be spread much more globally to cope with threats to sea trade.

Chapter Four

The Problem of Overland Transportation: Canals, Rivers, and Railroads

A train departed roaring. Before midnight it would be leagues away boring through the Great Northwest, carrying Trade—the lifeblood of nations—into communities of which Laura had never heard . . . Suddenly the meaning and significance of it all dawned upon [her]. The Great Grey City, brooking no rival, imposed its dominion upon a reach of country larger than many a kingdom in the Old World . . . Her force turned the wheels of harvester and seeder . . . Her force spun the screws . . . of innumerable squadrons of lake steamers . . . For her and because of her all the Central States, all the Great Northwest roared with traffic and industry . . . It was Empire, the resistless subjugation of all this central world of the lakes and the prairies.

Frank Norris
The Pit. A Story of Chicago

Whereas the ship made bulk transport of goods possible, it did so only for places close to the sea: littorals and archipelagos. Great trading cities were located on rivers at the head of tidal navigation, the distance a ship could come inland on the rising tide against the flow of the river. The great interiors of the continents remained as far outside the world's growing trade networks as they had ever been until the coming of the canal, the steamboat, and the railroad.

EOTECHNICS

Land transportation in the Eotechnic, in particular in hilly regions and for bulk goods, was slow, expensive, and unreliable. Humans and animals were the only sources of energy to move goods or vehicles. As von Thünen pointed out (1966), the friction of distance imposed very real costs in overland movement. Highly perishable or bulky, frequently used items such as wood for fires were moved the shortest distance to market. Such nonperishable agricultural goods as wheat could be transported much further, and self-propelled agricultural goods such as animals, whether for fiber or food, could simply be driven to market. Von Thünen's classic formulation appeared at the close of the Eotechnic, but it summarizes perfectly the period's problems.

Wheeled vehicles are extremely inefficient on anything other than hard-surfaced roads. On soft ground the wheels sink in and the vehicle is always moving uphill. There was always some awareness of this problem, and such "solutions" as the broad-wheeled wagon were mandated in some parts of Europe. The only real solution, however, was the hard-surfaced road, which Europe and Anglo-America lacked the political system to produce much before 1900. The Roman Empire built excellent hard-surfaced roads, but they were primarily of a military nature, designed to rush troops to the frontier. Europe and Anglo-America lacked the necessary combination of central direction and strength of military purpose from feudal times through the late 1800s. Roads were built and maintained at the local level, meaning that they served local needs first. They meandered along property boundaries, around the best agricultural land, from farm to farm and village to village. To modern eyes this organic pattern now seems haphazard. Landowners were legally required to supply a specified number of days of labor each year to build and maintain the roads that served them. This forced labor or corvée system produced roads suited to local landowners, but not to folk who wanted to move quickly through the region in a carriage.

In the Eotechnic, therefore, agricultural production for any place other than the isolated state modeled by von Thünen meant direct access to a seacoast by a navigable river. Sailing ships could penetrate inland only by riding in on the incoming tide. The head of navigation on a river was marked by the distance tides penetrated inland or the presence of some break in the slope of a river which induced falls or rapids. In southern England, London was the head of navigation in the fertile Thames River basin. In northern England, York grew up as far inland as ships could navigate the river Ouse. Paris played the same role on the river Seine, and Orléans on the Loire. Bruges began its career as the "Venice of the North" at the head of a tidal estuary that later silted up, requiring a canal to be constructed to the new port of Damme. Ghent, the great rival to Bruges, was the head of tidewater on the river Scheldt.

The southern coast of the Baltic offered considerable inland penetration for seagoing ships, hence its importance as a major wheat- and wood-producing region in the medieval economy. As a condition of membership the Hanseatic League "prescribed that cities should be situated on the sea-coast, at the estuaries of rivers or on the banks of navigable streams" (East, 1950, 339). Towns were allowed to canalize their rivers to provide the required access, but canalization never included locks: for example, Brunswick and Hannover connected themselves to the natural head of navigation on the river Weser at Bremen (East, 1950, 339) by improving the flow and course of existing streams, not by making true canals.

As Adam Smith pointed out, overland transportation was not economically feasible at this period. Sea transport was more efficient. As a result most thinking about land transport concluded that the proper thing to do was to recreate a maritime environment on land. Riparian canalizing of existing streams, as some Hansa cities had done, was the first step. Development of the pound lock to impound water on level stretches came second and allowed the later emergence of the third phase of development, the watershed canal. The first European use of a canal system to integrate a region instead of merely to improve the movement of goods was in the Po River valley of Italy. Milan's rise to power was based first on the riparian canalization of surrounding streams in the mid-1200s, then on the construction of pound locks in the mid-1400s to give Milan access to the Po (Vance, 1990, 41–52).

Something of the regional impact of the canalization of the area around Milan was achieved by the reopening of the Old Roman Foss-dyke between the English rivers Trent and Witham in the early 1100s (Dyos & Aldcroft, 1969, 38). This simple and very early canal across a watershed allowed goods to be shipped up the Witham from the port of Boston on the Wash to the river Trent. The Trent drained north into the river Humber and thus allowed shipping to enter the Ouse and move upstream on the tide to York. The Roman Fossdyke, however, had not been designed to unite a regional economy, but merely to speed troop movement to the northern capital of Eboracum, now York, and avoid the dangerous sea passage between the Wash and the mouth of the Humber.

A much more impressive watershed canal than the Fossdyke was produced by Hansa merchants to link the North Sea to the Baltic and the river Elbe upstream from Hamburg to the river Stecknitz and the Baltic port of Lübeck. The Stecknitz Canal had the first pound lock to raise and lower elevation, but its water supply was inadequate to replenish locks by groundwater seepage. Locks could be operated only once every two or three days (Vance, 1990, 44). The poor water supply was compounded by leaky locks, and it was not until Leonardo da Vinci developed the miter gate in 1497 that the problem of seepage through lock gates was solved (Vance, 1990, 48–49).

The first attempt to use a system of canals to integrate a polity was in France. The portage between the basins of the rivers Seine and Loire was canalized with the Briare Canal between 1604 and 1642. Forty locks were required to cross a summit of 281 meters. The state provided some six thousand troops to accomplish the task, but internal political problems delayed completion of the canal for about twenty-eight years (Vance, 1990, 54–57).

The success of the Briare encouraged Louis XIV to support the Canal du Midi, built between 1665 and 1681. This 240-kilometer cut with 101 locks was a true ship canal designed to offer an alternative route between the Mediterranean and the Atlantic. It more than paid back its cost in the Anglo-French wars of the 1700s. British ships could blockade the Strait of Gibraltar and close the sea route between the Mediterranean and the Atlantic coasts of France, but they could do nothing about trade on the Midi (Vance, 1990, 57–64). The success of the canal encouraged the French to propose an ambitious national network (fig. 4-1). Yet progress was limited by a system that preferred state funding to venture capitalization. As table 4-1 shows, the increase in canal mileage was slow, in particular compared to Britain.

Britain was the first European economy to develop an effective inland navigation system crossing a complex series of watersheds. The British network owed nothing to political decisions and was driven entirely by economics. Entrepreneurs strove to connect coal production regions with the growing number of consuming regions. Consuming regions were focused around other mineral deposits: clay for the production of pottery and iron ore for the production of iron. Canal mileage increased

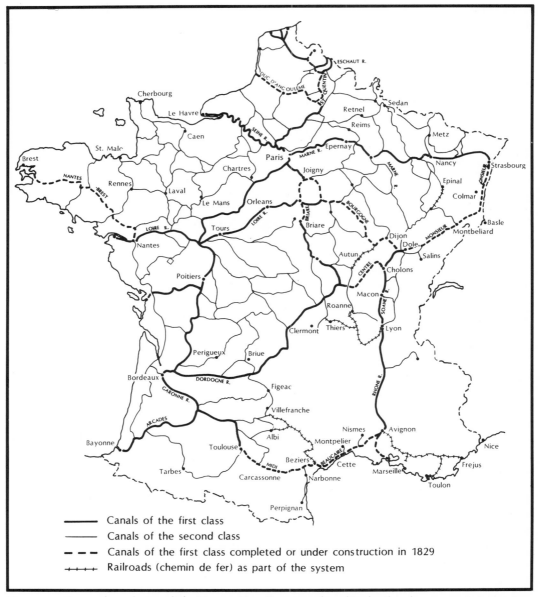

Figure 4-1 Proposed canal network for France, 1829. France sought political rather than economic integration. French canals were built on a lavish scale, compared to the much narrower canals favored elsewhere, and were financed by the state rather than by private enterprise. Vance 1990.

in Britain far faster than in France. By 1790 canals linked the country's four major estuaries: the Severn and the Humber estuaries were linked via the river Trent in 1772, the Trent was connected to the Mersey in 1777, and the Thames was plugged into the system at Oxford in 1790. The Thames and Severn Canal linked London directly to Bristol in 1789, and 1790 saw the completion of the Forth and Clyde Canal in Scotland, avoiding the long and hazardous sea route around northern Scotland. Britain's reduced rate of increase between 1830 and 1850 was accompanied by a rapid rise in railroad mileage (Porteous, 1977, 14).

Compared to French canals, British canals were narrow. The earliest British canals catered to 30-ton boats just under 2 meters in the beam, later ones to boats of about 60 tons and 3-meter beams (Hadfield, 1986, 58). The Midi catered to vessels of near 6-meter beams, thus was more of a true ship canal than a boat canal (Vance, 1990, 59). Narrow canals were cheap to build and operate, thus much better suited to the needs of an emerging industrial economy that was short on capital. British canals reached their capacity more rapidly than French ones, but by 1790 they had integrated the numerous regions that made up Britain's industrial economy, whereas the French system united only the two major agricultural regions, the river basins of the Seine and the Loire, and provided secure passage from the Mediterranean to the Atlantic.

Canals were also significant in the new American republic. Vance cogently accounts for the drive inland that after 1783 consumed American merchants stung by the Navigation Acts, which excluded ships of the new republic from free trade with British ports (1990, 110). Before the Revolution the colonial cities of the northeastern seaboard—Boston, New York, Philadelphia, and Baltimore—along with a whole coterie of smaller cities, had prospered as centers of the colonial merchant marine. Merchants in these now misplaced cities concluded that inland expansion and demographic increase were needed to replace markets lost abroad. Canals were the obvious solution to the need for inland penetration.

To turn inland the merchants of Boston, Philadelphia, Baltimore, and

Table 4-1 Canals and canalized rivers, 1650–1850, selected polities (kilometers)

Year[a]	Britain	France[c]	United States
1650		90	
1681		333	
1730	c. 1,770[a]		
1760	2,250[b]		
1790	3,594[b]	1,094	
1800	4,949		32[d]
1810	5,562		628[d]
1815		1,287	
1830	6,238[b]	2,216	1,598[d]
1850	6,474[b]	3,664	2,857[e]

Sources: [a]Porteous, 1977; [b]Hadfield, 1968; [c]Vance, 1990; [d]Brown, 1948; [e]Buley, 1950.

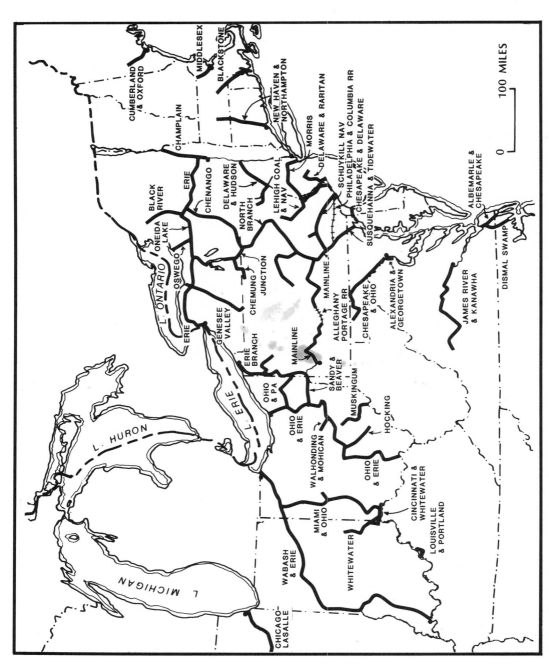

Figure 4-2 The canal network of the republic of the North, 1800–1860. U.S. canals were for the same narrow boats that were used in Britain, but they were built with public rather than private money. The only economically successful canal, the Erie, was designed to correct the geographic problems of the misplaced colonial city of New York when it needed to turn to inland trade after the Revolution. Vance 1990.

New York had to face the hard facts of physical geography. The two natural lines of access to the continental interior were, to the north, the St. Lawrence, which was controlled by Britain, and to the south, the Mississippi, held by Spain, then France, until 1803. Only New York's route up the Hudson and across the Mohawk into the Great Lakes made much sense topographically. Baltimore's attempts to canalize the Susquehanna were doomed because the taxpayers of Pennsylvania did not wish to improve Baltimore's prospects over those of Philadelphia, which at least had the Rube Goldberg contraption of the Pennsylvania Mainline Canal. Boston was hopelessly misplaced with no chance of a trans-Appalachian canal across the Berkshires (Vance, 1990, 114).

New York State followed the French model, using public rather than private funds, but built a narrow, British-style canal to carry boats of no more than one hundred tons (Vance, 1990, 125). The state elected to build a canal to Lake Erie above Niagara Falls rather than enter Lake Ontario at Oswego; hence the name, Erie Canal (Meinig, 1966, 160). Interests in the western part of the state opposed the Lake Ontario route because it would have required expensive lockage at Niagara Falls to lower boats from Lake Erie into Lake Ontario. The Lake Ontario route also would have allowed Montreal to compete with New York on an even footing. With the route chosen, boats progressed east via Buffalo and Rochester to Syracuse, thence to Rome, Utica, and the Mohawk Valley, finally entering the Hudson at Albany. This route reduced competition from Montreal, gave excellent access to the canal for farmers over much of the Lake Ontario shore plain, as opposed to only those close to the lake, and ensured military security. When the Erie was being planned, memories were still rife in upstate New York of the damage done by the British to U.S. ships on Lake Ontario in the War of 1812. Not until August of 1814, almost a year after Lake Erie was secured, did the United States regain control of Lake Ontario (Ellis et al., 1967, 141).

The Erie was opened in 1825, and its success prompted New York to spend more monies on a network of feeder canals (fig. 4-2). None of these was terribly profitable, but the Erie made so much profit that "canal fever" abounded (Meinig, 1966, 161). By the mid-1800s the Erie system had provided New York with an integrated regional economy and excellent access to the Great Lakes. New York's title, "the Empire State," was a direct reflection of the belief of East Coast merchants that, indeed, "westward the star of Empire takes its course." The Erie made it possible for the influence of New York City to penetrate more than 1,900 kilometers into the continental interior. Only relatively minor improvements were required to canalize the Detroit River between Lake Erie and Lakes Huron and Michigan and to construct lockage at Sault Ste. Marie to give access from Lake Huron into Lake Superior.

The Erie gave the American republic the finest inland transport system in the world and produced a vital subset of the U.S. polity, the republic of the North. Although the republic of the North had only a brief legal existence, when the southern states seceded during the Civil War, it was the single most important unit in the U.S. polity. Fourteen states were central to the republic of the North, states either on the Atlantic seaboard or connected to it at least in part by the Erie Canal and the Great Lakes system (fig. 4-2). Together these states occupied an area of nearly 110,000 square kilometers, roughly twice the size of the largest European polity, France, or of imperial Germany at its height in 1914.

PALEOTECHNICS

From a geopolitical perspective, the history of the 1800s is the history of the penetration of continental interiors made possible by the application of steam power. British hegemony was challenged by two land-based states in the late nineteenth and early twentieth centuries, albeit land-based states with significant maritime interests. In both Germany and the United States the railroad was a critical element of political integration, although in the latter it was augmented by the comprehensive use of steamboats on the river systems of the southern periphery.

Steamboats

Early steam vessels were fragile. They needed a constant supply of fresh water and used a huge amount of fuel. Engines could break loose in a heavy sea and crash through the hull. Salt water fouled boilers quickly. Inefficient engines meant that few ships could travel far without refueling. The "natural environment" for the first steam vessels was therefore the great rivers of the world, where they allowed penetration of the continental interiors. Rivers meant no heavy seas to cope with. Fresh water could be taken directly from the river. Fuel, at least in the early American republic, could be cut at any landing.

Steamboats emerged early in the U.S. South, with its fine network of rivers centered on the Mississippi (Watson, 1985). Without any spending on infrastructure other than the investment required by the boat itself, steamboats could penetrate the entire continental interior of North America. The trip from New Orleans to Minneapolis on the Mississippi could be accomplished with one portage around the falls at St. Louis. The Ohio offered upstream movement as far as Pittsburgh, obstructed only by the falls at Louisville.

Many people pioneered steamboats in the new American republic (Flexner, 1978). Fulton was the first to use his boats in a proper transportation system, to move people up the Hudson from New York to Albany. Fulton spent much of his life in England and France. In England he was friends with Rumsey, who was trying to use reaction propulsion in a steamboat. Fulton began his own experiments in 1793, but did not begin serious work on a practical steamboat until 1802 (Flexner, 1978, 219, 281). In the meantime he moved to France and attempted to sell Napoleon his ideas for submersibles to attack the British fleet. The emperor was more interested in steam tugs to pull an invasion fleet across the Channel and encouraged Fulton to build a full-size steamboat on the Seine. Fulton promised 26 kilometers per hour but delivered no more than a walking pace (Flexner, 1978, 292–93). Napoleon lost interest and Fulton moved back to England, where he attempted to sell the Royal Navy his idea for attacking ships with mines below the water-line. Just as he seemed to be having some success, the battle of Trafalgar brought his sales effort to an abrupt end; with the French fleet in ruins, there was no need for unconventional warfare. Fulton returned to the United States, but the British repaid him with a rare export license for one of Boulton and Watt's steam engines. This engine he promptly installed in a suitable hull and, with the powerful political support of Chancellor Robert Livingston, who held a monopoly granted by the state of New York for steamboat traffic on the Hudson, began service between New York and Albany in 1807.

Recognizing the crucial importance of western waters, Fulton entered into partnership with Nicholas Roosevelt to build at Pittsburgh a boat named *New Orleans* with the aim of navigating the Ohio and the Mississippi to that city (Dohan, 1981, 13–16). Agricultural surplus had long been floated down these rivers to the Gulf Coast. Shallow-draft keelboats could pass the falls at Louisville only at high water and with numerous oarsmen aboard to power them ahead of the current and retain the ability to steer. Roosevelt showed that steam could replace the oarsmen and that a steamboat could shoot the rapids under its own power (Dohan, 1981, 85–89). Yet the *New Orleans* was not the right boat for the Mississippi and Ohio river system. It drew too much water with its V-shaped hull, and it was too underpowered to proceed upstream easily against even the normal flow of the rivers.

One of the many to inspect *New Orleans* in Louisville was Henry Shreve, an important merchant boat builder and keelboat captain (Dohan, 1981, 71). Keelboats usually drew less than two feet loaded and had a sturdy keel, up to six inches wide and twelve high, which bore the brunt of frequent underwater collisions (McCall, 1984, 19). Keelboats were propelled upstream by crews of up to forty, and the normal time from New Orleans to Louisville was some three months (McCall, 1984, 21). After experiments with a smaller steamer built on keelboat lines and called *Enterprise,* Shreve recognized the need for a radical departure. He became convinced by the work of another early U.S. engineer, Oliver Evans, that high-pressure engines offered far more promise than the safer, low-pressure designs of Watt. A high-pressure engine operating at 100 pounds per square inch would weigh a twentieth of the low-pressure engines in the Fulton boats (Dorsey, 1941, 108) and would be four times more powerful (Dorsey, 1941, 111; McCall, 1984, 87). Power was needed to move upstream; lightness was also needed so that the engine could be mounted in the almost flat-bottomed hull he proposed without making his boat top-heavy. Shreve favored paddlewheel drive and a large boat, some 42.6 meters long and 7.6 meters in the beam. He then added passenger cabins in a second story, producing *Washington,* the first of a radical new breed of riverboats (McCall, 1984, 137–40).

After a legal battle in which Fulton tried to enforce his supposed monopoly of steamboat travel on western waters, *Washington* made its first paying trip upriver in March 1817, twenty-four days from New Orleans to St. Louis. Shreve's boats halved transportation costs on the Mississippi. His prediction of ten-day trips came true in less than ten years, and by 1853 the record was less than four and a half days for the 2,400 kilometer trip (McCall, 1984, 148–54, 177). Because Shreve refused to patent his ideas, some sixty Shreve-style boats were at work by 1819 (Meyer, 1948, 108). In its first two trips *Washington* was paid for and made $1,700 profit. In 1824 Shreve produced *George Washington* to demonstrate all he had learned. This side-wheeler had two powerful engines, one for each wheel to facilitate steering, and three decks instead of two. Other designers also innovated. On the shallower, faster-flowing Missouri, smaller, stern-wheeled boats fared better than the large side-wheelers that plied the Ohio and Mississippi. Side-wheels limited beam dimensions relative to length more than stern-wheelers did but were preferred where boats had to make open-ocean voyages, however brief.

New Orleans suddenly had access to some twenty-four thousand kilometers of navigable waterway (Dorsey, 1941, 129, 130). Between

The steamboat "Lizzie," loading cotton on Buffalo Bayou, Houston, Texas, 1900. The rivers of the U.S. South were superb natural highways for the export of the region's cotton crop before 1860, requiring only steamboats and landings. Thereafter, continued expansion brought into cultivation land accessible only by railroads. Cities like Houston grew as interfaces between steamboat and railroad technologies. Houston Public Library.

1819 and 1841 the number of steamboat arrivals at the port city went from 191 a year to 1,958, the weight of goods received rose almost fourfold to 551,211 tons, and the value of produce rose almost threefold to just under $50 million (North & Thomas, 1968, 200). New Orleans directly controlled some 777,000 square kilometers. The city's indirect hinterland, served through Saint Louis, Cincinnati, and Louisville, amounted to another 1,554,000 square kilometers (North & Thomas, 1968, 204). Other Gulf Coast ports, such as Galveston and Mobile, had to accumulate cargoes from the hinterlands of several rivers, so that their steamboats had to cope with the waters of the Gulf of Mexico as well as those of Gulf Coast rivers.

Although about sixty boats were at work in 1819, tonnage increased very slowly thereafter because of a lack of snag boats to clear the channels. With Shreve's development of a working snag boat in 1829, expansion was again rapid through 1850 (table 4-2). The great era of western riverboats was in decline by 1860 with competition from railroads. Many riverboats were still lost to snags and to explosions of high-pressure boilers. The average life of a boat was about five years.

The driving force of expansion in the Gulf Coast and up the Mississippi was cotton. The American upland form of cotton, which allowed expansion as far north as Cairo, Illinois, where there are only about two hundred frost-free days a year, was only just emerging in 1817, contemporary with Shreve's boat. Unlike the long-staple cottons of the Caribbean cotton economy, hybrid American upland cotton could tolerate cool conditions yet return a spinnable length of fiber and be productive. Cotton thus began to flow much more easily between the plantations and the Gulf Coast ports that put it into Atlantic trade. The growth of the cotton kingdom was also helped by the clearance of the Creek Indians from the southern end of the Appalachians by Andrew Jackson, at the head of, first, the Tennessee militia and, later, regular U.S. forces. Both production and export of cotton built rapidly as Shreve's riverboats allowed fast inland expansion, but also as British industrialization, and thus demand for raw cotton, accelerated between the end of the Napoleonic Wars and 1860.

Table 4-2 Merchant shipping on western
U.S. rivers, 1816–1860

Year	Tonnage
1816	9,930
1817	12,946
1818	24,512
1819	25,192
1820	27,269
1830	32,664
1840	117,952
1850	302,829
1860	167,739

Source: U.S. Bureau of the Census, 1975.
"western" = west of the Appalachians.

During this forty-five-year period the cotton South was clearly in Britain's periphery. Seventy-two percent of U.S. cotton was exported between 1815 and 1860, and between 1831 and 1860 Britain took almost 70 percent of these exports. After 1870 the percentage of exports began a long, inexorable downward trend, while their actual weight continued to rise. Domestic consumption saw the real gain, roughly doubling between 1870 and 1880, 1880 and 1900, and 1900 and 1920.

The history of the cotton South between 1817 and 1860 is the history of expansion to the western limit of nonirrigated cotton agriculture, the thirty-inch isohyet. This limit was in Texas, approximately along the line of the Balcones Escarpment. In Texas, however, King Cotton came up against a more serious immediate limit to continued expansion. Texas's rivers are singularly unsuited to steamboats. Although the counties on the Arkansas border were reasonably well served by the Caddo and Red rivers, that trade flowed to New Orleans. The Sabine River was partially navigable, the Trinity less so. Only the Brazos offered much inland penetration, and that only as far as Washington-on-the-Brazos, above which Hidalgo Falls prevented upstream movement. This was enough,

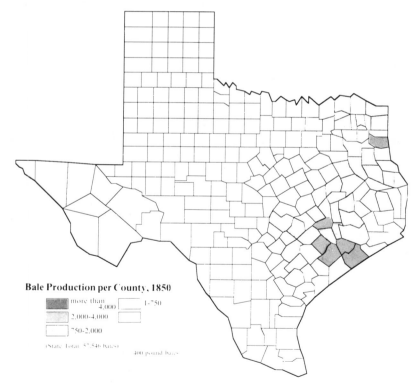

Bale Production per County, 1850

more than 4,000

2,000–4,000

750–2,000

1–750

(State Total 57,546 bales)

400 pound bales

Texas cotton producing counties in 1850

Figure 4-3 Texas cotton production in 1850. East Texas is the westernmost extension of the United States' humid South. To begin with, cotton was produced along the coast and rather restricted navigable sections of rivers, down which steamboats carried cotton to Galveston and ships bound for British textile mills.

however, to make Washington, today nothing more than a state museum, the first capital of the Texas Republic and the site of the signing of the Texas Declaration of Independence from Mexico (Puryear & Winfield, 1976, 12–13). The map of Texas cotton production in 1850 shows this riverine system, with cotton moving to Galveston from the riverine counties (fig. 4-3). Much land north of Washington-on-the-Brazos and east of the Balcones Escarpment was perfectly well suited to cotton but rendered inaccessible by a lack of navigable rivers.

Railroads

The railroad took the place of navigable rivers. By 1899 railroads allowed Texas cotton production to be pushed north along the Brazos River valley in an almost unbroken line to the Oklahoma border (fig. 4-4). Railroads such as the Houston and Texas Central, which came north from Houston via Waco to Dallas, opened prime cotton lands and shifted production dominance away from the riverine counties.

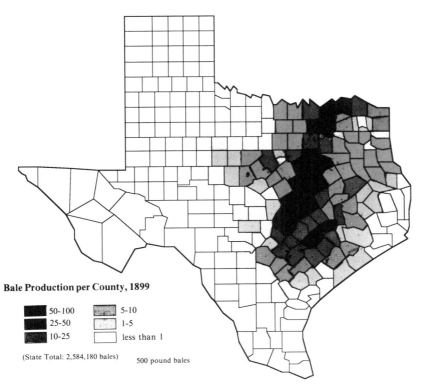

Bale Production per County, 1899

■ 50-100	▓ 5-10	
■ 25-50	░ 1-5	
■ 10-25	□ less than 1	

(State Total: 2,584,180 bales)   500 pound bales

Texas cotton producing counties in 1899

Figure 4-4   Texas cotton production in 1899. Railroads allowed cotton production to expand inland from the head of steamboat navigation and up onto the high plains around Lubbock. In East Texas rainfall cultivation was the norm, but on the high plains irrigation was needed, thus gasoline- or electricity-powered pumps. Railroads warehoused cotton in Houston rather than Galveston.

Yet the real impact of railroads in the penetration of the continental interiors was not so much in the U.S. South as in the full integration of the republic of the North. Other countries with similar problems in the political and economic integration of large, land-based polities also turned to the railroad, most notably France, Germany, and Russia—only later.

In Germany, for example, the growth of railroads was slowed by political problems. Until 1871 the German states were still independent, and most had their own railroads. The Prussian and Bavarian systems were the most significant, but they were planned for the particular needs of the state and not to facilitate through traffic when the barriers to easy movement and commerce were finally removed. Unlike the U.S. system, which grew alongside first the republic of the North, then the U.S. republic, imperial Germany had to piece together a network from pre-existing state systems.

Railroads were not conceived as a means of integrating land-based polities but to provide inland transportation for British coal mines in regions where canals were rendered prohibitively expensive by topography. Such mines had long used primitive iron railroads within the mines, with wagons moved by humans or animals. The Newcastle area coalfield possessed a major locational advantage in the Eotechnic, in that coal could be loaded easily onto sailing ships. Even after the mid-1700s, when the more easily accessible coal had been used up, the trip to the coast was short and downhill. Trains of loaded coal wagons were allowed to roll downhill, then horses pulled the emptied wagons back up. By the early 1800s, however, coal was being worked behind the range of coastal hills, and loaded cars had to be pulled up an incline as well as allowed to roll down. This led mine engineers to think of applying the steam power that drained their mines to the problem of hauling coal to the coast for shipment.

Of many pioneers George Stephenson was the first to understand the advantages and limitations of steam-powered railroads. Others, notably Richard Trevithick, built better locomotives, but no one conceived of an integrated system of steam-powered locomotives running on properly constructed roadbeds until Stephenson. His conservatism in laying out railroads to minimize variations in topography has been justly criticized (Vance, 1990, 199), but Stephenson turned railroads from the pipe dreams of mine engineers into a new and durable means of hauling goods and people over highly variable topography and long distances quickly and with low transfer costs.

On the strength of his work at the Killingworth colliery Stephenson was appointed engineer of the world's first public railroad, the Stockton & Darlington. This 20-kilometer line finished in 1825 was an alternative to a canal for moving coal from Darlington to port at Stockton. Passengers were a sideline and Stephenson's locomotives were designed to run at only 8 to 10 kilometers per hour (Ferneyhough, 1975, 60–62). The second public railroad in Britain, the Liverpool & Manchester, was thus the first one built for passengers. Stephenson was appointed to lay out the line, but public trials were held at Rainhill in 1829 to determine the best locomotive. The winner, Rocket, was designed mostly by George Stephenson's son, Robert, who improved on his father's pioneering use of steam blast by adding a more efficient multitube boiler. In his 1830 design, Northumbrian, he corrected remaining flaws to produce the first modern steam locomotive (Hollingsworth, 1982, 18).

The Liverpool & Manchester Railway was intended as a rapid route between the two cities. At 49.6 kilometers it was only 1.6 kilometers more than the airline distance. The railroad also cut 8 kilometers off the turnpike distance and 19, 24, and 43 kilometers off the distances covered by the three existing canals (Ferneyhough, 1980, 14). Speed was emphasized. The winner of the Rainhill trials was required to achieve a minimum average speed of 16 kilometers per hour over some 110 kilometers (Ferneyhough, 1980, 55). Rocket delivered all that was asked and more. Its average speed for 110 kilometers was 22.5 kilometers per hour, and it occasionally more than doubled that. Rocket was just as competent at hauling wagons of heavy goods. It performed with utter reliability. Within two years trains were averaging 45 kilometers per hour on the line. The success of the Liverpool & Manchester began a long, sustained rail boom in Britain which brought British railroad technology into temporary dominance. It was exported, licensed, or copied in the United States, Belgium, and Germany.

It was in the United States that radical changes in British technology were made, first in the locomotives, then in the trackage, then in the entire system. In the British locomotives, the axle was fixed through the frame so that there was only a small amount of side play and only one driven axle to minimize friction. Low power output limited the number of cars that could be pulled and the ability to climb hills. Tracks had to be as level as possible. Most of the capital cost of railroad construction is in laying track and preparing track bed. The Stephenson system was too expensive for the impoverished young American republic, which had to pay the soldiers in its Revolutionary Army in land and had to sell three million acres of western New York to Dutch investors to pay for the damage the British did to Boston in the same war.

U.S. distances were also vastly greater than British ones. The South Carolina railroad of 1834 ran nearly 250 kilometers through rich cotton lands from the port of Charleston to the Savannah River opposite Augusta, Georgia. By 1840 there were nearly twice as many kilometers of track in the United States as in Britain, and by 1860 there were three times as many (table 4-3). To save on scarce capital U.S. railroads had to be cheaply built. The Camden & Amboy Railroad in New Jersey called for a capital investment of $3.2 million for nearly 100 kilometers, of which $500,000 went to iron rails imported from England. "Almost as much went to preparing the roadbed, and $370,000 went for real estate" (Hindle & Lubar, 1986, 137). The Camden & Amboy was built in the Stephenson manner, with round-bottomed rail carried on iron chairs bolted to stone blocks set deep in the ground. Charles Vignoles, who had worked with Stephenson on the Liverpool & Manchester

Railway, spiked flat-bottomed iron rail directly to wooden crossties. This system lowered construction costs and time needed to build track. Vignoles communicated regularly with U.S. engineers, and his track design was first used extensively on the Baltimore & Ohio Railroad in 1832 (Ferneyhough, 1975, 246; 1980, 155).

With tracks laid in this way, British-style locomotives derailed frequently, especially so when American engineers skimped on ballast under the crossties. By 1837 Joseph Harrison of Philadelphia had developed three-point compensated springing, which gave locomotives the inherent stability of a tripod on even the roughest tracks. Harrison's Hercules of 1837 led directly to the American standard 4-4-0 locomotive, "of which 25,000 were built for the USA alone" (Hollingsworth,

Table 4-3 Railroad track by population and area, 1830–1960, selected polities (kilometers)

| Year | Per 1,000 people | | | | |
	Germany	Britain	United States	Japan	Russia/USSR
1830	—	0.01	—	—	—
1840	0.01	0.13	0.27	—	—
1850	0.89	0.48	0.62	—	—
1860	0.26	0.64	1.56	—	—
1870	0.54	0.84	2.54	—	—
1880	0.75	0.85	2.99	—	—
1890	0.87	0.85	4.26	0.05	—
1891					0.24
1900	0.92	0.82	5.47	0.14	
1910	0.94	0.80	6.13	0.18	
1920	0.94	0.77	6.15	0.25	
1926					0.52
1930	0.90	0.73	5.62	0.39	
1939					0.51
1940	0.89	0.69	5.02	0.45	
1950	0.73	0.65	4.21	0.33	
1959					0.59
1960	0.67	0.61	3.64	0.31	

| | Per 1,000 square kilometers | | | | | |
	Germany	Britain	United States	Japan	France	Canada
1830		0.50	0.01		0.06	
1840		7.61	0.98		0.75	
1850		31.18	1.87		5.34	
1860		46.48	6.30		16.85	
1870		70.33	13.60	0.08	28.57	
1880	62.57	79.76	19.17	0.41	43.04	1.18
1890	79.27	88.57	34.27	4.84	62.04	2.29
1900	95.56	95.73	53.20	16.50	71.05	2.91
1910	113.18	102.43	72.32	22.69	75.47	5.13
1920	161.24	134.05	83.59	35.74	70.22	8.35
1930	163.00	133.74	88.38	56.56	77.94	9.51
1940	173.55	131.87	83.95	65.82	74.63	9.51
1950	148.65	128.43	81.49	71.77	75.92	9.75
1960	145.17	125.09	79.56	72.79	71.69	9.95

Sources: Statesman's Year-books.

1982, 5, 24–25). Having two driven axles reduced the load on each axle, and allowed lighter, cheaper rails to be used. This design made possible the vigorous expansion of U.S. railroads between 1840 and 1860. Trackage trebled from 1840 to 1850, trebled again from 1850 to 1860, then again from 1860 to 1880 (table 4-3). After 1880 the increased weight of traffic caused the 4-4-0 to give way to larger locomotives running on steel rather than iron rails.

The U.S. system of route construction was also far cheaper than the British. Britain's 180-kilometer London & Birmingham Railway was built in the late 1830s at a cost of just under $165,000 per kilometer (Vance, 1990, 207). Thirty years later the Union Pacific's 1,600 kilometers from Omaha, Nebraska, to Promontory Point, Utah, cost only $16,284 per kilometer (Vance, 1990, 307). Very little tunneling, earth movement, or bridge construction was done in the U.S. system. Either of the two main cuttings on the London and Birmingham Railway "represented the movement of more spoil than was required in the building of the . . . Union Pacific" (Vance, 1990, 208).

The impact of railroads in the United States was markedly different by region. Between 1840 and 1860 most trackage was laid in the republic of the North. The excellent inland waterway system was joined by an equally fine railroad network that remedied many of the former's problems. The first of these was crossing the Appalachian barrier. The merchants of Baltimore, Philadelphia, and Boston, unable to develop

DeWitt Clinton. This 1831 design for the Mohawk & Hudson Railroad was on British lines. Without a pony truck or three-point compensated springing, such 0-4-0 locomotives were unsuited to the development of railroads in America, where distances were great and the heavy capital investment needed for well-laid track was avoided. Ford Museum, by the author.

canals to compete with the Erie, turned with relief to the railroad. Even so, New York had a virtual monopoly of access to the continental interior from the completion of the Erie Canal in 1825 to the almost simultaneous completion between 1851 and 1852 of five railroads between eastern ports and the Great Lakes (Vance, 1990, 286).

The republic of the North was an emerging core area with two hinge points. New York was the hinge to the Atlantic world and the British core. Chicago was the hinge to the internal peripheries of the West and South. The U.S. economic historian, Robert W. Fogel, has argued that Americans misdirected investment monies into railroads by the mid-1800s and that canals would have served the economy of the country better until the late 1880s (Fogel, 1964). Fogel regards canals and railroads as interchangeable, whereas in fact they are complementary. Railroads carried passengers and perishable goods far more efficiently than canals, which the railroads recast in their proper role as carriers of

Hercules 4-4-0 locomotive, Beaver Meadows Railroad, 1837 (replica). This locomotive, designed by John Harrison of Garrett & Eastwick of Philadelphia, was the first to use three-point compensated springing. Each pair of driven wheels was pivoted around a center point that was connected to the mainframe via a large leaf spring. The third point was the pivot of the small-wheeled pony truck at the front. In effect the locomotive "stood" on three "legs" of constantly varying length, and thus could move at speed along inexpensive track. The early American republic simply lacked the capital to build substantial mileages of high-grade track, and without three-point compensated springing the under-populated continental interiors would have been much harder to open for trade. Baltimore & Ohio Railroad Museum, by the author.

nonperishable bulk commodities. Fogel's analysis hinges on the social savings that would have accrued had less scarce capital been used on canals rather than railroads. Yet social savings are hard to measure, and Fogel did not weigh them against the time savings that accrued to the economy from the use of railroads, which are much faster than canals.

Time savings were important to the rational management of an economy spread over as much area as the republic of the North. Even the earliest railroads could average 30 kilometers per hour over long distances compared to 5 kilometers per hour for canal packet boats. The New York & Erie averaged nearly 32 kilometers per hour between New York City and Dunkirk on Lake Erie in 1851. In the same year the Michigan Central managed 30 kilometers per hour between Detroit and New Buffalo on the eastern shore of Lake Michigan. Lake steamers provided the connecting links to bring New York within three days of Chicago (Mayer and Wade, 1969, 39). Direct railroad connection in the mid-1850s shortened the trip to just over forty hours.

Thus was created an efficient market over the entire republic of the North, and Chicago became a focus for the railroads that would extend that republic's hegemony west and south. Between January 1 and September 27, 1873, Chicago passed on from the West and South just over 1.5 million barrels of flour, 6,606 tons of wheat, 11,983 tons of corn, 5,327 tons of oats, 104,106 tons of cured meats, 28,321 tons of tallow, and 11,700 tons of wool (Chamberlin, 1874, 281). Chicago had a direct sea link to the British core via the Great Lakes and the Erie Canal.

American class 4-4-0 locomotive. Such locomotives, derived from Hercules, were built in huge numbers between the 1850s and about 1900. They could pull considerable loads at reasonable speeds on inexpensively laid track. Baltimore & Ohio Railroad Museum, by the author.

But Chicago also saw itself as a core and the West and South as its natural periphery:

> The time has already come when the arrest of developing manufacturers in California, by the opening of the Pacific railroads, which exposed them to the competing woes of lower paid labor in the East, will engage the pecuniary sagacity of Chicago in preparing her to be the great Shop for supplying the infinite demand of the far West, in time, for manufactured goods. Alaska itself is not extolled as a fur trader without implying Chicago as the future purchaser. The like remark may be made of Texas, and of Mexico. (Chamberlin, 1874, 102–3)

The success of Chicago in the 1850s created the necessary geographic conditions for the diversion north of the agricultural surplus of the southern states once the Civil War was over. After 1865 the railroad net focused upon Chicago detached the cotton states from the periphery of the British maritime world system and attached them to the periphery of the continental world system of the republic of the North. This fundamental fact of transportation geography was recognized by all states that aspired to be land-based polities in the wake of U.S. success.

By 1860 the republic of the North accounted for nearly two-thirds of the 49,286 kilometers of U.S. trackage (Miller, 1969, 27) (table 4-3, fig. 4-5). By 1870 the Great Lakes had been surrounded and the first line pushed west across the continent from Omaha to Oakland (fig. 4-6). By 1880 six U.S. and one Canadian transcontinental lines served the U.S. republic rather than just the republic of the North.

The U.S. system and the American 4-4-0 locomotive served the railroads of the developing world well until the late 1800s. European countries, with adequate capital to finance expensive trackage, tended to prefer the speed and freedom of running enjoyed by engines with only two large driven wheels, such as 4-2-2s. As long as train weights were low, such engines were adequate. European railroads used lightweight, wooden, four- or six-wheeled passenger cars well into the 1890s. U.S. railroads developed much larger, heavier, and better-equipped passenger cars in the 1840s to ensure passenger comfort on the long distances traveled. Such larger cars were carried on two four-wheeled swiveling bogies to allow them to negotiate sharp curves. The 4-4-0 was a response to high U.S. train weights as much as to topography and bad track geometry.

By the late 1830s the basic locomotive problems had been solved and emphasis in railroad operation returned to where Stephenson had rightly put it: on the operation of a complete system. This meant planning operating procedures, economics, research and development, and safety. In the 1840s U.S. railroads enthusiastically embraced the telegraph so that they could continually update the positions of trains in a largely single-tracked system. By the 1870s Saxby's interlocking signal frame ensured that drivers could receive accurate information without the use of vast numbers of signal operators. Although Saxby was British his system was standardized in Europe and the United States rather than

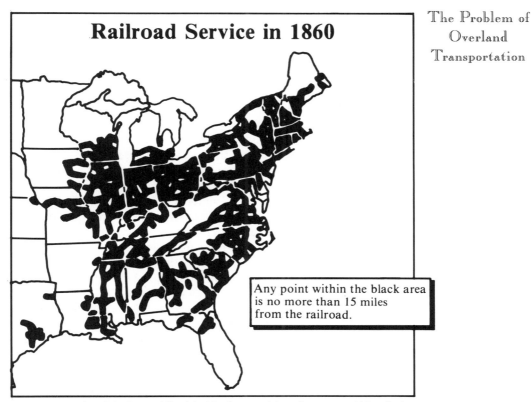

Railroad Service in 1860

Any point within the black area
is no more than 15 miles
from the railroad.

Figure 4-5 Railroads in the republic of the North, 1860. Most places were within a day's round trip of a railroad by oxcart, and railroads provided political and economic integration. Steamboats served the South well as long as most trade was with Britain and regional political integration was not sought. Redrawn from Miller 1969.

Britain (AAR, 1986–87, 103). Also in the 1870s came the first use of Janney's automatic train coupler, which for a long time was used only by U.S. railroads (AAR, 1986–87, 5).

Civil engineers played as vital a role in railroad development as did mechanical engineers. Lowering construction costs by using cheaper bridges or line layout was important. Like most of the first generation of railroad engineers, John Edgar Thomson spent time with George Stephenson before returning to his native United States, where he eventually headed the Pennsylvania Railroad. His design for the horseshoe curve allowed the Pennsylvania Railroad to cross the Allegheny Mountains on a reasonable grade, and he led his line to become the "standard of the world" (AAR, 1984–85, 17). The career of Albert Fink, who early on designed several very advanced bridges, was crowned in the 1870s by his work on railroad economics and by his achievement of standardized freight rates in the U.S. South (AAR, 1984–85, 87).

Organized research and development characterized the career of George Westinghouse. Unlike other pioneer engineers, Westinghouse worked on a contract basis rather than for a specific railroad. His greatest achievements were in the field of safety, especially the automatic air brake of the late 1860s (AAR, 1986–87, 99).

It was France, where supplies of coal were limited, that made innovations in locomotives. Like the United States, France operated a basically continental economy, but the need for power and speed had to be tempered by efficiency. The Paris, Lyons, & Mediterranean Railroad had to cross difficult topography, for which its 4-2-0 locomotives were unsuitable by the early 1870s. Its 2-4-2 design of 1876 gave better pulling power, and more important, the small undriven wheels at the rear allowed a large, efficient firebox, thus better steam production. The PLM class 121 was continually improved. Its 1888 rebuilding marked a watershed in steam locomotive development and the beginning of French technical ascendancy in steam locomotive design. Three major improvements were responsible: Walschaert's valve gear, which was more complex but more efficient than Stephenson's; very high boiler pressure; and compounding, admitting high-pressure steam to one set of cylinders, then allowing it to expand further in a second set of cylinders operating at lower pressure (Hollingsworth, 1982, 40–41). After 1888 French locomotives, although requiring high levels of technical skill to build and drive, used fuel more efficiently than any others. The French had an understandable concern for efficient fuel use: their coal reserves were the poorest of any major European nation, and production per capita fell sharply after the loss of the Alsace-Lorraine coalfield to the Germans in 1870.

Railroad Service in 1880

Any point within the black area is no more than 15 miles from the railroad.

Figure 4-6 Railroads in the American republic, 1880. Railroad growth in the South was rapid as cotton production expanded to fill the needs of the mills of the U.S. north as well as those of Britain. The Pacific coast was linked to the Atlantic, but for political rather than economic reasons, and the link was heavily subsidized. Redrawn from Miller 1969.

After 1888 there were clearly two schools of thought on steam loco-motive design: French and other. U.S., British, and German designers were more concerned with power output and reliability than with fuel consumption, especially where highly trained workers and locomotive engineers were not available. To achieve high power output, U.S. de-signers simply added another set of driven wheels to the 4-4-0 to pro-duce the simple (noncompounded) 4-6-0. This locomotive had the vir-tues of low axle loading and ease of construction, maintenance, and driving, but its firebox was constricted by being over the last driven axle. The Pennsylvania Railroad favored 4-4-2 Atlantic designs with large fireboxes, not least because it had to cope with the worst topography of any line between the East Coast and the Great Lakes. The 4-4-2's steam-ing capacity in the mountains outweighed its disadvantages in pulling power and high axle loading. Because of the latter the Pennsylvania had to lay very heavy steel rails that could withstand great loads on each axle.

The greatest Pennsylvania 4-4-2, the E-6 of 1914, was the most pow-erful 4-4-2 ever built. It used superheating as a much cheaper way of improving thermal efficiency than compounding. The idea was that of a German engineer, Wilhelm Schmidt, known appropriately to his inti-mates as Hot-Steam Willy. Schmidt recognized that loss of steam tem-perature and resulting condensation in the locomotive's cylinders were a primary cause of inefficiency. To prevent this he heated the steam after it left the boiler. In this way the thermal efficiency of steam locomotives could be raised between 10 and 30 percent (Westwood, 1977, 132). French compounds achieved good results in part because they divided the temperature drop between high- and low-pressure cylinders, which reduced the condensation loss. Superheating offered much of the in-creased power or decreased fuel consumption of compounding with far less complexity (Westwood, 1977, 133). Along with the Pennsylvania E-6, the 4-4-2 compound La France designed by Alfred de Glehn repre-sents one of the greatest of all steam locomotives. The de Glehn design weighed substantially less and had a much lower axle loading, yet did comparable work.

The ideas of compounding and superheating diffused rapidly, and others quickly refined them. In Britain the chief engineer of the Great Western Railway, George Churchward, imported a de Glehn 4-4-2 com-pound in 1903 but discovered that superheating and improved "breathing" from the use of long-travel valves and enlarged cylinder steam ports gave just as good performance with none of the construc-tion, operating, and maintenance problems of the compound (West-wood, 1977, 125).

The E-6 was a spectacular performer that still hauled the fastest scheduled train in the world in 1933 (Hollingsworth, 1982, 59). The price of this performance was high fuel consumption, albeit of relatively moderate-quality coal, and extremely heavy track. Outside the United States, railroads used relatively light rails that could not take heavy axle loadings. As demands grew for longer, more luxurious trains, thus for faster, more powerful locomotives, track capacity became a crucial bottleneck. During the age of steam no major system outside North America could carry an axle load greater than 22.5 tons. Low axle loadings meant that locomotives outside the United States had to stress thermal efficiency if they were to pull fast trains or climb rugged topography. By 1914 the Pennsylvania Railroad could accept 33-ton axle loadings, a loading that became normal in the United States.

The demand for rapid service in the republic of the North, in particular between its polar capitals, New York and Chicago, dominated the development of U.S. railroad technology around 1900. Better track and locomotives lowered the time between the two cities to less than twenty hours. Thereafter it was capacity and comfort that improved rather than outright speed. Heavier, all-steel passenger cars replaced wooden ones. Proper heating, cooling, and eating facilities were added, and the train became a thread of metropolitan amenities moving through the countryside (Stilgoe, 1983).

Heavier cars required more powerful locomotives. First to appear was the Pacific. The 4-6-2 wheel arrangement allowed, as with the 4-4-2, a generous firebox producing ample steam for speed or mountain climbing. Six driven wheels made for good traction, and spreading the weight over three axles kept axle loading within reason. The finest early Pacific was the Pennsylvania Railroad's superheated K-4 of 1914. This elegant and simple design was as admired and copied as the de Glehn 4-4-2 had been, especially in Britain. It led almost directly to the London & North Eastern Railway's A-4 of 1935, holder of the world speed record for steam locomotives at 203 kilometers per hour. Yet the K-4 reached a serious limit for steam power, that of fire stoking by human labor. The K-4 was the last powerful U.S. locomotive built without a mechanical stoker, although many engines were retrofitted because they were not capable of sustaining maximum power output without one. European locomotives had smaller grate areas and rarely resorted to mechanical stoking.

After World War I U.S. designers emphasized even more luxury, which put pulling power at a premium. In 1926 the New York Central introduced its legendary Hudson 4-6-4, a design that regularly hauled 1,270-ton trains of eighteen cars at 88 kilometers per hour. The four trailing wheels under the firebox allowed the generation of vast amounts of steam. Timing on the crack Twentieth Century Limited was cut from the 20 hours with which the service was established in 1902 to 18 hours by 1932. Better service came from all-around improvement of the system, not just from faster locomotives. Until 1936 the New York Central's tracks ran at street level through the center of Syracuse, New York, and trains were allowed to move at little more than a walking pace. Elevated tracks allowed a $16\frac{1}{2}$-hour New York to Chicago schedule by 1936 and further minor improvements cut it to 16 hours by 1938 (Hollingsworth, 1982, 125).

In the years between World Wars I and II U.S. railroads steadily improved in performance, but there was no substantial change in their infrastructure. Except on a few heavily trafficked coal lines, axle loadings remained between 30 and 33 tons. Given this limitation U.S. designers concentrated more and more on efficiency. William Woodard was greatly concerned that his locomotives should have large enough fireboxes to produce the steam needed to do their job. He favored four- and even six-wheeled trailing trucks to allow large fireboxes, thus wheel layouts such as 4-6-4 and 2-6-6. He also allowed steam to expand more than other designers did, limiting his engines to a maximum cutoff point for incoming steam at 60 percent of the piston stoke where others allowed 90 percent. Woodard used higher boiler pressures and lighter components than other designers, and he improved mechanical drive systems (Westwood, 1977, 149–50). All of these improvements combined in the Lima Locomotive Works Super-Power concept to provide some of the most powerful yet fuel-efficient simple locomotives ever built.

By the late 1930s the Association of American Railroads (AAR) had as its goal moving 1,000-ton trains at 100 miles (160 kilometers) per hour. Existing locomotives, however good, were not up to the task (Reed, 1972, 42). Among solutions offered were the classic American 4-8-4, one of the finest simple locomotives ever built, and the 4-4-4-4 of

The 4-8-4 locomotive, first introduced in 1926, saw steam out on most U.S. railroads. The last were phased out in the early 1960s. The four-wheel trailing truck beneath the cab allowed a huge firebox, thus the ability to generate lots of steam for high speed and mountain climbing. Baltimore & Ohio Railroad Museum, by the author.

the Pennsylvania Railroad, one of the most complex and innovative. The Norfolk & Western class J 4-8-4 of 1941 hauled a 1,000-ton test train at 176 kilometers per hour, yet was designed to be incredibly cheap and easy to maintain, running 24,000 kilometers per month and visiting the repair shop only once every year and a half (Hollingsworth, 1982, 172–73). The Norfolk & Western, headquartered in Roanoke, Virginia, was as much interested in pulling power on its mountainous routes through the Blue Ridge and Appalachians as it was in speed, and the J had the highest pulling power of any 4-8-4 built.

The innovative and complex Pennsylvania Railroad 4-4-4-4 of 1942 was never fully developed. Two separate sets of driven wheels, it was argued, would make for shorter, lighter connecting rods and more steam cylinder area. The Norfolk and Western Js had two cylinders measuring 686 by 813 millimeters, whereas the Pennsylvania T-1s had four of 501 by 600 millimeters each, some 18.5 percent greater swept volume. The T-1s had a phenomenal reputation for speed but tended to break traction on one or the other set of driven wheels in mountainous country, in particular in the rain. Such sudden losses of adhesion made them uncomfortable for crew and passengers (Reed, 1972, 281).

Despite its strengths, the 4-8-4 simple locomotive suffered from high fuel consumption. Lines such as the Norfolk and Western and the Pennsylvania carried heavy coal traffic from the Appalachian coalfields and paid pithead price for their fuel. They tended to trade "some extra (cheap) coal for less (expensive) work in the [maintenance] shops" (Hollingsworth, 1982, 101). These lines highlighted both the principal advantage and the principal drawback of the French compounds. Minimizing fuel consumption by compounding required excellent maintenance and highly skilled drivers.

Despite the remarkable achievements of such U.S. designers as Woodard, the greatest steam locomotive designer of all time was André Chapelon, who so greatly improved French compound locomotives built on de Glehn's principles that they amounted to new locomotives. In 1926 Chapelon rebuilt the 1907 Paris-Orléans railroad's 4500-class Pacifics to an unparalleled level of performance. Although no mean performers as designed, Chapelon increased their power output by 85 percent to 3,700 horsepower (Hollingsworth, 1982, 78). A much modified experimental Pennsylvania K-4 of 1937 made only 3,500 horsepower from a locomotive nearly twice as heavy and with nearly double the axle loading. Chapelon's great skill was internal streamlining, easing the flow of steam from boiler to cylinders. His work was greatly admired and copied by others. In Britain Gresley radically improved his A-1 Pacific by adopting Chapelon's techniques, increasing pulling power by more than 20 percent in the A-4 of 1935 and producing a locomotive that broke the magic 100 mile per hour barrier with consummate ease. In 1938 one A-4, the *Mallard,* established an official world steam record at 203 kilometers per hour. Ironically, the record improved by only 4 kilometers per hour on that of 1935 established by a class 05 4-6-4 locomotive of the German state railroad, a locomotive equally influenced by Chapelon's design principles.

For all Chapelon's influence and skill his own designs were complex and expensive to maintain. The rebuilt Paris-Orléans Pacifics might have been the most efficient locomotives ever built in terms of fuel usage, but high maintenance costs meant they cost more to run than "the fleet of simple rugged 2-8-2s—the 141R class—supplied from North America at the end of World War II" (Hollingsworth, 1982, 79). The 141R Liberation class "totally vindicated" American principles and were the last steam locomotives in mainline service on French railroads (Hollingsworth, 1984, 151).

Chapelon's final masterpiece, his 4-8-4 of 1946, produced only five hundred horsepower less than the Norfolk and Western class J with half the grate area and not much more than half the weight. By 1946, however, a national coal shortage caused the French government to instruct the railroad system to cut coal consumption, pushing the French to a policy of electrification. Chapelon's 4-8-4 could outperform any electric locomotive then existing and was so economical on fuel "as to nullify any potential coal saving through electrification" (Hollingsworth, 1982, 181, 186). Yet it required highly skilled drivers and highly skilled mechanics, which electric locomotives could do without. Like so many other Neotechnic innovations, electricity substituted massive capital investment in sophisticated technologies that could be run by unskilled operators for a lower level of investment in simpler technologies that required skilled operators. The steam era of the world's railroads thus drew to a close.

STEAM RAILROADS AND THE STATE

The development of steam-powered railroads is inextricably linked with the emergence of land-based polities after 1850. As table 4-3 shows, the amount of trackage constructed by each polity varied substantially. Measurements against population size give a very different picture than measurements against land area.

Britain achieved a high areal density early and was not overtaken by Germany until early in this century. At its height in 1930 the United States had a lower areal density than Britain in the 1890s or Germany at the turn of the century. As befits a maritime system, Japan has always had an even lower density. As the German geographer Wagner pointed out in 1903, however, the republic of the North had a railroad density only slightly less than a Europe defined by the Mediterranean, the Baltic, the North Sea, and Germany's eastern border. Europe had 70 kilometers of track per 1,000 square kilometers to the republic of the North's 66 kilometers per 1,000 square kilometers. Eastern Europe (Scandinavia, the area east of Austria-Hungary but north of the Mediterranean, and Russia) had a density of 10 kilometers per 1,000 square kilometers compared to the western United States's 14 kilometers per 1,000 square kilometers (Wagner, 1903, 835).

Per capita figures show a very different picture (table 4-3). The United States exceeded British per capita track density in the 1840s. In the 1860s it had more than twice as many kilometers of track per person as

Britain and six times as many as Germany. At the peak in 1910 the United States had nearly eight times as much trackage per head as Britain and retained a six-fold advantage over Germany. The rebuilding of U.S. railroads in the early 1900s with much heavier track also greatly increased the capacity of the system. The tonnage of freight traffic increased 76 percent in the United States in the decade 1900 to 1910 compared with a 60 percent increase in Germany (table 4-4). The amount of work done expressed in ton-kilometers increased even more, and from a higher base, in the United States (table 4-4). Part of this can be attributed to the sheer distances Americans had to work with, but a significant component of the 235 percent increase carried over the twenty years between 1890 and 1910 can be attributed to relaid track. By 1910 most U.S. trackage could handle 33-ton axle loadings, whereas German railroads could handle only 18-ton loadings.

Railroads were the key to economic and political integration of continental states and were seen as such by Germany, France, and Russia. By 1910 Germany had the highest areal density of railroads of any major polity, an event closely presaged by such geographic commentators as Wagner (1903). But German railroads tended to be concentrated in the major states, particularly Prussia and Bavaria, and linkages between states were poor in the 1800s. France developed a competent system centered on Paris and kept about even with Germany on a kilometers per capita basis. In a state of roughly similar area to imperial Germany, however, France had a much lower areal density of trackage. The other great land-based polity of the late 1800s was Russia. The development of this system was driven by a higher level of theoretical concern than that of any other nation, even including Prussia. "In Germany locomotive designers held, typically, Chairs in neighboring universities or high-

Table 4-4 Railroad freight carried, 1890–1955, selected polities

Year	Thousands of metric tons		
	Germany	Britain	United States
1900	360,165	426,514	642,641
1910	575,330	516,054	1,130,960
1920	337,200	323,158	1,502,435
1930	438,200	309,218	1,344,806

Year	Millions of ton-kilometers					
	Germany[a]	% change	United States	% change	Soviet Union	% change
1890	22,500		135,186			
1900	37,000	+64	251,183	+86	38,900	
1910	56,400	+52	452,383	+82	65,800	+ 69
1928	73,900	+31	714,065	+58	93,000	+ 41
1932	44,800	−39	385,304	−46	169,000	+ 82
1937	80,600	+80	594,087	+54	355,000	+110
1940	nd		614,643	+ 4	415,000	+ 17
1955	52,904		1,026,498	+67	971,000	+134

Sources: Mitchell, 1980; U.S. Bureau of the Census, 1975; Westwood, 1982.
[a] West Germany only after 1945.

er technical schools; professors designed and designers professed. This practice was even more marked in Russia, where new locomotive types were designed not so much by professors as by committees of professors" (Westwood, 1982, 3).

Russian designers pioneered line testing of their locomotives, but the mass of mathematical descriptions thus produced rarely described the real performance of a locomotive. The comment of a British designer to the Russian Lomonosov is apt: "Professor, you know that in this country trains are pulled by locomotives, not differential equations" (quoted in Westwood, 1982, 4). Nevertheless, locomotive testing was a good idea, as the remarkable success of the Pennsylvania Railroad's stationary testing plant at Altoona proved in 1904.

Russia's excessively theoretical approach was further weakened by revolution. By the end of the civil war in 1922, 63 percent of all locomotives were out of action, the track was so bad that the best locomotives had an axle loading too high for most lines, and the economy was crucially short of steel (Westwood, 1982, 9). Rebuilding took time, and the process was delayed by continued concern with theoretical levels of performance. Eventually the Party acted. It deemed that too much effort had been wasted on electric and diesel locomotives between 1922 and 1929. Steam traction was recognized as the prime motive power for the short term, and there was a general move to U.S.-style railroading. It was proposed to upgrade track from 17 to 27 tons axle loading, and locomotives were ordered from the United States (Westwood, 1982, 92–96). As late as the 1950s, however, axle loadings had improved only to 18.5 tons.

The importance of the development of land transportation, in particular to the development of the United States and Russia, the two great land-based states of this century, was not lost on geopolitical observers. Mackinder commented extensively on the emergence of the land-based states in his heartland theory (1904). Even Mahan, whose early career was based on applying the principles of sea power to geopolitics, accepted Mackinder's ideas (Blouet, 1988b). The only question seems to have been how the land-based polity of Eurasia would be structured, since the republic of the North had attained almost undisputed hegemony in North America after 1865.

Despite the buildup of the German high seas fleet that was so disturbing to the powers with maritime interests after 1900, Germans also dreamed of a central European customs union from the Pyrenees to the Polish frontier, bordered on the south by the Mediterranean and on the north by the Barents Sea (fig. 4-7). Contemporaries in Germany saw this Mitteleuropa as the basis for German resistance to "the Russian, English, American, and perhaps the Chinese world empires" (quoted in Fischer, 1967, 9). Other Germans suggested that a broader European power base was needed if Germany was to "achieve and hold an overseas empire" (Fischer, 1967, 9). Germans also recognized that the basis of Mitteleuropa was a central European economic union conceived to counteract high U.S. protective tariffs (Fischer, 1967, 10).

By 1910 the German economy was the best organized and most modern in the world, "regularly introducing every modern innovation and invention, fed by a network of technical academies organized exclusively to serve it, and manned by a disciplined, industrious and thrifty population . . . Economic expansion was the basis of Germany's political world diplomacy" (Fischer, 1967, 19–20). With the outbreak of World War I in 1914 the concept of Mitteleuropa took on new strength: by 1915 it dominated German policy. Poland, acting as a buffer state against Russia, whose "basically Byzantine-Oriental culture separates it from the Latin culture of the West," was to be the cornerstone of Mitteleuropa. Poland might be peopled by Slavs, but they lacked the "Russian's Mongol strain, and they were divided from the Orthodox Russians by their Catholic and Protestant religions" (Fischer, 1967, 203). The last great German war aims conference of July 1918 not only emphasized the importance of Poland but also the importance of its railroads to German security. Of the six demands made of Austria-Hungary at that conference, two were concerned with this issue: that "Poland, being the most important transit area to the east, must be economically dominated by Germany," and that Germany must therefore "possess a dominating influence over its railways" (Fischer, 1967, 531). As late as early September 1918 Germany was still demanding that Poland "come into the Central European customs Union" and that "a railway company under predominantly German influence" be founded (Fischer, 1967, 533).

"Mitteleuropa"

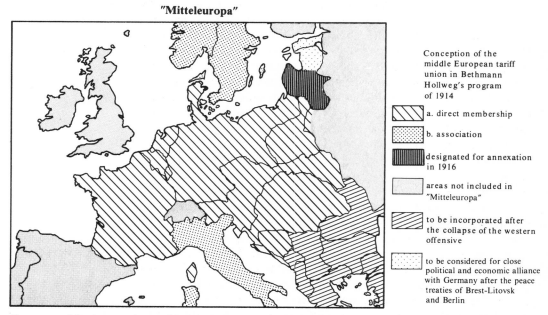

Conception of the middle European tariff union in Bethmann Hollweg's program of 1914

a. direct membership

b. association

designated for annexation in 1916

areas not included in "Mitteleuropa"

to be incorporated after the collapse of the western offensive

to be considered for close political and economic alliance with Germany after the peace treaties of Brest-Litovsk and Berlin

Figure 4-7 Mitteleuropa, imperial Germany's proposed European "superstate" from the Pyrenees to the river Elbe and a major German war aim in World War I. Redrawn from Fischer 1967.

The German aim of Mitteleuropa that dominated policy in World War I was pursued on an even grander scale in World War II. Hitler perceived Kaiser Wilhelm II's colonial aspirations outside Europe as a fundamental error. He conceived a Third Reich, a "Super-Germany" from the "Urals to the Pyrenees, eliminating the power of France and Russia, and supplied with raw materials from its colonial empire in Central Africa" (Herwig, 1976, 185–86). It was this power base that would allow the Third Reich, preferably in alliance with Britain, to expand out into the Atlantic and destroy the United States. The takeover of Austria and the invasions of Czechoslovakia and Poland were all part of this process, although Hitler did not count on Britain's declaring war upon his invasion of Poland. Failing to take Britain but having succeeded in France, Hitler was well on the way to his "Super-Germany" by late 1940, and Mitteleuropa was reality. In two years of relatively inexpensive warfare, Hitler succeeded where Wilhelm had failed. But in 1941 Hitler made the fatal error of pushing east into the Soviet Union, attempting to complete his "Super-Germany" to the Urals. From that point on the Third Reich was doomed to overextension and collapse.

The geostrategy of the Third Reich was based less on railroad integration of the spatial economy and polity than on the Neotechnic transportation technologies that emerged between 1900 and 1920. Roads and aircraft were far more significant than railroads. Russia, however, still saw the railroad as the key integrator. Whereas German railroad freight traffic fluctuated widely between the wars, in part because of the Depression and in part because of the rise of motor trucks, Russian railroad traffic experienced a steady increase. The Third Reich, like the American republic as early as the 1920s, was becoming a Neotechnic polity, whereas the Soviet Union remained a Paleotechnic one well into the 1960s. Even in the 1970s ton-kilometers of freight were still increasing steadily on Soviet railroads, whereas they peaked in the mid-1960s in Western Germany and in the early 1950s in America and Britain (U.S. Bureau of the Census, 1975, 431; Mitchell, 1980, 624, 626).

ELECTRICITY AND SURFACE TRANSPORTATION IN THE NEOTECHNIC

Neotechnic land transport falls into two categories: that which uses the surface of the land, and that which passes above the land in the air. Neotechnic surface transport also falls into two categories: one with fixed routes, and one with marked route flexibility. All these have had a massive impact on the world system, completing the process of continental penetration begun by the railroads while speeding it up and both reducing the effective scale of movement and improving economies of scale.

The common ancestor of most Neotechnic improvements in transport was the bicycle. Bicycle technology introduced humanity to the advantages of individual mobility at a speed level much higher than the unaided organism could achieve. In the 1880s bicycle technology forced the pace in light, strong steel tubing for frames; ball bearings and chain power transmission systems; suspension systems; and lightweight wheels. The bicycle makers' obsession with lightness and efficiency of power transmission makes the bicycle the key to the later development of automobiles and aircraft when the spark ignition engine began to mature in the 1890s.

Neotechnic improvements in land-based transport are also confused by the development of two independent technologies that had similar spatial impacts. The earlier of these was the use of electricity. The second, and in the long run more important, was the application of the internal combustion engine. Hybrid internal combustion–electric systems also developed, in particular for railroad use.

Streetcars

Electricity is not a form of energy per se, only a means of transmitting energy efficiently. It was first successfully used to power a vehicle in 1879 by Siemens und Halske in Berlin (Dunsheath, 1962, 131–33, 181). By 1883 a practical electric streetcar was running in Richmond, Virginia, picking up its power from a device that looked like a fisherman's troller rolling along the top of an overhead wire (Dunsheath, 1962, 185). *Troller* was soon modified to *trolley*, and a spring-loaded pole became used to hold a single wheel in contact with the underside of the power wire, the circuit being completed by return current passing through the metal rails.

Streetcar networks grew explosively in the United States after the mid-1880s during a period of rapid industrialization, urbanization, and immigration. The victory of the republic of the North in the Civil War guaranteed rapid industrialization, but a labor source was not available at short notice from internal demographic increase. With considerable land remaining to be settled in the West, most old-stock Americans preferred to own land and farm rather than work for wages. Labor for industry could only come from massive immigration from the industrializing nations of Europe, where agrarian populations were being alienated from the land.

The arrival in the United States of large numbers of such immigrants resulted in massive urbanization. Old-stock Americans did not want to live close to the new immigrants, who were increasingly different in language, culture, and religion toward 1900. Old-stock Americans thus used their higher incomes and the streetcar to move to the new suburbs that developed on land adjacent to the old, pedestrian-scale cities (Warner, 1968; Ward, 1971, chap. 5). This process was begun in the 1870s by the horse- or mule-drawn streetcar, but it was radically accelerated by the trolley after 1885. By 1890 the United States had 2,100 kilometers of electric streetcar track. From 1890 through 1913 some 3,200 kilometers were added each year (table 4-5).

The trolley was also seen as a solution to the pollution caused by horse-drawn vehicles. Many U.S. cities used wood-block streets to soak up horse urine, but solid waste still had to be cleared. By the late 1800s the germ theory of disease had awakened people to the problems posed by organic waste, whether animal or human, and cities were enthusiastically installing sewage systems and water treatment plants. Getting rid of the animal waste in the streets was the next logical move, and the electric trolley offered a way out. By 1890 Boston had already retired some nine thousand horses.

No other industrial nation adopted the trolley with the enthusiasm of the United States (table 4-5). Germany was a very distant second and Britain and France hardly placed. A subtle reason for the lack of trolleys in Britain was the massive emigration of people to the United States, Australia, Canada, New Zealand, and South Africa between 1880 and 1900, which markedly reduced demand for new housing and suburbanization (Byatt, 1979, 40–42). When demand turned up again, just after 1900, internal combustion–engined busses were available. They were thus enthusiastically adopted in Britain before they were common elsewhere.

U.S. development of streetcar technology proved very profitable to U.S. industry. Although the initial technology was German, it was in the United States that operational problems were first encountered and overcome. The first successful motors for streetcar use were designed by Frank Sprague for the Richmond system. This was no mean achievement. Streetcar motors were exposed to water, dirt, and vibration and had to occupy a small space (Dunsheath, 1962, 183–84). Despite the early German lead, German companies had to buy streetcar technology from the United States after 1890. Not only that, but by the mid-1890s U.S. companies had achieved economies of scale through mass production, and costs fell by more than 70 percent between 1891 and 1895 (Byatt, 1979, 32–33).

Streetcars had two great advantages over the steam-powered railroads that had been responsible for modest suburban growth from the 1840s. Because they were powered by energy transmitted electrically from a distant, stationary power plant, they had very low axle loadings. Vehicles were light and could run on cheap rails. Despite the high fixed capital costs of the power plant, the fixed capital costs per kilometer of route were lower than those for a steam-powered system. Large, efficient power plants could be amortized over a large route network, and sur-

Table 4-5 Electric trolley track, 1890–1913, selected polities

Year	United States	Germany	Britain	France
	kilometers of single track		*route kilometers*[a]	
1890	2,092	nd	13	nd
1895	16,677	1,363	72	97
1903	44,665	5,536	2,353	2,004
1913	72,425	nd	4,176	nd

Source: Byatt, 1979.
[a]Route kilometers usually equal about two-thirds of track kilometers.

plus power produced could be sold to nontransportation users, such as homeowners for lighting and industry for machinery.

The second advantage lay in the power available for acceleration. For all types of transport the power required to cruise is very much less than the power required to accelerate to cruising speed, in particular when high rates of acceleration are desired. Steam locomotives had some very desirable characteristics, such as developing maximum torque from rest, but they needed to carry water, fuel, and a huge firebox and boiler to make steam. An electric motor has the same desirable torque characteristics, yet is not encumbered with firebox, boiler, water, and fuel. A small, light electric streetcar can accelerate extremely fast because it draws energy from a large, efficient, distant generating plant. High acceleration rates meant that stops could be more frequent than with steam power and a similar or even better service schedule could be maintained. In the United States, steam railroad suburbs began 6 to 8 kilometers from the city center. Electric traction made it possible to fill the gap with suburbs beyond the economic reach of horsecars, yet too close in to be served by steam (Vance, 1990, 389).

Electric traction had further technical advantages. The simple steam locomotive, in particular the U.S. two-cylinder locomotive, was notorious for its "hammer blow" impact on the rails. Where locomotives were

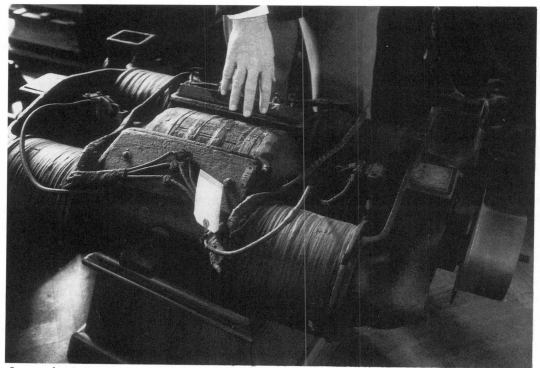

Sprague electric streetcar motor, 1887, 7.5 horsepower. Such immensely compact motors needed no fuel or auxiliaries, only wiring back to their generators. They could be fitted under the floor of a mobile vehicle or power machines in a factory. Ford Museum, by the author.

started in stations, rails had to be half as strong again as needed with the more balanced European three- or four-cylinder locomotives. Yet even three- or four-cylinder locomotives required skilled drivers, frequently slipped their wheels on starting, forced rails apart, and wore both the rail and their own steel tires. Rails in stations were sometimes replaced weekly. Electric traction avoided these problems. Power was delivered smoothly and progressively, with no hammer blow and little likelihood of slipping. Electric trolleys and locomotives were thus easier to drive than steam locomotives, an important consideration given the rapid increase in their number. Almost anyone could be taught to drive a streetcar. Given that a steam train also needed a crew of three—one to drive, one to stoke, and one to guard the rear—trolleys were also more labor efficient on a per vehicle basis.

In the United States this labor advantage was assisted by a simple fare system. All rides in a given city, no matter how far, cost a nickel—a fare that encouraged frequent patronage and was easily collected by the driver. European systems charged by distance and required an extra person to collect fares, issue tickets, and make change. In the U.S. system the only proof of payment needed was a transfer and it, too, was standardized across the transit region (Vance, 1990, 386–87).

Despite the advantages of the trolley, mass adoption of the automobile led to its rapid decline. Trackage peaked in 1917 at just under 67,000 kilometers, fell to 45,600 by 1934, and to 8,000 by 1957 (Vance, 1990, 391–92). However frequent the service, trolleys were never as convenient as automobiles, and routes were fixed.

Interurbans

Little has been written on long-distance rural electric traction. Vance has a useful section on "the farmer's railroad," but it concentrates on the experience in the North, where such lines usually served two or more towns and hence were "interurban" (1990, 392–95). In many ways the interurban presaged the motor truck, which strangled it almost at birth, and it pioneered technology combining the internal combustion engine with electric power transmission. Remarkably few miles of rural interurban railway were built, its heyday was perhaps 1905 to 1915, and it almost invariably lost vast amounts of money.

As we have seen, steam technology was a poor choice for relatively short distances around cities or where fixed costs had to be kept low. Electric streetcar systems had to amortize both the fixed and operating costs of a power plant over as much track as possible. Rural service required low fixed and operating costs. Internal combustion engines offered low operating costs but required complex power transmission systems to multiply their poor torque outputs. Gearboxes worked well with small engines but frequently broke when used with large ones. An internal combustion–engined streetcar had most of the advantages of electric and steam systems but had difficulty transmitting power from its large engine smoothly and reliably to the driven wheels. The problem was solved by using electrical transmission within the vehicle.

The true interurban was thus powered by electricity generated by a large, on-board, spark-ignited internal combustion engine, not a distant power station. Manufacturers simply closed off one end of a regular streetcar and installed a gasoline engine and generator. Gas-electrics, as they were called, developed in the 1890s in Europe and were first marketed in the United States in 1904. The first really successful design came from General Electric in 1909. Gas-electrics were usually sold as "traffic generators" to get traffic up to the point where capital could be raised to build a fixed power plant and install overhead transmission wires (Hanley & Corley, 1973, 39–43).

Most such lines can be researched only locally, but at this level they are very revealing (Kniseley, 1987). In 1900 Bryan, Texas, was a typical courthouse town in the cotton South, with good railroad connections to Houston and Galveston. The interurban promised access to potential cotton-producing areas that could not be affordably served by steam railroads and that were too far from the cotton gins and other services of the country towns for ox carts. There was also an element of competition. Bryan put up money for a 44-kilometer gas-electric interurban into the cotton lands of the fertile Brazos River valley because it feared that courthouse towns to the north and west might do it first and preempt Bryan's trade (Kniseley, 1987, 13).

Electric interurbans pioneered another critical piece of electrical transportation technology, the multiple-unit controller developed by Frank Sprague in 1895. This technology allowed one driver to control a multiple-car set (Dunsheath, 1962, 186). Electric locomotives powerful enough to move large trains of regular cars had axle loadings little lower than those of comparable steam locomotives. The multiple-unit controller allowed a light motor on every axle. Adding more cars meant adding more power, with no change in axle loading. The multiple-unit controller thus maintained the streetcar advantage of low track weight and cost while allowing large trains to be run on suburban commuter lines.

Some 29,000 of the 67,000 kilometers of streetcar lines in America in 1917 were interurbans (Middleton, 1961, 12). These lines developed in three main areas of the United States: the republic of the North, southern California, and northern central Texas. The single largest integrated system was the Pacific Electric, which provided transportation throughout the Los Angeles basin at average speeds higher than on today's freeways, most of which follow the old interurban rights of way. The Texas Electric system was focused on Dallas–Fort Worth, with continuous extensions as far north as Denton and as far south as Waco and Marlin, with outliers beyond that. The republic of the North was covered by a welter of unrelated local systems, often very large and focused on individual cities, which it connected to their neighbors.

All of these systems were mainly used locally. Traveling salespeople enlarged their sales areas. Farmers moved perishable goods to town more easily. People took trolley vacations, or at least spent a day at the amusement and country parks that sprang up at the ends of most lines.

The New York–Chicago tour became one of the most written about, if more rarely experienced. In 1915 the trip could be made in thirty-one to forty-five hours, depending on connections, and encompassed twenty-three electric and one steam system (Middleton, 1961, 43–44). Certain sections paralleled steam railroads or even used their tracks. Between Rochester and Canandaigua, New York, for example, Rochester & Eastern interurbans raced and beat crack New York Central steam expresses (Middleton, 1961, 95).

In Europe Sprague's multiple-unit technology allowed suburban electrification of existing railroads as well as development of commuter lines reaching underground into the heart of the world's then largest city, London. The great advantage of suburban electrification was increased frequency of service, thus better amortization of fixed capital in rails and terminals. Typical of London's suburban steam services were the Great Eastern and the Southern railways. The Great Eastern served mainly working-class areas east of London. The Southern served mid-dle- and upper middle–class areas south of London. Suburban London railways had reached bottleneck conditions in their terminals by the late 1800s. Contemporaries estimated that electrification could double the capacity of existing terminals. Steam trains took six minutes of platform time to turn around, whereas electric multiple-units could do so in two. Average speeds rose between 30 and 150 percent on lines that electrified.

The London, Brighton, & South Coast Railway, a major component of the Southern, let a contract to Allgemeine Elektrische Gesellschaft (A.E.G.) of Berlin in 1906 to supply overhead electrification equipment through British licensees. A trip of 14.4 kilometers with nine intermediate stops that took thirty-six minutes with steam locomotives fell to twenty-four minutes with electricity. Traffic also recovered. In 1902 the South London line of the railway had carried more than 8 million passengers. Competition from electric streetcars and internal combustion–engined busses reduced that figure to less than 4 million by 1909. Electricity had it up to 7.5 million in 1910, and with a 10 percent return on capital outlay (Moody, 1979, 7–9). Another component of the Southern, the South Eastern & Chatham Railway, estimated gains from electricity at 40 percent in speed and 60 percent in frequency of service (Moody, 1979, 21).

By 1924 the Southern found itself with three different and incompatible systems of transmitting power. After consultation, in particular with the Pennsylvania Railroad, continued overhead electrification using A.E.G. technology was rejected because of high construction and maintenance costs (Moody, 1979, 23). Infrastructural considerations loomed very large in this decision. Most British suburban lines were crisscrossed with road bridges that would have to be raised to give adequate clearance for overhead installation. The Southern therefore standardized on direct-current, 600-volt, third-rail power transmission, which one of their component companies had bought from the Pennsylvania Railroad. Despite their infrastructural advantages, such third-rail systems could carry only low voltages without electricity arcing to ground across the air gap in wet weather. This limited them to relatively low power systems with generating stations relatively nearby.

London's other major suburban system was the Great Eastern, which did not electrify until the 1950s, although it considered doing so in the teens. Instead the Great Eastern brought in Henry Worth Thornton from the United States's Long Island Railroad and worked its Liverpool Street terminal to very tight turnaround times (AAR, 1984–85, 79). This required careful management as well as a class of steam locomotives designed to maximize acceleration (White, 1963, 183). Despite these locomotives it is clear that the Great Eastern chose a software rather than a hardware solution to the problem of high-frequency service.

Sprague's multiple-unit technology was also at the heart of the development of the London underground early in this century. So was U.S. investment capital. The underground began in the 1860s as a "cut-and-cover" system with trains pulled by steam locomotives. Although these locomotives were supposed to "consume their own smoke," the Circle Line was usually referred to as the "sewer" line (Byatt, 1979, 49). The City & South London Railway opened in 1890 was the first to use electric locomotives underground. Locomotives were provided by Siemens Brothers of Berlin. Although it was clearly better than the old steam-hauled underground, the use of electric locomotives limited the system, and there were all the teething faults typical of a pioneer technology.

After 1901 the various existing and proposed underground railways in London were brought up by Charles T. Yerkes, a Chicago financier who also brought in New York and Boston financiers to capitalize the system. The Yerkes tubes, the basis for almost all the modern systems, were quickly constructed using tunneling shields and "equipped and run on standard American lines" (Byatt, 1979, 51). The system made little profit to begin with, its capacity having been pushed well ahead of demand. High, thus profitable, load densities were not reached until the suburban boom that followed World War I. Internal combustion–engined busses were strong competitors. They began to appear in large numbers after 1906 and carried 288 million passengers by 1911, at which time the Yerkes tubes carried 370 million and the streetcars 631 million (Byatt, 1979, 52). Nevertheless, Yerkes was quite clear-sighted about what his tubes would do to the spatial structure of London. As early as 1896, when his system was merely a pipe dream, he claimed that "a generation hence, London will be completely transformed, so that some people will think nothing of living 20 or more miles from town owing to these electric railways" (Edwards & Pigram, 1986, 10). Yerkes died in 1905, probably fortuitously for the London underground. His elaborate financial dealings in Chicago streetcar lines in the 1880s and 1890s were designed to enrich him, not serve the public, and he was run out of that city in 1897. He never had time to milk the London system as he did that of Chicago and thus left as his monument the world's first great integrated subway system (Bobrick, 1981, 129–30). Remarkably, no new lines were added to the system planned by Yerkes until 1969, although the Yerkes tubes were extended (Howson, 1981, 5).

Mainline System Electrification

System electrification and electric locomotives seemed to offer better possibilities of long-distance high-speed service than steam early in this century. Naive but serious promoters offered the possibility of ten-hour service between Chicago and New York at a cost of $10 on the Chicago–New York Electric Air Line Railroad (Middleton, 1961, 27–29). Such service would have required cruising at 160 kilometers per hour, and no steam locomotive of 1900 was capable of speed remotely like that. On long trips steam locomotives had to be changed several times because no firebox could hold the accumulated ash of sustained high-speed working with coal. As early as 1901 high-speed electric working was being explored in Germany. The Study Group for Electric High-Speed Railways formed by Siemens und Halske, the Allgemeine Elektrische Gesellschaft, and the Prussian state government ran a highly modified, experimental interurban car at 210 kilometers per hour (Hollingsworth & Cook, 1983, 26).

Long-distance electrification occurred first in countries with no coal, or with mountainous areas, or both. Italy and Switzerland had plentiful falls of water to generate electricity, and little or no coal. Italians turned to electricity first in 1908 on the line connecting the port of Genoa to its hinterland over a substantial climb (Hollingsworth & Cook, 1983, 30). Pushing railroads across the steep grades of the Swiss Alps became far easier with the power outputs that could be achieved by hydroelectric generating stations.

The first U.S. use of electric power to drive locomotives long distances came in the American Rockies. The "Milwaukee Road" (the Chicago, Milwaukee, St. Paul and Pacific Railroad) had electrified 1,075 kilometers of track in two sections by 1919: 370 kilometers across the Cascades from its Tacoma terminal to Othello, Washington, and 705 kilometers over the Rockies from Avery, Idaho, to Harlowton, Montana (Hollingsworth & Cook, 1983, 38). The General Electric "bipolar" locomotives built for this run were so good that they were in service through the late 1950s. Their pulling power well exceeded that of the most powerful steam locomotive ever built for passenger trains, Union Pacific's Challenger class (table 4-6).

Electric locomotives and multiple-unit controllers offered speed as well as pulling power. Concern with speed rose in the 1930s because of budding competition from airlines and more seriously in the 1950s, when airlines finally proved their potential. High-speed electric railroads have been most accepted in Europe and Japan, where distances between cities are shorter and where population density and land prices are high. Trains averaging 160 to 200 kilometers per hour between the centers of cities easily beat aircraft on stage lengths up to 650 kilometers, especially when airports have to be built well out of town.

Electric traction also offered countries with water power freedom from imported fuel oil, or a reduction in the use of native coal with its attendant pollution problems. Japan's Shin-kansen network realizes these advantages and will stretch when complete from Kagoshima to Sapporo. The first services began in 1964, and the system quickly realized the potential suggested by the German experiments of 1901. Shin-kansen trains have every axle powered, attain 210 kilometers per hour, and average 176.5 kilometers per hour over the 1,176 kilometers from Tokyo to Fukuoka. Such a service would put Chicago 8¼ hours from New York (Hollingsworth & Cook, 1983, 142–43). In a mountainous country like Japan, however, high speed has been bought only with

Table 4-6   Selected steam locomotive statistics

Year	Polity	Railroad	Locomotive name, type, class	Wheel arrangement	Compound (c) or simple (s)
1830	GB	Liverpool & Manchester	Northumbria	0-2-2	s
1832	USA	Mohawk & Hudson	Brother Jonathan	4-2-0	s
1837	USA	Philadelphia, Germantown, Norristown	Campbell	4-4-0	s
1837	USA	Beaver Meadows	Hercules	4-4-0	s
1855	USA	Western & Atlantic	American	4-4-0	s
1870	GB	Great Northern	Stirling 8' single	4-2-2	s
1876	F	Paris, Lyons & Mediterranean	121	2-4-2	s
1884	I	Upper Italy	Vittorio Emanuele	4-6-0	s
1900	USA	Lake Shore & Michigan State	I-1	4-6-0	s
1900	F	Nord	Atlantic de Glehn	4-4-2	c
1901	USA	Pennsylvania	Atlantic E3 sd	4-4-2	s
1902	USA	Chesapeake & Ohio	Pacific F1S	4-6-2	s
1906	D	Prussian State	P8	4-6-0	s
1907	F	Paris-Orléans	Pacific 4500	4-6-2	c
1908	D	Bavarian State	Pacific 53/6	4-6-2	c
1914	USA	Pennsylvania	Pacific K4	4-6-2	s
1926	USA	Northern Pacific	Niagara A	4-8-4	s
1926	USA	New York Central	Hudson J3A	4-6-4	s
1935	GB	London & North Eastern	Pacific A4	4-6-2	s
1935	D	German State	Hudson O5	4-6-4	s
1941	USA	Norfolk & Western	Niagara J	4-8-4	s
1942	USA	Pennsylvania	Duplexii T1	4-4-4-4	s
1942	USA	Union Pacific	Challenger	4-6-6-4	s
1946	F	French National	Niagara 242 A1	4-8-4	c

Sources: Hollingsworth, 1982; Hollingsworth, 1984.
nd = no data.
[a] Adhesive weight is the weight applied to all the driven axles.
[b] Engine only, without tender.

expensive route leveling. Even more spectacular than the Shin-kansen is France's Train à Grande Vitesse (TGV) of 1981. As with its canal system of the early 1800s, the French aim is to provide unparalleled national integration. The Paris-Lyons service began in 1983 at two hours for 426 kilometers, 40 of which are on low-speed existing track (Hollingsworth & Cook, 1983, 200–201). The Shin-kansen and the TGV require special track to accommodate their high speeds and, in the case of the Shin-kansen, also because normal Japanese railroads are built to 3-feet, 6-inch gauge rather than the standard 4 feet, 8½ inches considered minimal for high-speed working.

Steam pressure (kg/cm^2)	Superheater (yes or no)	Grate area (m^2)	No. cylinders	Weight (tons)	Adhesive weight[a] (tons)	Axle loading (tons)	Maximal pulling power (kg)
3.50	no	0.75	2	11.5	3.0	3.00	720
3.50	no	nd	2	6.4[b]	3.2	3.20	464
6.35	no	1.1	2	nd	7.25	3.60	1,984
6.35	no	nd	2	14.0	9.0	4.50	2,045
6.35	no	1.35	2	41.0	19.5	9.75	3,123
9.80	no	1.64	2	66.0	15.5	15.50	5,101
9.00	no	2.20	2	49.7	27.5	14.00	5,545
10.00	no	2.25	2	83.7	48.0	16.00	6,958
14.10	no	3.10	2	136.0	61.0	20.30	10,800
16.00	yes	2.75	4	120.0	35.6	17.80	10,921
14.40	yes	5.20	2	165.0	58.0	29.30	12,400
12.7	no	4.40	2	185.0	71.5	24.00	14,696
12.0	yes	2.58	2	78.5	52.0	17.75	12,140
16.0	yes	4.27	4	136.5	53.0	17.50	nd
16.0	yes	4.50	4	149.0	53.0	18.00	nd
14.4	yes	6.50	2	242.0	96.0	33.00	20,170
15.8	yes	10.70	2	335.0	118.0	29.50	27,950
18.6	yes	7.60	2	350.0	91.5	30.50	19,000
17.5	yes	3.80	3	170.0	67.0	22.50	16,086
20.0	yes	4.71	3	213.0	56.0	19.50	14,870
21.0	yes	10.00	2	396.0	131.0	33.00	36,287
21.1	yes	8.75	2+2	432.7	124.0	31.50	29,300
19.7	yes	12.30	2+2	486.0	184.3	31.00	44,100
20.4	yes	5.00	4	225.0	84.0	21.00	nd

"Internal" Electrification of Locomotives

System electrification worked well on routes where high traffic density justified high fixed costs. On less densely traveled lines the principle of "internal" electrical transmission of power was borrowed from the interurban gas-electrics, but was applied with much more efficient compression-ignited (diesel) engines. Diesel-electrics now dominate the world's nonelectrified railroads. There have been attempts to produce locomotives with electric transmission of power from steam or gas turbines, and there have been attempts to transmit power from a diesel engine mechanically, usually with hydraulics. Yet diesel-electrics have consistently won out, not least because electric transmission allows them to be worked in multiple units using Sprague's controller and one crew. Large trains can otherwise only be worked by powerful and expensive locomotives, by large numbers of locomotives, each with its own crew, or by complex and trouble-prone control systems. Diesel-electrics can be mass-produced to a simple, standardized design, and the requisite number can be simply coupled together.

These advantages were first realized in the United States in the 1930s. Earlier attempts in Prussia, the United States, Russia, and Canada foundered because the diesel-electric was conceived only as a traction unit, not part of a complete system from manufacturer to user. Success came in 1937, when General Motors' Electro-Motive Division introduced its E-series locomotives. GM brought to the railroads the same advantages they brought to the automobile: low price and easy parts availability because of mass production and standardized design. The

Pennsylvania Railroad GG1 electric locomotive, 1934. Electric locomotives were ideal for densely traveled routes with frequent stops because of their rapid acceleration. They were also good in mountainous terrain, despite the high costs of route electrification, because they were not limited to the power they could generate from the fuel they carried. Baltimore & Ohio Railroad Museum, by the author.

reliability of these units and their frequency of service compared to steam quickly became legendary. One set covered 454,000 kilometers at an average scheduled speed of 90 kilometers per hour over 365 continuous days of service (Hollingsworth & Cook, 1983, 78–79). No steam locomotive could hope to approach such levels of availability. At the end of World War II the U.S. steam locomotive fleet was severely rundown, and most railroads converted to diesel-electrics rather than continue with steam.

The Soviet Union also experimented with diesels in the 1920s. Lenin gave improved transport very high priority and strongly supported both dieselization and electrification (Westwood, 1982, 18, 21). One argument for diesels was that they were less easily interrupted by sabotage or war than electrified lines. Similar arguments had been advanced in Prussia before World War I, and Kaiser Wilhelm had rejected electrification in favor of steam on the basis of military security (Westwood, 1982, 57). Numerous Soviet experiments with diesels in the 1920s came to nothing, and the country returned to steam traction in the 1930s. System electrification began after World War II. By 1975 some 44,800 out of 138,000 kilometers of track were electrified (Hollingsworth & Cook, 1983, 148) and nearly 60 percent of the country's ton-kilometers of work were being hauled electrically (Hollingsworth, 1984, 202). Mod-

American Locomotive Company diesel-electric locomotive, 1924. The first diesel-powered locomotive to use electric transmission of power. Diesel-electric locomotives did far less damage to tracks than steam locomotives of equal power because they delivered their power smoothly instead of in "hammer blows." Since they needed no water other than that in their closed cooling systems and never needed ash cleaned from nonexistent fireboxes, they were far easier to look after than steam locomotives. Baltimore & Ohio Railroad Museum, by the author.

Winton 201 diesel engine, 600 horsepower, 1934. In 1930 General Motors bought Winton for its lightweight, high-speed diesel engine technology, and the Electro-Motive Company for its railcar technology. GM's Electro-Motive Division is still the world's largest producer of diesel electric locomotives. Baltimore & Ohio Railroad Museum, by the author.

ern Soviet diesel-electrics date from 1945, when designers began to copy U.S. locomotives (Hollingsworth & Cook, 1983, 120).

The other major countries to prefer diesels to system electrification since World War II have been Germany and Britain. German engineers, however, tended to prefer hydraulic transmissions because they were more efficient, wheelslip was less, and they eliminated the problems that plagued electrical transmission. These claimed advantages resulted in the only modern sale of foreign-built locomotives in the United States. In 1961 the Denver and Rio Grande Western Railroad and the Southern Pacific Railroad bought a total of twenty-one diesel-hydraulics from Krauss-Maffei of Munich, a company long famous for powerful and innovative steam locomotives (table 4-6; the Bavarian 53/6 was built by Maffei). At 4,000 horsepower the Krauss-Maffeis offered 1,600 horsepower more than U.S. diesel-electrics, and hydraulic transmission greatly reduced wheel slip. These locomotives were ideal freight haulers in the Rockies, where they were usually used in threes. Although most downtime on diesel-electrics is caused by electrical problems, the diesel-hydraulics proved little more reliable, and "complexity of construction and inaccessibility for repairs" caused Southern Pacific to withdraw them in 1968. Similar experience with Krauss-Maffei diesel-hydraulics built under license in Britain, the Great Western region's Warship class of 1962 based on the German federal railroad's V200 class, restricted future use of diesel-hydraulics to Germany, where higher standards of maintenance seem to have been available (Hollingsworth & Cook, 1983, 138–39, 114–15).

European diesel-electrics have generally ignored the lessons of standardized design and multiple-unit control systems learned in the United States. The British have insisted on building specialized locomotives for specialized work, making for unnecessary complication. The U.S. concept of a simple range of mass-produced standard designs that can be coupled together for increased power under the control of a single crew is far more efficient. The British reasons for the switch to diesel-electrics were for other than the U.S. advantages: avoidance of the high capital expense of system electrification except on very heavily trafficked lines; the deteriorating quality of available coal in the early 1950s; and the increasing difficulty of getting labor to perform dirty, disagreeable routine maintenance of steam locomotives (Hollingsworth & Cook, 1983, 10, 118). A reason for preferring single to multiple units may lie in fears of the power of labor unions at the time of dieselization. The Baldwin Locomotive Works in the United States, for example, produced as its first diesel a 6,000-horsepower unit because of uncertainty whether labor unions would demand a crew on each unit (Hollingsworth, 1984, 164).

High-Speed Diesel-Electric Sets

Because they promised higher speeds on lightly trafficked lines than were believed possible with steam, high-speed diesel-electric sets were pioneered in Germany and America in the early 1930s. In 1933 German state railroads put into service the "Flying Hamburger" two-car diesel-electric sets. These averaged 125 kilometers per hour for the 287 kilometers from Berlin to Hamburg (Allen, 1978, 60). Many German population centers were within three hours of Berlin with the "Flying" services. Business people and government officials could take an early train to almost anywhere in Germany, work half a day, and return home that evening. With departures at 7 A.M. and 6 P.M., it is possible to spend eight hours at a destination three hours away. Given Berlin's location east of Germany's population center, not all cities were within such easy reach: at 1,157 kilometers, Cologne was just over ten hours from Berlin, and at 1,369 kilometers Munich was just over thirteen hours away (Allen, 1978, 64). Even so, Germany in the 1930s was well on the way to a level of integration in the national economy only achieved elsewhere in the 1960s and 1970s. The German services had the same impact as the U.S. airline industry in the late 1930s: to minimize overnight business trips and to produce an integrated national economy.

In America it was the need for long-distance high-speed operation that made high-speed diesel-electric sets attractive: they allowed long hauls without stops to change locomotives. In 1934 a General Motors M-10001 six-car train set of the Union Pacific Railroad lowered the record speed for the 5,216 kilometers between Oakland, California, and New York City from 71.5 hours to a still unbroken 57 hours, a remarkable start-to-stop average of 91.5 kilometers per hour (Hollingsworth & Cook, 1983, 64–65). Yet despite the success of high-speed sets in Germany and the United States and serious consideration of their adoption in Britain, steam had one final fling in the 1930s.

Both Germany and Britain were reluctant to use expensive imported petroleum. Chapelon's work on improving the efficiency of steam locomotives offered performance as good as that of the diesel-electric sets by the mid-1930s. Richard Wagner in Germany and Nigel Gresley and William Stanier in Britain used Chapelon's work to refine their locomotives (Allen, 1978, 27). By 1936 Wagner's 05 was in limited service, providing more comfortable accommodations than the "Flying" sets but averaging only 5 kilometers per hour less between Berlin and Hamburg. Nigel Gresley's A-4 locomotive of 1935, built as an alternative to importing "Flying" sets from Germany, was scheduled to take six hours for the 608 kilometers between London and Edinburgh (Allen, 1978, 21). By 1937 it was averaging 116 kilometers per hour for the 303 kilometers between London and York (Nock, 1969, 151).

In the absence of centralized, national administration of Britain's railroads, however, accelerated service was as restricted as it was in the United States. Without system acceleration of the sort pioneered by the German "Flying" sets, such high-speed services running on regular track had a serious negative impact on normal operations. To allow one high-speed train a day between London and Edinburgh, "the track had to be swept of other traffic long before she was due; and in her wake she left a sizable vacuum" (Allen, 1978, 23). Limited use of high-speed trains requires expensive dedicated trackage, whereas system acceleration can use existing fixed capital more thoroughly.

The basic ecological concept behind high-speed trains was to improve the temporal efficiency of high-priced executive labor in advanced industrial societies. Such labor only became common in the 1920s after the managerial revolution wrought by Taylor, Gilbreth, and Ford in the teens. Acceleration of railroad schedules began in the late teens in a United States where World War I meant higher profits and rapid economic growth rather than higher expenses and economic turmoil. Yet high-speed technology through system electrification required far too much capital, as the Milwaukee Road demonstrated. The Depression slowed the adoption of accelerated services, but as the industrial economies recovered, new diesel-electric and improved steam technologies offered the possibility of productivity increases for executive labor when overnight business stays could be reduced. The trend of the thirties was greatly accelerated after World War II, and the lessons learned in the

1930s were applied with some care. Electric propulsion took over completely, whether via system electrification or within diesel-electric train sets. At the same time, either dedicated high-speed systems were built or systems were accelerated.

Before World War II European accelerations occurred on a country-by-country, even company-by-company basis. After World War II, as a more genuinely European economy emerged, in particular with the formation of the European Economic Community (EEC), system problems recurred. The first serious attempts at European service had to be developed with diesel-electric train sets similar to Germany's "Flying" sets. The first trans-Europe express (TEE) services began between Amsterdam and Zurich in 1957, covering the 1,050 kilometers with thirteen stops at an average of 110 kilometers per hour. The use of diesel-electric sets was required despite the fact that nearly all the line was electrified by 1957. Each country along the route—Holland, Belgium, France, and Switzerland—had a totally different electrical operating current and frequency (Hollingsworth & Cook, 1983, 124–25). After 1964 solid-state electronics allowed diesel-electric sets to be replaced with electric trains that can run on any of the four different currents (Hollingsworth & Cook, 1983, 144–45). The ambitious proposed TEE network for 220 kilometer per hour running all over Europe never materialized because competition from airlines increased.

In the area of spatial integration of the national economy, British developments of the 1980s have been significant. Japan and France chose high-speed technologies that require expensive, restricted routes. The British accelerated the entire system. British engineers used their considerable experience with powerful individual locomotives to produce the Inter-City 125 high-speed train (HST) sets that entered service in 1978. The name derives from a design speed of 125 miles (200 kilometers) per hour, comparable with that of Japan's Shin-kansen. This allows averages of 160 kilometers per hour from city center to city center, bringing almost all major cities in England within three hours of London, and Scottish cities such as Edinburgh within little more than four hours.

Part of the reason for system acceleration is scheduling. If long-distance, high-speed trains have to coexist with slower trains, they may overrun them. Careful scheduling can avoid some of this, but accelerating all services is the only reasonable alternative to constructing specialized high-speed systems—the Japanese and French solution. System acceleration is expensive but, so it would seem in the British solution, less expensive than new route construction in densely traveled corridors. In the British case total system acceleration has not yet been achieved. For example, in the evening rush west of London from Paddington Station, eight HSTs and eighteen slower trains must leave in a two-hour period. Fourteen kilometers west of Paddington HSTs have accelerated to 200 kilometers per hour and regular trains to 140 kilometers per hour, and at the 28-kilometers mark the HST has to be diverted to pass a regular train that left some eight minutes earlier (Nock, 1980, 24–25).

Of the major industrial nations only the United States has abandoned high-speed surface transportation. Over the much longer stage lengths between major U.S. cities, aircraft offered significant advantages even by the late 1930s. The U.S. population is also geographically more mobile than that of Europe, and fixed route corridors between major population centers are less common as well as less densely populated. Rail services need fixed population centers, not fluctuating markets. Fluctuating markets and populations are better served by more route-flexible transport systems.

Chapter Five

The Return to Overland Route Flexibility: Bicycles, Cars, Trucks, and Busses

By means of the motor car the upland areas where . . . the railroad enters at a considerable disadvantage can be thrown open to commerce, industry, and population. These uplands are likewise often the most salubrious seat of living, with their fine scenery, their bracing ionized air, their range of recreation . . . Here is . . . the special habitat of neotechnic civilization, as the low coastal areas were for the eotechnic phase, and the valley bottoms and coal beds were for the paleotechnic period.

Lewis Mumford
Technics and Civilization

Overland transportation change in the Neotechnic has been so dramatic that it warrants special attention. In spite of remarkable increases in organizational efficiency and economy of scale in sea transportation and remarkable acceleration of schedule along fixed land routes, the real triumph of the Neotechnic has been elsewhere. Neotechnic technology combined the old route flexibility of the Eotechnic with the speed and scale of movement made possible in the Paleotechnic. It then provided remarkable acceleration of service, beyond even that attainable by high-speed trains, in the form of aviation, which eventually could ignore all topographic barriers. The flexibility of route choice that the streetcar and interurban only hinted at, and the speeds they attained, were achieved first by the bicycle, then by internal combustion—engined vehicles. The process begun by the bicycle was continued by the automobile, the bus, and the truck. An associated device, the tractor, also deserves mention. A vehicle that owes its origins to both the truck and the tractor, the tank, revolutionized land-based warfare toward the end of World War I by bringing the advantages of mobility unencumbered by the need for any roads at all.

THE BICYCLE

A return to Eotechnic characteristics of transportation, where individual rather than group mobility was the norm, came with the bicycle. In the industrial world, in particular North America and Europe, the bicycle has been a recreational or child's vehicle for a generation or more. In other parts of the world, such as Japan, it has only been superseded as the major form of transportation by motor vehicles in the past twenty years or so. In developing industrial and many nonindustrial societies, it is still the most significant vehicle of individual mobility.

The bicycle was also a key technology forcer of the late nineteenth century. It was one of the first complex technologies to reach a mass market: the middle classes in North America and Europe in the 1890s, and the working classes just after the turn of the century. Bicycles also helped break down the relatively rigid distinctions between the sexes that characterized Paleotechnic society. They allowed middle-class women to leave the confines of the home unchaperoned.

Although simple hobby horses were available in the early nineteenth century, only in 1861 was a crank axle added to the front wheel (Ritchie, 1975, 54). As early as 1864 an American commentator saw the advantages of the bicycle over the horse with remarkable clarity:

> It costs little to produce, and still less to keep. It does not eat vast loads of hay, and does not wax fat and kick. It is easy to handle. It never rears up. It won't bite. It needs no check or rein or halter, or any unnatural restraint. It is little and light, let alone it will lean lovingly against the nearest support. It never flies off at a target unless badly managed, and under no circumstances will it shy at anything. (Ritchie, 1975, 53)

Changes in the early bicycle quickly negated at least some of these advantages. The "front wheel grew larger to gain greater speed, while the back wheel grew smaller to save weight" (Ritchie, 1975, 77). "Ordinary" bicycles of the 1870s had front wheels up to 160 centimeters in diameter for high speed. The rider sat high and almost over the front wheel. Even a small obstruction in the road caused a serious tumble. "Ordinaries" were sporting devices for athletic young males, never practical daily transportation. Even so the contemporary enthusiast press claimed that the "ordinary" enabled working men to move to "cheap and healthy homes at a moderate distance from their town" (Ritchie, 1975, 90). Commuting by "ordinary" was probably little more common than the feat accomplished by Thomas Stevens, who rode one around the world east from San Francisco between April 1884 and January 1887 (Stevens, 1889).

The potential of chain drive to the rear wheel was understood, but chain, gear, and bearing technology were not up to it in the 1870s. By the mid-1880s technology had improved and the "safety" bicycle became possible. The rider sat between two wheels of moderate and equal size, with the crank axle attached to the rear wheel by a chain transmission that geared up the rider's pedaling force to give the speed advantage of a big wheel without its disadvantages in ridability and safety. Women and older people could ride safety bicycles as readily as athletic young men. By 1890 the bicycle had assumed its lasting shape (Sharp, 1896, 153–58). The diamond frame of cold-rolled steel tube was light yet strong and resisted bending in all planes, whereas the simple curved frame of the "ordinary" bent very easily indeed (Sharp, 1896, chap. 23).

The need to transmit power by chain forced rapid improvements in the design of chains and in the heat treatment of bearings to resist wear. Plain gave way to ball or roller bearings to cope with the high torque that could be exerted on the crank axle and to give freer running. Ball bearings required metallurgy not needed in previous technologies. Heat treatment to produce stronger, more durable steel was advanced as much by the demands of the bicycle industry as by those of the armament industry for better steel for large naval gun barrels and for armor plate for battleships.

Wheels were also improved. The heavy, compression-spoke wire wheels of early "ordinaries" (Sharp, 1896, 336–38) were replaced by

Columbia safety bicycle, 1889. The high-wheeled "ordinaries" to the rear were suited only to athletic young men. The safety bicycle, introduced in the mid-1880s, could be ridden by anyone. It was the first Neotechnic transportation device to allow individual mobility beyond the few miles a person could walk each hour. Ford Museum, by the author.

tension-spoke wheels about 1870 (Bartleet, 1931, 18). The new wheels were light and strong, and they smoothed out bumps much better than compression-spoke wheels. Tension-spoke wheels made it possible to ride long distances on the iron- or solid-rubber–tired "ordinary" because the size of the front wheel gave adequate insulation from bumps. The smaller wheels of the safety bicycle required additional suspension. Sprung frames failed because they weakened the diamond frame by requiring it to be open and hinged at some point (Sharp, 1896, 295–97). Sprung wheels added weight and complexity to a device that was already complex and of which weight was an enemy. Fortunately the pneumatic rubber tire improved ride comfort even above that of the "ordinary." It also had far less rolling resistance on an irregular surface than a steel or solid rubber tire (Sharp, 1896, 487–91).

Pneumatic tires were invented twice: once by Englishman Robert Thomson for horse-drawn carriages in the 1840s, then again by the Scottish veterinary surgeon John Boyd Dunlop in 1888. Dunlop's young son complained that he and his solid-tired bicycle were being shaken to pieces by streetcar tracks and rectangular stone paving blocks. Cavity rubber tires with air imprisoned in them were known, but Dunlop inflated his tires to a pressure of several atmospheres. Much of the reason for Dunlop's success and Thomson's failure is that the environment was right for adoption of pneumatic tires in the late 1880s: the safety bicycle was rapidly replacing the "ordinary," and there was increasing demand for a better ride. Inventors other than Dunlop also worked out solutions to a crucial problem: how to hold the tire to the wheel rim. The first wire bead patent was granted in 1890 and was quickly bought by Dunlop. It was a crucial step in holding the inflated rubber tire securely on the metal rim of a tension-spoked wheel (Burton, 1954, 17–22).

Bicycle technology was useful beyond providing geographic mobility. By the turn of the century the pages of *Scientific American* and other such journals were filled with bicycle-based inventions, some bizarre, some fanciful, but many practical: pedaled boats, exercise devices, bicycle-operated showers, spiked wheels and ski attachments for snow and ice, bicycles with flanged wheels for railroad use (Calif, 1983, 38, 50–59). The military also used bicycles. The British used them to lay wire quickly for field communications in the Boer War. Bicycle cavalry moved at least as quickly as horse cavalry over the open South African veldt. Troops were issued folding bicycles weighing only eighteen pounds that could be carried on a soldier's back. The U.S. military mounted Colt machine guns on bicycles, and the Germans and Americans both used portable field radios powered by bicycle generators (Calif, 1983, chap. 9). Three- and four-wheeled cycles were adapted as urban delivery wagons for fresh foods or popular newspapers, which required rapid dissemination if their news was to stay "fresh."

The safety bicycle was a far more practical vehicle than the "ordinary." Just how much so is shown by three young Englishmen who cycled nearly 31,000 kilometers around the world between July 1896 and August 1898. "We took this trip . . . because we are more or less conceited, like to be talked about, and see our names in the newspapers. We didn't go into training. We took things easy" (Frazer, 1899, v). They rode "good, sturdy roadsters painted black. In the diamond frames were leather bags stuffed with repairing materials. Over the rear wheels had been fixed luggage carriers, and to these were strapped bags containing underclothing" (Frazer, 1899, 1). Just how much British society had been transformed by the bicycle may be gleaned from a comment about entering Transylvania out of relatively civilized Europe for the first time on their run east from London: "We felt we were slipping away from civilization. The legend, 'Good accommodation for cyclists,' that adorns nearly every wayside cottage in England, was unknown in that part of the world. Indeed, there was neither good accommodation for cyclists nor for anybody else" (Frazer, 1899, 20–21).

Bicycles allowed women to leave their homes relatively easily without the assistance of men to handle horses and without being restricted to a distance they could walk. In 1896 in one of H. G. Wells's lesser novels, *Wheels of Chance,* a draper's shop assistant, Hoopdriver, takes his annual holiday by bicycle. He meets a young, unchaperoned, upper-middle-class woman cycling along the road, where he can pass as almost her social equal. Although in the end he does not win her, and there is much comment about the danger to her reputation of wandering around the countryside in such a fashion, Wells's message is clear: the bicycle will liberate women, and it will lessen the distance between the sexes and the social classes.

Wells wrote at a time when annual holidays were becoming normal for the lower-middle classes that Hoopdriver represented and when inexpensive, mass-produced clothing could disguise social origins with relative ease. Hoopdriver wore "his new brown cycling suit—a handsome Norfolk jacket thing for 30 shillings" (Wells, 1984, 13). Dressed thus, on vacation, and mounted on a secondhand bicycle, "the draper Hoopdriver . . . vanished from existence. Instead was a gentleman, a man of pleasure, with a five-pound note, two sovereigns, and some silver at various convenient points on his person" (Wells, 1984, 17). Hoopdriver poses a problem for the young woman after he has rescued her from an appropriately mustachio-twirling cad. Because she has no experience of men below her own class, she explains his relative lack of manners and education by guessing, "You come from one of the colonies? . . . You were educated up country" (Wells, 1984, 123).

Thus the bicycle introduced conundrums as well as solutions. It freed Hoopdriver and his ilk from the confines of city and class, but it required a reordering of society. Neotechnic transport brought increased intersection of the orbits of people from different social worlds (Strauss, 1961, 65–67). In Paleotechnic society these worlds were geographically isolated, whereas in the Eotechnic the lack of land transport had caused landowner and peasant, millowner and proletarian to live in almost constant contact. Paleotechnic increase in wealth and improvements in relatively long-distance but fixed-route transport systems allowed them to live apart. Spatial segregation by class was well under way when Engels commented in 1845 that Manchester mill owners had removed themselves from the polluted unpleasantness of the Paleotechnic city: "They enjoy healthy country air and live in luxurious and comfortable dwellings which are linked to the centre of Manchester by omnibuses which run every fifteen or thirty minutes" (Engels, 1968, 55). Others used steam-powered trains to commute fifteen or twenty miles. As the railroads penetrated London in the 1830s and 1840s, stations were planted in empty fields that "were in time heaped up with houses for which a free first-class season ticket for the first year was not infrequently a house agent's, if not a railway company's, bait" (Dyos & Aldcroft, 1969, 231).

The bicycle and factory-produced clothing began the Neotechnic reordering of the social classes, although not their spatial reintegration. In Eotechnic society public order was a matter of appearances (Lofland, 1973, 27). Since the social classes lived in fairly constant contact, cues to the identity of strangers were based on their appearance: style and quality of clothing in the main; bodily grooming to a lesser extent. Sumptuary laws often controlled who was entitled to wear what clothing, a principle nearly gone in modern societies, except for police and military officials and, rather briefly, college graduates. In the Paleotechnic the social order became increasingly spatial. Classes were geographically separated on the basis of their ability to pay for transport. The bicycle offered a return to the relative spatial "democracy" of the Eotechnic but was unable to achieve a reintegration before the more expensive automobile appeared. The automobile allowed the Paleotechnic spatial ordering of society to be retained. As José Antonio Viero-Gallo, the assistant secretary for Justice in the government of President Allende of Chile, remarked of his country's failed experiment in democratic socialism, "Socialism can only come riding on a bicycle" (Ritchie, 1975, 165).

The bicycle's achievements have been obscured in the industrial world by the success of the streetcar, bus, and automobile, yet it was the first transportation device to improve individual mobility. On good roads bicycles could carry people anywhere at reasonable speeds and with true route flexibility. Early in this century the upper working class around London used bicycles to commute to railroad stations (Hugill, 1983). They thus enjoyed greater residential flexibility and less dependence on streetcars, which were expensive in any case. Many kept a bicycle at each end of their railroad commute, and season tickets on many of London's commuter lines included bicycle parking. In the United States the policy of a single, low fare for the streetcars and free transfers from one line to another precluded such developments. The more acute winters of the industrial cities in the U.S. North may also have discouraged cycling.

Advanced industrial nations other than the United States used bicycles for commuting through the late 1950s, and some continue to do so today. Holland claims to have some five million bicycles in regular use among a total population of thirteen million people (Ritchie, 1975, 174). Even in wealthy industrialized nations bicycles still have many uses. Switzerland maintains three military bicycle regiments, and many soldiers keep their bicycles at home with their rifles. The Swiss claim that bicycle soldiers are "the quickest regiment in the army" over 30 kilometers and that no other force could adequately defend the country's vast network of forest trails (*Wall Street Journal,* Europe, June 16, 1988). In many countries bicycles remain unregistered and untaxed, thus much more difficult to count than automobiles. Outside the industrial world bicycles still dominate, although they are rarely acknowledged because polities aspire to the motorization of the industrial nations (McGurn, 1987, 186). China produced some 24.2 million bicycles in 1982, and in the same year it was believed that some 150 million were in use. That year's production represented a 38 percent increase over 1981 (Scherer, 1984, 76, 204).

Theoretical Maximum Size and Shape of the One Hour Commuter City

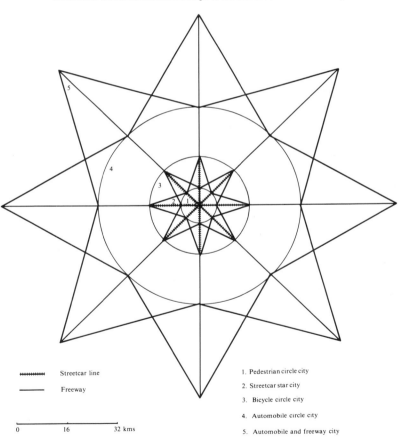

Streetcar line

Freeway

0   16   32 kms

1. Pedestrian circle city
2. Streetcar star city
3. Bicycle circle city
4. Automobile circle city
5. Automobile and freeway city

Figure 5-1 Theoretical maximum size and shape of the one-hour commuter city using different modes of transportation: streetcars, bicycles, automobiles on surface roads, and automobiles on freeways.

An average person can walk comfortably at just under 5 kilometers per hour, expending .1 horsepower. Since people are willing to commute about an hour, a pedestrian city could have a maximal diameter of 10 kilometers if all people needed to commute to its center. Streetcars could average about 16 kilometers per hour but were restricted to fixed lines. A city built around fixed lines and on a homogeneous plain would have looked like a star with a diameter of 32 kilometers (fig. 5-1). People close to town could afford more time to walk to the streetcar than those farther out, hence the star pattern. If expended on a bicycle the one-tenth horsepower of the pedestrian would propel a rider at about 16 kilometers per hour because the bicycle uses human muscles more efficiently (Wilson, 1973). The bicycle allowed our theoretical city to become 32 kilometers in diameter, thus greatly increasing its potential area.

One-hour pedestrian, eight-line streetcar, and bicycle cities have respective potential maximum areas of 250, 860, and 2,500 square kilometers. The bicycle thus had marked theoretical advantages over fixed transport lines operating at the same route speed because it made more land available for business and residential use. It therefore decreased the cost of land, a major component of the fixed costs of production.

THE AUTOMOBILE

The first successful automobiles were clearly derived from bicycle technology in chassis design, suspension, and power transmission. Benz's 1885 vehicle was a large tricycle equipped with a crude, spark-ignition internal combustion engine. Its frame was cold-rolled steel tube, it had tension-spoked cycle wheels, it transmitted power to the rear wheels by chain, and it used ball bearings and heat-treated components in large number. With only .75 horsepower the Benz needed weight-saving bicycle technology, as did most other automobiles before the turn of the century. Benz, like most makers, adopted pneumatic tires only in the late 1890s, early tires having problems carrying heavier loads than the bicycle could manage (Georgano, 1982, 85–86).

If the bicycle could increase the potential area of a city tenfold, then the automobile had even greater impact. A city 64 kilometers in diameter is four times as large as the bicycle city and more than forty times as large as the pedestrian city. A modern automobile can travel much more than 32 kilometers in an hour, but congestion and parking are more limiting for automobiles than for bicycles. Houston is probably at the maximum practical size for a city that depends almost totally on automobiles for transportation. The freeway has reintroduced the star-shaped city. Houston has seven divided highways reaching out into the countryside, four of them freeways (fig. 5-2). Commuting is feasible out to about 65 kilometers. The city now engulfs farm to market roads that once marked its limit. If we accept 32 kilometers as the practical maximum commuting distance with no freeways and 65 kilometers as the limit with them, the star-shaped, eight-freeway city gives a theoretical maximum area of around 23,000 square kilometers, compared to the automobile city's 10,000.

Real-world studies of commuting time are rare. Forer's 1978 article on Christchurch, New Zealand, points out that the time-area of Christchurch declined 7 percent between 1880 and 1970, whereas the physical area rose 673 percent. The article includes the complexities of the real world: waiting time for streetcars; walking time to and from parking lots and streetcar lines; and the impact of elevators on vertical time-distance.

A further advantage of the internal combustion–engined automobile needs mention. Automobiles could use steam engines, but they would have to carry heavy loads of fuel and water. Steam vehicles built early in the twentieth century were trucks or tractor units designed to haul several road trailers in a road train; they usually wore steel or, at best, solid rubber tires. Neither the tires nor the engines wore much, but the trucks were expensive capital items that returned profit only over many years. And like railroad locomotives, they required skilled drivers. Inter-

Major Commuting Highways, Houston, Texas 1985

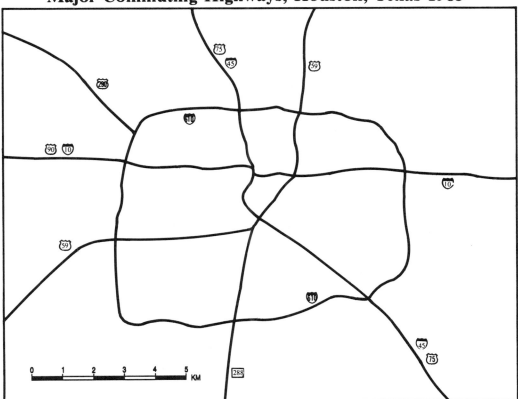

Figure 5-2 Major commuting highways, Houston, Texas, 1985. Houston is as near as a city can come to being uninterrupted in its growth by human interference or topography. Land use zoning is unknown, and only the Houston ship channel requires bridging.

nal combustion—engined vehicles require as much if not more capital, but distribute it differently. The engines and pneumatic tires may last only years instead of the decades that steam engines and steel tires last, but they concentrate capital in machine tools in the factory because precision is far more important than for steam engines. Such tools are only profitable when used intensively in mass production, which lowers the cost of the vehicle and encourages innovation, in the form of improvements in the next generation of tools and vehicles. Internal combustion—engined vehicles thus "fit" better with advanced capitalist economies than steam engines. They are inexpensive to buy and inexpensive and easy to use, and they require concentrated skill and capital only in manufacturing. In the absence of tariff barriers, the automobile provides a natural product for multinational enterprise, valuable relative to its weight. Ford, who recognized this advantage early, had begun the world's first major multinational corporation by 1913 (Maxcy, 1981, 66).

Ford's Model T production line was inflexible, capable of turning out only one product. In nineteen years, changes were few despite the millions of vehicles produced and the vast number of lessons presumably learned. Fordism was not the key to mass production. It was Alfred Sloan at General Motors (GM) who worked out the system of flexible mass production still used today, tying major model changes to a three-year life for machine tools. Sloan is more famous for "planned obsolescence," which merely alters body panels, than for mass production. He deserves more credit. Sloanism, mandatory innovation tied to the expectable life of working capital, has been a far more powerful agent for technical change in the Neotechnic than the simple mass production represented by Fordism (Flink, 1988, 234).

Only the Japanese have altered the production process much since Sloan, and once again the automobile industry has taken the lead. To Sloanism has been added the important notion of *jidoka,* or worker responsibility. To both Ford and Sloan, workers were merely cogs in the machinery, as they had been in any Paleotechnic factory. Popular movies of the 1920s, such as Charlie Chaplin's *City Lights* and Fritz Lang's *Metropolis,* illustrate their attitude perfectly. Sloanism required a slightly more competent and flexible work force than Fordism, and thus gave power to labor unions in the U.S. automobile industry, the only highly unionized segment of U.S. industrial society. Yet Sloanism retained an adversarial relationship between management and labor that owes much to the Paleotechnic factory, with its notion of labor as expendable units, just like machinery. It was left to Japan, a society with almost no Paleotechnic history, to reinstitute the notion of individual worker responsibility. In Japanese automobile factories, any worker has the right to stop the assembly line if a fault is seen. Under Fordism or Sloanism halting assembly was a major disaster brought about only by wildcat strikers. The Japanese worker might still be paid much less than management and have much lower social status, but he or she is not treated as a disposable part in an assembly line or a labor union contract.

A second Japanese innovation is the *kanban,* or "just-in-time system" of parts supply. Ford attempted a similar practice in the early 1920s but abandoned it in favor of vertical integration and complete control over component supply. The capital-poor Japanese adopted *kanban* in the 1950s partly because it reduces inventory costs for mass producers and loads at least some of the costs of capital investment onto parts suppliers. In this sense it is like Detroit before Ford, when producers merely assembled parts bought from others. *Kanban* ties parts suppliers closely to assemblers because only enough parts for a few hours of production are kept on hand. The system depends on reliable transport to move components from the place of manufacture to the place of assembly and on freedom from industrial stoppages. Given the relatively poor transportation of the 1920s and the strained relations between management and labor, vertical integration made more sense.

In a major report, *The Future of the Automobile,* the Japanese *kanban-jidoka* system is described as "a new standard of best practice for the world. It supplements the old Ford system in the plant and the Sloan formula for co-ordinating the production chain as the recipe for . . . success" (Altshuler et al., 1984, 161).

Variations in Adoption Rate

The adoption of the automobile proceeded far faster in some industrial societies than in others. In market research, the adoption of an innovation by one household in three marks it as a necessity. On such a measure the United States and Canada were automobile-oriented societies by 1921 (table 5-1). The core polities of Western Europe—Britain, France, and Germany—did not reach the same threshold until the late 1950s. Italy caught up shortly thereafter, and noncore states seem to be following suit. By 1975 Western Europe had arrived at a ratio of one automobile for every four people or fewer, the mark of an almost totally automobile-oriented society. Polities outside the core, such as Spain, are experiencing increases in adoption rates, as are some Eastern European polities, notably Czechoslovakia. Adoption rates in Japan increased in the late 1960s and early 1970s to the point of being comparable with those of the United States in the teens.

Such differences must be accounted for. Exact measures of gross national product for early in this century are lacking, and before World War I government spending as a proportion of GNP was higher in Britain than in the United States. Nevertheless Britain's per capita GNP was not markedly lower than that of the United States (table 5-2). It was higher until roughly 1905, when the United States drew slightly ahead. Before and again after World War I Britain did not tax gasoline, and until 1910 it levied no heavy taxes on vehicles. The already marked difference in adoption rates for 1910 cannot be explained in crude economic terms.

After 1910 British use tax, based on the diameter of the piston, was designed to restrict engine size. This tax system was strengthened in 1921 (Church, 1979, 79). As a result vehicles tended to be small and to

Table 5-1 Persons per registered private automobile, 1900–1988, selected polities

Year	United States	Britain	France	Germany[a]	Italy	Spain	Czechoslovakia	East Germany	Poland	Japan
1900	9,512									
1910	202	763	13,248	2,048	1,156		2,545			
1921	12	176	729	1,019	222		359		1,589	6,762
1930	5.3	42	197	133	151				1,849	1,105
1938	5.2	24	37	54	137		172			1,195
1950	3.8	22	23	98	55	314	93	243	680	1,733
1955	3.2	14	25	31	25	221	55	153	680	565
1960	2.9	9.2	14	12	9.5	105	34	58	255	212
1965		5.9	8.2	6.3	5.3	40	17	25	127	63
1970	2.3	4.7	5.0	4.4	3.7	14	9.8	15	68	15
1975	2.0	4.0	3.9	3.5	3.2	7.4	6.7	8.9	32	7.6
1980	1.9	3.6	3.4	2.7	2.6	4.9	5.9	6.2	15	5.4
1985	1.8	3.4	2.9	2.4	2.4	4.2	5.2	5.0	10	4.7
1988	1.8	3.1	2.6	2.1	2.1	3.8		4.6	8.4	4.3

Sources: Europa World Year Books; Eurostat: Eurostat Review, 1972–81; Mitchell, 1980; Statistical Abstracts of the U.S.; Statistical Handbooks of Japan; U.S. Bureau of the Census, 1975.
[a]West Germany only after 1945.

be powered by small engines, but Herbert Austin demonstrated in his Austin 7 of 1922 that small vehicles could perform reasonably well. The use tax on an automobile rated at 7 taxable horsepower was not excessive, and fuel taxes were removed once World War I was over, not to return until 1929 (Church, 1979, 111). The horsepower tax tended to freeze out large-engined U.S. vehicles. Ford's Model T was built in Britain both to reduce freight costs and avoid the McKenna import duties, introduced in 1915, of 33⅓ percent: yet Model Ts had to pay so much use tax on taxable horsepower that sales declined badly after 1921 (Church, 1979, 80).

One cannot argue that the British were slow to adopt the automobile because public transport was better developed in Britain than the United States in the period 1900–1930. The single-fare U.S. trolley system, usually run as a local monopoly, was markedly superior to the mix of small-scale private and public transport bus systems that passed for local transport in British cities. The heyday of British urban public transport came in the late 1930s, with the development of local government monopolies modeled on the London Passenger Transport Board of 1933 and thus on U.S. practice. Before this, at least in the London region, no single company had been "prepared or able to put up the capital necessary" to provide the coordination and extension of a system grown piecemeal (Howson, 1981, 131; Jackson, 1973, 213). Under the post–World War II Socialist government local government monopolies prospered and most British cities came to enjoy well-organized bus systems, albeit without the U.S. advantages of single fares and generous transfers. In recent years the Thatcher government encouraged a return to a more competitive system with monopolies discontinued, although local governments retain their bus systems.

Why then did the automobile prosper in the United States earlier than elsewhere in the industrial world? The key to the answer lies in the technical nature of the American automobile in the critical years 1900–

Table 5-2 GNP or NNP, 1880–1930, selected polities (U.S. dollars per capita)

Year	United States (GNP)	Britain (GNP)	Germany[a] (NNP)
1880	186	237	89
1885	204	220	94
1890	210	240	114
1895	199	236	114
1900	231	281	137
1905	294	283	153
1910	349	297	168
1920	835[b]	762[c]	
1930	740[b]	549[c]	

Sources: Mitchell, 1980; U.S. Bureau of the Census, 1960; *World Almanac & Encyclopedia,* applicable years.
1 pound sterling converted to 5 U.S. dollars.
[a] 1 Reichmark = 23.84 cents in 1900.
[b] 1957 value.
[c] 1975 value.

1930. American automobiles in this period were simply much more competent than European automobiles. Not only were they inexpensive to buy and pay taxes on (the various states levied low use taxes), but their overall running costs were low despite high fuel consumption. Fuel was inexpensive but, except in oil-producing states, not much less expensive than in Britain. In the late 1920s it was certainly less costly to run an Austin 7, which burned fuel at a rate of a liter every 18.5 kilometers, than a Chevrolet or Ford sedan, which burned a liter every 8.5 kilometers. It was not, however, less costly to maintain the small engines used by British and other European manufacturers. Poor oils and engine bearings made for high wear factors (Hugill, 1988a, 118). The gap between European and U.S. manufacturers widened as autos with closed bodies were adopted in the 1920s. As weights rose, U.S. designers increased engine and bearing size; European makers, saddled with punitive use taxes, did not. The wear factor of the 1938 Austin 7 was nearly ten times that of the 1938 U.S. Ford V8. Admittedly part of the gain came from the generous bearings surfaces of Ford's V8, but Chevrolet's more conventional 1938 straight 6 had a wear factor more than seven times better than the Austin's (Hugill, 1988a, 118).

High power and low weight are critical in heavy urban traffic; they allow vehicles to accelerate safely through or into traffic. It is hard to define a safe minimum because traffic density and driver expectations are critical, but power-to-weight ratios around 18 kilograms per horsepower seem reasonable for modern traffic. American middle-class automobiles, such as the Buick, achieved such figures in the late 1920s. Less expensive Fords and Chevrolets followed in the mid-1930s. European vehicles did not achieve similar figures until the late 1950s; interestingly, adoption rates rose rapidly thereafter.

Part of the reason for the failure of European power-to-weight ratios to improve in the 1930s was the switch to closed bodies. Part was use taxes based on engine displacement or gasoline consumption, or both. How constraining the British tax system was is evidenced by the 36 horsepower put out by the Italian Fiat Tipo 508s of 1934; the British Riley 9 put out 29 horsepower. Yet Fiat was a mass producer that had hotted up one of its sedan engines. Riley was a well-respected specialist in sports cars and performance sedans, rather like BMW today. To compound the disparity, the Riley engine was nearly 10 percent larger than that of the Fiat. Fiat's Tipo 508 can also be compared to Ford's Model Y 8, introduced in Britain in 1933. The Ford engine was slightly smaller than the Fiat and it had side valves. It cannot be fully compared with the overhead-valve Fiat and Riley engines, but it turned out only a very modest 22 horsepower. With the constraints of the technically uninformed British use tax removed, Ford demonstrated in its next generation of engines that it too could produce fine power units. The Anglia 105 E engine, which replaced the Model Y engine and its derivatives, was almost exactly the same size as the Fiat engine of 1934 and turned out 39 horsepower, three more than the Fiat (Culshaw & Horrobin, 1974, 131). Unfortunately it was 1959 before Ford produced such an engine, twenty-five years after Fiat. Ford had too much capital tied up in the 1933 unit to abandon it before 1959.

The means of giving European automobiles a high power-to-weight ratio was at hand in the 1930s, but it was a capital-intensive one. The conventional automobile had a heavy chassis. Designers such as the Italian Lancia, who patented the idea in 1919 (Tubbs, 1969, 230), and the German Rumpler, who came up with it later the same year (Borgeson, 1984, 270), recognized that the chassis could be abandoned if the body were designed to carry the loads imposed by suspension and drivetrain. But because such unit construction was extremely expensive to tool up for, it could only be applied to long production runs. It also raised problems for the vehicle's occupants in terms of noise, vibration, and harshness.

Despite capital cost and technical problems, unit construction began to be adopted in Europe in the late 1930s, but not at European hands. Ford was heavily involved in Europe, having completed what was intended to be Europe's largest vertically integrated automobile factory outside Detroit at Dagenham, near London, in 1929 (Hugill, 1988a, 128). The Depression brought protectionism back into vogue, and Ford had to add plants in France and Germany. When Dagenham switched over in 1933 from large American Fords (Models A and B) to a small Ford, Model Y, designed in Detroit for the British market, one result was to spread the iniquities of engines based on misguided British use taxes far and wide. The German and French Ford factories built Model Y, and when World War II was over the Japanese firm Toyota used the engine of the prewar British Ford Model C, itself an updated Y, as the basis for its first engine in the 1-liter class (Toyota, 1967, 249–52).

Ford dominated the European market of the early 1930s to a remarkable extent. In Europe in the 1930s, as in the United States in the 1920s, GM strove to catch and surpass Ford. It bought two European companies: British Vauxhall in 1925 and German Opel in 1929 (Hugill, 1988a, 127–28). Because its production capacity alone was not enough to put it ahead of Ford, GM innovated, introducing the world's first mass-produced, unit-construction sedan through its German subsidiary as the Opel Olympia of 1935. Unit construction gave a weight reduction of just over 20 percent compared to similar separate-chassis GM cars (Hugill, 1988a, 129). This, together with the adoption of independent front suspension, which allowed shorter, thus lighter, bodies, was the key to a rapid improvement in power-to-weight ratios in the European market. Britain's Morris, France's Renault, and Italy's Fiat, which had routinely copied Ford, switched wholesale to copying GM, so that GM's technical leadership did not result in any marked increase in market share. British Ford, much less conservative than the U.S. company and relatively independent until the 1960s, followed GM's lead shortly after World War II. By the early 1950s all the major European producers except Volkswagen had moved to unit construction, thus preparing the way for the rapid increase in automobile adoption of the late 1950s.

Volkswagen followed a central European design philosophy established by Edward Rumpler and Hans Ledwinka in the early 1920s. These designers lowered weight by, respectively, moving the engine to the rear and pioneering the backbone chassis. These developments lowered vehicle weight nearly as much as unit construction did, and they did not need high capital investments. When Hitler propounded Motorisierungpolitik as a prescription for economic recovery before rearmament (Overy, 1975), he needed a vehicle to run on his proposed autobahns (Hugill, 1988a, 131). Rumpler was Jewish, and Ledwinka, who had apprenticed under Rumpler, was a Czech working for Tatra, thus a non-Aryan. Porsche, who worked for Volkswagen, was familiar with the ideas of both and, unlike them, was acceptable to Hitler. Porsche stole Ledwinka's ideas so blatantly that Volkswagen settled out of court for $2,130,000 when Tatra finally sued for patent infringement in 1967 (Hugill, 1988a, 123). The Ledwinka system produced a vehicle that, while excellent at low speeds on bad roads, suffered handling peculiarities at high speed on good ones. Numerous attempts to copy the Volkswagen failed, usually from misunderstandings of the dynamic nature of a vehicle with its weight concentrated at the rear. Cost cutting with regard to a critical suspension component ruined the reputation of the best copy, Chevrolet's Corvair, and brought on Ralph Nader's crusading book, *Unsafe at Any Speed* (1972). This marked the end of the rear-engine automobile.

Although unit construction began the improvements in power-to-weight ratio necessary for European mass adoption of the automobile in the late 1950s, it was European modification of an American idea that completed Europe's transition to automobility. The Greek-born British engineer, Alec Issigonis, modified a system pioneered as early as 1906 by Walter Christie of New York. At the onset of fuel rationing in the Suez crisis of 1956, Issigonis's employer, Austin, sought a fuel-efficient vehicle. Issigonis thus had to reduce vehicle weight more than was possible with unit construction. Like Christie he turned the engine sideways and had it drive the front wheels. Issigonis succeeded where Christie failed because of improvements made in constant-velocity joints to avoid the sudden changes in rotational speed when steered wheels were driven by conventional U-joints.

Austin brought Issigonis's tour de force, the Mini, to market in 1959. The vehicle was a revelation in terms of performance and compactness, and all the major European companies set about copying and improving upon it. The Mini offered a better power-to-weight ratio, 16.9 kilograms per horsepower, as opposed to 18.9 for the Ford's conventional front-engine, rear-drive Anglia 105 E, also introduced in 1959 (Hugill, 1988a, 118). It seated four in approximately the same discomfort as the Anglia, got about 20 percent better gas mileage, and most notably, was far easier to park in a London then becoming insanely congested (Hugill, 1988a, 132). Mounting the engine across the vehicle allowed the Mini to be only 305 centimeters in length to the Anglia's 390. In a congested city the Mini had far more parking opportunities than Ford's otherwise excellent unit-construction vehicle.

Since 1959 all the world's mass-market manufacturers have followed Issigonis's lead. The impetus in the United States, where the automobile industry had grown hidebound since World War II, did not come until 1973. Until the OPEC-induced oil price crisis, U.S. manufacturers had concerned themselves increasingly with their domestic market. In the prewar years, on the other hand, the United States was the world's dominant producer and exporter. Some blame attaches to the strong dollar after World War II, which made U.S. exports very expensive. But competent subsidiaries, especially in Europe, failed to consolidate U.S. economic power, probably mostly because they stopped treating Europe as a distinct market with special needs. In the late 1950s GM in particular, but also Ford, began to treat European customers as junior Americans. In his autobiography Maurice Platt, long-time chief designer at Vauxhall, comments on a much heavier-handed U.S. presence (1980). American designs for excessively large, heavy, fuel-hungry vehicles were forced onto European subsidiaries at precisely the time the Mini was being born.

Infrastructural Considerations

The automobile cannot be considered apart from its infrastructure, especially highways and parking. If we assume that all vehicles are parked at home each evening and at work each morning, every vehicle needs two parking spaces. Let us also assume that drivers park in a space one and a half times as long as their vehicle and that the standard U.S. traffic lane width is 3 meters. Vehicles also need a mobile space, which varies in length because drivers allow about one vehicle length as braking distance for each increment of 16 kilometers per hour in speed. If we take 32 kilometers per hour as the average speed in our one-hour commuter city, each vehicle requires an area three times its parked area both traveling to work and homebound, since few cities use roads with reversible flows. Thus each vehicle needs an area at least nine times the product of its length and the standard lane width.

The extreme range of normal four-passenger automobile lengths on this planet has been from the 6.1-meter-long American vehicles of the late 1960s to the 3.05 meters of the Mini. A 6.1-meter vehicle theoretically needs at least 164.7 square meters of urban space, compared to 82.4 square meters for a vehicle half as long. An average American three-bedroom single-family house occupies about 150 square meters, and the average European home is somewhat smaller. In terms of overall space demands on the city, we may say that the automobile is as demanding as the residence in Neotechnic societies. In America this demand is compounded by the fact that there are nearly two automobiles in the average household, whereas most of Europe averages only about one per household and Japan slightly less than that.

The problems of road construction were far more intransigent than those of railroad construction. Railroads, which are dedicated rights-of-way owned by a single company, are easily financed by market forces.

Competing companies serving the same cities each built their own right-of-way. Financing roads, especially long-distance roads, is much harder. The very flexibility of the system defeated market forces, even in industrial societies such as Britain, which had developed the market system to a high degree, and in such British-origin societies as North America. Only statist societies seemed easily able to command the construction of a long-distance road network. The *routes nationales* built in the France of Louis XIV and added to by Napoleon Bonaparte are excellent examples of a road system built for reasons of administration and national defense in the Roman manner well before the Neotechnic. Upon the advent of the Neotechnic, the value of these roads was considerable.

> Good roads in France encouraged bicycle manufacturing and touring. They made it easier for businessmen there to envisage a substantial market for cars (in contrast to Germany, where both Daimler and Benz found it very difficult to convince their associates to take the risk of automobile manufacturing). They inspired early producers to publicize their cars by intercity demonstrations and races. And they made cars more practical for residents of rural areas and small towns. (Laux, 1976, 8)

It was in the statist societies of Europe that long-distance roads developed earliest and best. Most landowners in market societies saw the advantages of local roads from farm to marketplace, but few saw any reason for high-speed roads when railroads sufficed. Thus it was Fascist Italy and Nazi Germany that emulated eighteenth-century France and began national systems in the period before World War II. In market societies such as Britain and North America long-distance roads languished. Even local roads were initially a problem in market societies. Road construction was vastly expensive, and the corvée labor system used to build most local roads was totally unsuited to the task. The problem was most acute in North America, where high gains could be expected from road construction but both rural population density and willingness to pay taxes were low (Hugill, 1982).

The solution was taxation on an unprecedented scale. In the United States taxes were levied first at the state level, later at the national level. Although Americans were willing to pay taxes for military protection, until 1945, they were levied in serious amounts only in times of war. Taxation for roads, once begun, became continuous. Part of the solution was to tax gasoline, thus taxing the road user specifically, but the United States had an automobile for every twelve persons as early as 1921 and state taxes were not levied on gasoline until 1923 (Wixom, 1975, 87). Taxation began at the state level, but not via gasoline. Progressive states began to raise their own Good Roads funds: New York state approved the issue of $50 million in bonds in 1904, enough for some 8,000 kilometers of water-bound macadam road at a cost of about $6,250 per kilometer. New York's actions were based on those of Massachusetts, which began to issue road bonds in the late 1890s (Hugill, 1982).

In neither New York nor Massachusetts were automobiles much considered. The main user lobbies were the League of American Wheel-

Grover's Mill Road, Mercer County, Before Improvement.

Grover's Mill Road, Mercer County, After Improvement.

New Jersey highway, 1905, before and after improvement. Eastern U.S. states saw the advantage of highway improvement early. *Twelfth Annual Report of the Commissioner of Public Roads,* New Jersey, 1905.

men, representing the bicyclists, the post office, which wanted to develop rural free delivery (RFD), and the farmers' granges. The last saw two advantages. RFD brought them at least some of the benefits of town, as the growth of mail-order retail and farm service companies quickly attested. But good roads were also seen as the key to international competition in farm produce. Cheap, virgin lands in the U.S. West, Canada, Australia, New Zealand, and other newly opened agricultural regions were producing food more cheaply than U.S. farms, in particular those in the East, where the fixed costs of land were high. Eastern farmers accepted the argument that a substantial fraction of their transport costs to Europe lay in the farm-to-railhead section and that good roads would lower those costs as much as 60 percent (Hugill, 1982, 329).

Once this state road–building period was under way the growing automobile lobby, usually the very wealthiest people in the early years of the adoption of the automobile, added its pressures to those of the farmers. By 1911 New York had doubled its bond issue to $100 million. The total of state-issued bonds for the whole country by 1912 was $136,878,000, with a further $155,633,955 issued at the county and township levels (Hugill, 1982, 339). This made for $3.07 of debt per head of population. The expenditure on "major national security" that year was $2.97 per head, and the principal on the national debt amounted to $12.52 per head.

Localized public spending was all on water-bound macadam roads, which were ideal for horse-drawn vehicles and bicycles but not well suited to automobiles. The torque exerted by driven wheels required a much harder surface than the simple compacted gravel of water-bound macadam. Bitumen-bound macadam and concrete roads, which could stand automotive traffic, were three to three and a half times as expensive. This brought in the federal government, tentatively in 1916, then extensively with the Highway Act of 1921. By 1924 the government was providing about half the cost of new roads, about $15,000 for every mile built (Hugill, 1982, 342). State governments raised the other $15,000 per mile from gasoline taxes. The local road problem in the United States was solved, and the highway system boomed in the 1920s. Even the Depression failed to slow its growth seriously. The New Deal concentrated much effort on infrastructure, especially in the rural South, where road conditions were unusually bad, but whence much of Roosevelt's support had come.

An unintended consequence of a fine network of local roads was the demise of Ford's Model T, which had a suspension system that was well adapted to the roads of the first two decades of this century, as was the entire vehicle. When introduced in 1909 Model T had a power-to-weight ratio of about 36 kilograms per horsepower, as good as such prestige vehicles as Peerless, Pierce-Arrow, and Cadillac (Hugill, 1990). At 725 kilograms to a Cadillac 30's 1,089, Model T was lighter because Ford used sophisticated, strong, yet light steels in his chassis. Model T could thus drag itself clear on bad roads more easily than heavier cars with the same power-to-weight ratio. The Ford also had a simple, two-speed, epicyclic, foot-operated transmission that the unschooled drivers of the day found much easier to operate than the manual H-pattern, three-speed gearshifts of other vehicles. Not until GM offered the syncromesh gearbox on the Cadillac in 1919 did conventional transmissions become as easy to use as that of the Model T.

The real brilliance of Model T was its suspension system. Until the development of Model T, U.S. manufacturers copied the German-built Mercedes of 1901, a car built by Daimler for its lucrative French market, but Model T had a suspension for American roads. The Mercedes-style suspension was fine on the hard-surfaced roads of France. Ford's Model T suspension was superb in the mud holes that passed for roads in America. It allowed both axles to achieve high and different angles relative to the body, thus keeping all wheels on the ground. Its flaw was that, at speeds much over 32 kilometers per hour, and on a good road, its rate of roll was high. Ford made things worse by refusing to install suspension dampers; the essentially similar suspension of Ford's Model A of 1927 behaved well when damped. The high rate of roll induced seasickness on hard-surfaced roads. Ford's mass-market competitors, Chevrolet and Dodge, had no such problems.

Ford Model B, 1905. By 1905 the technical problems of designing and building a working automobile had been solved, and the technology had diffused from Europe to the United States. Like its contemporaries, Ford's Model B was derived from the great Mercedes of 1901, and its suspension was less than ideal on bad U.S. roads. The problems that remained—suspension, cost, longevity, and the availability of parts and service—were solved by the Model T of 1909 and Ford's creation of the first great system of mass marketing as well as mass production. Ford Museum, by the author.

Iowa highway, 1918. Better roads were vital to the continued penetration of the continental interiors. The truck is a Ford Model T, Ford being as important for pioneering trucks as for automobiles and tractors. *Report of the State Highway Commission*, Iowa, 1918.

Suspension systems and ride criteria remained a critical limiting component of the U.S. automobile until relatively recently. GM in particular persisted with separate-chassis vehicles in its U.S. market despite the marked weight reduction the company had pioneered in its unit-construction designs for its European subsidiaries, Opel and Vauxhall. Unit construction posed at least three problems. First, it was not amenable to the annual model change GM had so successfully used to increase sales since company president Alfred Sloan introduced it in the 1920s (Sloan, 1964, 238–47). Second, unit-construction vehicles were prone to rusting, and it took some time to work out techniques to promote reasonable vehicle life in climates where a considerable amount of salt was used on the roads. Third, early unit-construction vehicles tended to be poor in noise, vibration, and harshness on all but the smoothest roads. A considerable amount of time and money had to be spent on suspension design and development to overcome these problems. In separate-chassis vehicles it was possible to isolate the passenger compartment from the chassis and suspension components at relatively low cost. Modern suspension design has resolved these problems, as proper antirust treatment before painting has resolved the problem of body life expectancy. Thus unit-construction vehicles have become the norm in all the world markets, not just where roads are smooth and the weight-saving advantages of unit construction are needed to promote fuel efficiency.

If the local road network was expensive, a national system of high-speed roads was even more so. High-speed roads also suffered political problems in market economies that practiced local democracy. Large and small landowners frequently combined to restrict road development, in particular in long-settled areas, where land-holding units tend to be small because of inheritance patterns. Even in cultures that practice primogeniture (land goes to the eldest male child), enough families are childless or have only female children for increasingly complex land-holding patterns to emerge over time. Partible inheritance, where land is divided among all the children, produces complex patterns in a few generations. Although democratic governments can buy land under the right of eminent domain, they must expend considerable time as well as legal and financial resources to do it. Statist societies have less trouble than market societies with strong local democratic traditions. The success of U.S. freeway development thus seems anomalous unless we take account of settlement history. Americans have generally practiced partible inheritance, but in a new country with a short history, exercising eminent domain was relatively easy. To this must be added the unquantifiable American belief in "progress" and in roads as the key to economic development. Compared with Britain, also a market society with strong local democratic traditions, U.S. roadbuilders found fewer heirs to deal with to get rights-of-way and more acceptance of roads as valuable in themselves.

Only in the statist societies of Fascist Italy and Nazi Germany did high-speed road development occur faster than in the United States. Britain, on the other hand, was by far the most laggardly of all the industrial nations. France, which had a fine system of long-distance roads as a legacy from its history as a statist society, invested in freeways slowly and only after the *routes nationales* became lethally overcrowded.

High-speed roads could only be achieved by limiting access to them. At first sight limiting access seems to violate the principle of route flexibility, but as vehicle speeds rose in the 1920s route flexibility had to be balanced against safety considerations. Because Ford's Model T cruised at only about 30 kilometers per hour, limited access highways were hardly needed. In the late 1920s and early 1930s leaded fuels and larger engines allowed such vehicles as Chevrolet's Master Six and Ford's V8 to cruise nearly three times as fast, and limited access highways began to be seen as a necessity. By the late 1930s much smaller-engined European vehicles were using leaded fuels, unit construction or other weight-saving techniques, and improved aerodynamics to approach similar speeds. Higher-octane leaded fuels, better oils, and better bearings allowed European vehicles to equal U.S. vehicle cruising speeds in the late 1940s.

Such parity in the performance of European and U.S. vehicles was presaged in the 1930s by four designs. The first was the unit-construction, front-drive French Citroën 7cv (for chevaux or horsepower) of 1934. Second was the unit-construction, independent front suspension, Detroit-designed German Opel Olympia of 1935. Third was Porsche's backbone chassis, rear-engine, rear-drive Volkswagen. Fourth was the DKW, an important but little remarked upon German car of 1931 that was in many ways the ancestor of the Mini of 1959.

France and Britain ran neck and neck as European car markets in the teens and twenties, with Germany a poor third. Because they reacted to the Depression differently, Britain and Germany began to move ahead of France in the 1930s (Fridenson, 1982). Britain suffered less from the Depression than the United States or the rest of Europe. Britain was also a more market-oriented society, and British makers turned quickly to low-priced cars rather than sell nothing. Hitler's statist society pushed a radical building program for high-speed roads in Germany, then an inexpensive people's car to run on them, both as a way out of the Depression. France, a centralized democracy, cut all forms of spending, public and private, and did nothing. Even the market-oriented French automakers could not escape a tight money policy.

In 1934 André Citroën introduced his radical 7cv on the French market in a vain attempt to save his company from the Depression and his own gambling habits. He lost his company to its major creditor, the Michelin tire company, but Michelin continued production. The design of the 7cv was inspired by a prototype produced in 1931 by the Budd Company of Philadelphia. Citroën held the French license for Budd's unit-construction technology. Budd, a major innovator in the U.S. auto industry, is usually neglected by automobile historians, perhaps because the company never made a vehicle that carried its own name. Yet it made numerous prototypes. In 1926 Budd built a front-engine, front-drive prototype that four years later entered very limited production as the Ruxton. Budd used the Miller front-drive system with the engine carried in line behind the front axle to lower the body strikingly because no drivetrain to the rear wheels was needed (Grayson, 1978, 360). In 1931 Budd built a smaller front-drive prototype that took the logical step of removing the chassis and going to unit construction. André Citroën almost certainly saw this prototype on a visit to Philadelphia that year (Owen, 1975, 199). Citroën designed a similar vehicle in France but produced it with Budd technology (Owen, 1975, 199, 202).

The front-drive Citroën was a success, but it failed to close the gap that was apparent by 1938 between British and French adoption rates. The Citroën may have been a fine technical answer, but too few people were asking the question. Money was in too short supply in France in the 1930s to encourage spending on Citroën's relatively expensive vehicles. Costs as well as weight had to fall before Europe could move to mass adoption.

The drive for cost reduction began in the most market-oriented European society, Britain. The history of the second successful design, the Opel Olympia, belongs in Britain rather than Germany, where the design entered production. British automakers had long sought to build inexpensive, light vehicles. Opel in Germany and Citroën in France had pursued the same goal, but nowhere near so successfully as Britain's Austin. Herbert Austin, born and raised in Australia, was as much concerned with democratizing the automobile as Henry Ford. In 1922 he produced the Austin 7, the best of all the light cars of the 1920s. It was licensed in France as the Rosengart; in Germany as the Dixi, the backbone of BMW; and after threats from Austin over patent infringements, by Nissan in Japan, where it was called the Datson (later Datsun) (Hugill, 1988a, 126).

Ford and GM preferred to build regular American vehicles in the British market in the teens and twenties. This changed with the Depression. Ford, who had taken the precaution of routinely shipping European light cars to Detroit for inspection, came up with a design based on the best European practice, Model Y of 1931. Model Y was introduced at the new British Ford plant at Dagenham in early 1932. Ford could have obliterated its British competitors had it used Dagenham's 15 to 20 percent cost advantages to build a car "to standards of design and quality similar to those of its most immediate British competitors" (Rhys, 1976, 247–48). Ford chose to make the car reliable and profitable instead, allowing companies like Morris and Austin to thrive. Morris blatantly copied Model Y (Edwards, 1983, 195–96) and was in turn equally blatantly copied.

But Model Y was a miniature American Ford Model B, not a radical departure. It was GM that took the radical step in an attempt to catch Ford in Europe. GM was keyed more to the German market because its subsidiary there, Opel, was a genuine mass producer, whereas its British subsidiary, Vauxhall, had started as a producer of high-priced luxury sports cars. As early as 1926 U.S. business was projecting Germany as the "Best Potential Automobile Mass Market in Europe" (Heinze, 1926). Despite GM's help, Opel seemed unable to catch German Ford, which produced Model Y as the Köln from 1933 and took Dagenham's modified Y, the Model C, as the Eifel in 1935 (Oswald, 1977, 112, 115). In 1935 Opel introduced the Detroit-designed, unit-construction, independent front suspension Olympia (Oswald, 1977, 325–27). The British version of the Olympia, the Vauxhall 12 of 1938, was the first European middle-class car to have a power-to-weight ratio almost equal to that of American family cars of the late 1920s (Hugill, 1988a, 118–19). The Olympia could have been, but the combined effect of restrictive horsepower and gasoline taxes kept engine capacity lower in Germany than Britain at the time.

British manufacturers copied the Olympia as avidly as they had copied Model Y. Morris replaced its blatant copy of Model Y with a more freehand copy of the Olympia in 1938 (Edwards, 1983, 238). Morris bought unit-construction technology from Pressed Steel, Limited, a Budd licensee. All major European companies recognized the worth of unit construction and either had adopted it or were on the verge of adopting it in 1939. GM followed the Olympia in 1937 with the smaller Kadett, which was even more popularly copied. Although Fiat's chief designer claimed his 700 was based on the Olympia, it turned out looking remarkably like a Kadett (Giacosa, 1979, 52–58). Renault's Juvaquatre was a blatant copy (Fridenson, 1972, 284). In Britain, Hillman produced a very Kadett-like prototype in 1938 but never proceeded (Olyslager, 1973, 58). Hillman preferred to stay in the 10- to 12 taxable horsepower class and introduced unit construction on the Minx of 1940 (Allen, 1985, 81).

The third radical technology was introduced in Germany by state

rather than market action. Hitler admired Ford and saw the motorization of Germany as a tool for economic recovery before rearmament.
Hitler wanted a German vehicle that would be superior to the imported
U.S. technology represented by Ford and GM and inexpensive enough
for blue-collar workers. His target price was 990 reichsmarks, well
below the RM 2,100 asked for the Opel Kadett of 1937, the inexpensive
car nearest to the Volkswagen in specifications (Oswald, 1977, 314,
326).

Europeans needed a light, strong vehicle with a small engine and a
suspension that could cope equally with good and bad roads. Experiments in the teens and twenties indicated that lightness could be best
achieved by unit construction or a backbone, tube chassis. The first was
plagued by high capital costs. The backbone chassis cost less and was
lighter and more rigid than the American perimeter frame, but it posed
problems for attaching body and suspension components. Part of the
solution was to use swing-axle rear suspension mounted directly to the
final drive. Rumpler had patented this suspension as early as 1903 and
featured it on his midengine Tropfen-Auto of 1920 (Borgeson, 1984). In
1923 Rumpler's disciple, Hans Ledwinka, mounted a twin-cylinder,
horizontally opposed, air-cooled engine and gearbox to the front of a
larger tube and took the driveshaft through the tube to the final drive at
the rear, which had swing-axle suspension. The Czech Tatra type 11 was
remarkably successful; Hitler used one extensively as he campaigned
throughout Germany in the late 1920s.

Although a backbone chassis was light, Central European designers
began to think of removing the weight of the driveshaft by putting the
engine next to the driven wheels. Thus came into being the series of rear-
engine, rear-drive cars that led to the Volkswagen. Ledwinka led in the
latter technology because he recognized that the central engine mounting of Rumpler's Tropfen-Auto of 1920 worked well with unit construction but could not be used with a backbone chassis. Ledwinka mounted
an engine at the rear of a backbone chassis, behind the rear axle, where it
drove the rear wheels. His Model 77 Tatra of 1934 was a remarkable
vehicle with a powerful V8 engine, highly streamlined and capable of
160 kilometers per hour (Georgano, 1982, 610). The combination of
swing-axle suspension and a large engine carried behind the rear axle
made the Tatra deadly to unskilled drivers. So many German officers
were killed in commandeered Tatras in occupied Czechoslovakia in
World War II that it became known as the Czech secret weapon.

Swing-axle rear suspension gave exemplary ride comfort on bad
roads at low speed. If the car was cornered hard on smooth roads, the
lightly loaded outside wheel could "tuck under," reducing the rear-
wheel track. A vehicle emerging from a bend in such a condition rolled
over to the outside. Any vehicle with swing-axle rear suspension will do
this, but a concentration of weight behind the rear axle exaggerates the
tendency. Ledwinka understood this and substituted a much lighter flat-
four engine in his Tatra 97 of 1937.

The Nazis initially selected Josef Ganz to design the Volkswagen.

Ganz had promoted rear-engined, backbone chassis designs in the 1920s, but he was Jewish. Rumpler could have led the design team, but he was Jewish too. Ledwinka was a non-Aryan Czech, a subhuman species in Nazi demonology. Porsche was Austrian, as was Hitler, and apolitical. And it was Hitler, according to Borgeson (1980), who drove the whole people's car concept, even down to basic design, after a meeting with Porsche in 1934.

The Nazi state committed vast resources to the Volkswagen. An innovative financing system allowed workers to elect "union" payroll deductions until they accumulated enough to pay for their automobile. Their deductions helped finance the vast new factory laid out at what is today Wolfsburg. Initially the town was named Stadt des KdF, just as the people's car was named KdFwagen after *Kraft durch Freude* (Strength through Joy), the slogan of the Deutsches Arbeitsfront, the "union" to which all German workers perforce belonged. Although the KdFwagen was not mass-produced before World War II, it was after the war, the name reverting to Volkswagen (Sloniger, 1980). Porsche's design had a top and cruising speed of 100 kilometers per hour even in 1938, with minimal fuel consumption.

Although many companies copied Porsche's design after the war none was successful, usually because their cars handled badly and generated customer resistance (to the point of lawsuits in the case of GM's Corvair [Hugill, 1988a, 124]). Rear-engined cars seemed sensible for the cramped infrastructure of Europe, but although it was easy to make them light, it was dangerous to make them powerful if the crude, swing-axle rear suspension was retained. By the early 1960s suspension technology had improved to the point where rear-engined cars were safe, as Hillman's Imp of 1963 and the radically redesigned GM Corvair of 1965 demonstrated. But by the early 1960s a fourth design for light cars, far superior to these three, had proved its worth and was diffusing rapidly through Europe's automakers.

The fourth design was based on the 1931 German DKW, which mounted a twin-cylinder engine across the front of the vehicle, driving the front wheels in the manner patented in 1908 by Christie. The Miller/Budd/Citroën front-drive system mounted the engine in line behind the front axle to achieve similar ends. The primitive constant-velocity joints of the early 1930s worked only with long drive shafts, which limited the 1931 DKW to a two-cylinder engine and miserable performance. When a larger, three-cylinder DKW was built in 1940, the engine was mounted in line and ahead of the driven front wheels. This allowed long drive shafts but lost the advantages of compactness and light weight enjoyed by the two-cylinder design. The lack of modern constant-velocity joints also meant that both DKW designs had poor turning circles.

Despite their disadvantages both DKW designs had long lives, in part because the development costs and patent rights vanished along with the Third Reich in 1945. Both designs were used by the Swedish company, Saab, to enter automobile manufacturing after World War II, as well as by the Japanese company Suzuki. The DKW company was re-constituted in West Germany after the war to build a developed three cylinder, but the tooling remained in the East Bloc. The three cylinder remained in production as the East German Wartburg until 1988 (*Cars International 1989, 448*), and the twin cylinder lived on as the East German Trabant until 1990. The three-cylinder layout, highly developed and now using four-stroke engines and four, five, six, and even eight cylinders is the basis of the modern German Audi, the successor company to DKW in the West. Despite the success of these layouts in Audi's hands, neither one has been the basis for ongoing designs from any other major company.

In 1959 Issigonis married a transversely mounted four-cylinder engine placed above the gearbox and transmission to a sophisticated constant-velocity joint to create the Mini. In less than 305 centimeters of length the Mini accommodated four passengers at standards of comfort better than those of the Volkswagen Beetle (407 centimeters) and no worse than the unit-construction Ford Anglia 105E of 1959 (390 centimeters). The Mini had leg room equal to or better than that of the average U.S. car of the time, which was usually about 500 centimeters in length. The Citroën and later DKW three-cylinder front drives gave good passenger accommodation but in vehicles somewhat longer than the Volkswagen and with very poor turning circles. The Mini did everything right. It was short, and it had a turning circle of 9.45 meters, far better than the 11.68 meters of the DKW design. The Mini could thus be parallel-parked in a space only 46 centimeters longer than the car itself, a valuable ability in a crowded European city where in the 1950s most parking was still on the street. The Mini also had as good a power-to-weight ratio as a prewar U.S. Ford V8.

With important refinements, principally the removal by Fiat's chief designer, Giacosa, of the transmission from below to beside the engine (Giacosa, 1979, 250), the Christie system as developed by Issigonis has become the standard technology of all the world's mass-market automakers. The Issigonis/Christie system allowed mass automobility to develop outside the United States. It is probable that weight and size reduction of four-person vehicles peaked with the Mini in 1959. Only 20 percent of the Mini's length was devoted to mechanicals, and no subsequent design has done quite as well. By transforming even long-settled market societies such as Britain into accepters as well as adopters of the automobile, the Mini and its copies brought on better roads and parking facilities, in particular in American-style shopping centers, which have ironically somewhat reduced the need for the radical space saving of the Mini. The currently most popular vehicles on the European market are "superminis," typified by 1983's Fiat Uno of 364 centimeters length, nearly as long as the 1959 Anglia and with far more interior room.

The adoption of the automobile worldwide has proceeded at very different rates because technology, infrastructure, and societal values have interacted differently in different places. In the case of societal values, political, economic, and social decision making have also interacted differently. In the United States demographics and land-holding patterns allowed a market solution to work at a low technological level. Much more sophisticated technology was needed in a market society such as Britain, where infrastructure development was delayed by historical social conditions. In France in the 1930s, which had an adequate infrastructure, economic problems prevented mass adoption of a competent technology. In Nazi Germany a statist society overrode social problems in infrastructure development and reaped positive economic benefit from public spending in the Keynesian manner. German automobile technology was also forced, albeit in an inappropriate direction.

Nazi Germany is an example of how statist societies can introduce new theories of social organization more easily than market societies. The major impact of the automobile has been to decentralize urban systems, making them independent of fixed transport lines. Erich Koch, the Nazi gauleiter of East Prussia, strongly advocated such decentralization in the 1930s, and it became part of the Nazi program (Mullin, 1982).

To the Nazis the attractions of decentralization were military—the reduction of strategically vulnerable points of concentration. Yet the main advantages of automobile-induced decentralization were really economic. In a document in the Ford Archives dated 1920 and attributed to Henry Ford, but with only a typewritten signature, Ford pointed to the problems of urban congestion as insoluble. "It is not the advantages of the city that are doomed, but the disadvantages—the congestion, the inequality which reigns even in the matter of air and sunlight and groundspace" (Village Industries, Vertical File). Ford suggested that the automobile would decentralize the city and fuse the social advantages of city life with the organic advantages of country life. In a later article he added that big industry would eventually have to move "to the country where labor is steady and overhead costs low" (Pearson, 1924). Ford regarded an ideal society as one in which men worked not only in industry, but also to grow food on small farms. In an interview of 1921 he stated that he saw no reason why the extensive periods of underproductivity of both rural and urban workers should continue (*New York Tribune*, Feb. 20, 1921, part 8:1).

Ford's suggestions of the early 1920s are like those of the Russian anarchist, Petr Kropotkin (1899), who argued that the problems with capitalism lay not in the capitalist mode of production but with its inhuman scale, an argument strongly advanced by Lewis Mumford (1934). Kropotkin argued a geographic solution, reducing the scale of enterprise to the village level. Mumford saw that electrical and internal combustion engine technologies could accomplish Kropotkin's goals without losing advantages of scale. Ford echoed that thought when he responded to an interviewer that his back-to-the-country scheme would not cause smaller industry.

On the contrary . . . industry in the future is going to be organized on a big scale—somewhat along the lines of the vertical trust . . . In our own Highland Park plant we first cut down on the cost of production by taking the work on an endless chain to the man. Now we go one step further. Instead of having the man come to the city we take the work out to him in the country. Improved transportation methods have made that possible. (Pearson, 1924)

It was sentiments like this that made Lenin, in his drive to "put an end to the division between town and country," pay attention to Ford. Lenin was convinced that technology held the key to the success of the Russian revolution. He borrowed a colleague's phrase: "The age of steam is the age of the bourgeoisie, the age of electricity is the age of socialism" (1920, 52). His reasoning was simple. Monopoly power lay with the landowners who owned the coal mines that made Paleotechnic society work. With early 1920s technology, electrification was not subject to monopoly power, and most electricity was generated locally by small streams. At this scale a flexible land transport system was a must. Stalin concluded a deal with Ford in 1929 to build a large manufacturing plant in Russia for Model A cars and trucks (Wilkins & Hill, 1964, 218).

The Russian scheme never came to much, although the Russians clearly learned mass-production skills at the Ford plant and turned out, by Russian standards, a reasonable number of vehicles. What is significant in all this is the convergence between 1920 and 1939 of thought and action about the shape of Neotechnic society. Between those dates the seeds were sown in the United States for a society that lowered its fixed costs by moving enterprise away from congested cities, even if only to new suburbs. This radically improved both productivity and quality of life for first white-collar, then blue-collar workers. The middle classes moved out of the congested cities to larger, better-appointed houses in the suburbs, and were followed by the skilled working classes. After World War II suburbanization began to extend to all elements of U.S. society, first to the regular working class as car ownership accelerated in the 1950s, and finally to the poorest ethnic minority, southern blacks, by the early 1970s.

Continued road improvement was important in achieving suburbanization. After World War II GM involved itself with the "road gang" that persuaded the Federal government to fund the interstate highway program in 1956 (Flink, 1988, 371). The automakers in the "road gang" concentrated on the development of urban freeways for suburban commuters rather than interstate highways. It was the trucking and oil interests in the "road gang" who pushed strongly for interstates.

This pattern is now being repeated in Europe and Japan. Their middle classes were heavy adopters of the automobile in the 1960s as higher wages, superior technology, and declining mass transit made a combined impact. The U.S. pattern is broken only where large, old central cities impose vast costs through congestion. In such areas automobiles provide access to the unexploited countryside and commuter transport for those who work in new businesses located on cheap land away from fixed transport systems. Even the complicating factor of British "Green

Belt" policy, which held farmland immune from development, has recently been swept away by the Thatcher government.

Europe's "Golden Triangle," defined by the cities of London, Paris, and Bonn, now looks increasingly like the United States. Road transport is dominant away from the great city centers. Even rail commuters use automobiles to reach their station, where they either park them or use the "kiss-and-ride" system. An example of automobile-centered growth in this region is Southend-on-Sea, some thirty-five miles east of London, which has long been a dormitory town for rail commuters. Since the early 1960s Southend has closed its old downtown to through traffic and made it into a gigantic pedestrian mall accessed by a series of ring roads and feeder highways. Considerable growth has occurred in data processing and other white-collar businesses as firms seek to leave the congestion of central London. Several large shopping centers have been built around Southend, in particular along the main road to London.

Observers such as the U.S. writer Flink have questioned whether Europe has yet attained an automobile "culture." Flink is bemused by a viewpoint generated in southern California, where multiple vehicle ownership is a necessity and local economic factors confuse the issue. He makes numerous errors of fact and judgment with regard to the development of Europe, concluding that "automobility is incredibly expensive" and calculating the cost of owning a vehicle in simple dollars without calculating the concomitant benefits in access to inexpensive land and inexpensive housing. Southern California's notorious housing prices are driven by an excessive regional in-migration. Without the automobile the regional economy would collapse because its transport needs require route flexibility in the journey to work. This is increasingly true elsewhere than Southern California. A worker in a given urban area needs to be able to commute by automobile almost anywhere within that urban area without the expensive process of moving house. At least this is the case until diseconomies of scale set in, as they seem to have done in southern California.

As Figure 5-1 shows even the automobile city has a theoretical maximum size. Cities such as New York or London can exceed it because they still have good fixed-route mass-transit systems from the early, electrically driven Neotechnic. Cities such as Los Angeles have reached the stage of diseconomies of scale that Henry Ford foresaw early in this century. Cities such as Houston are close to the theoretical limit. Closure below the theoretical limit or a revolution in the transportation of things, people, and information would be well advised. Given that we seem to be becoming an "information society," the notion of using homes as computer work stations makes sense, just as Ford suggested in the 1920s that work should be taken to people rather than people to work.

TRUCKS, TRACTORS, AND TANKS

Although the automobile has been central to the development of Neotechnic society, trucks, tractors, and tanks have also played a part.

Table 5-3 Persons per commercial vehicle (light trucks, trucks, and busses), 1910–1988, selected polities

Year	United States	Britain	France	Germany[a]	Italy	Spain	Czechoslovakia	East Germany	Poland	Japan
1910	9,128	749								
1921	85	203	418	2,038	1,556		6,423			2,068
1930	33	99	100	383	655		682		1,621	497
1938	29	79	91	179	522		695		2,072	277
1950	17	47	56	137	204	337		193	539	192
1955	16	41	35	88	132	286	124	181	374	71
1960	15	34	28	79	110	191	108	140	248	33
1965	13	31	22	65	80	83	92	107	171	19
1970	11	32	18	58	60	46	66	84	126	15
1975	8.2	30	14	56	50	34	52	65	80	13
1980	6.6	28	20	39	34	27	41	42	52	13
1985	6.0	24	17		30	25	38	40	43	13
1988	5.8	21	13		29	15	34	38	37	13

Sources: CSO Annual Abstracts; Europa World Year Books; Eurostat; Eurostat Review; Mitchell, 1980; Statistical Abstracts of the U.S.; Statistical Handbooks of Japan; U.S. Bureau of the Census, 1975.
[a] West Germany only after 1945.

Trucks made it feasible for companies to locate freely within cities. They have also proven useful for long-distance movement of perishable, relatively high-value commodities. Tractors mechanized agriculture and bolstered the huge increase in productivity brought by synthesized fertilizers. Tanks made truly mechanized mobile warfare possible.

Add to these the remarkable spread of specialized automotive power machinery and the depths of the automotive revolution become apparent. Construction machinery is now remarkably sophisticated and complex. Earthmovers, scrapers, and shovels have allowed us to literally reshape society by making buildings and infrastructure alike much less expensive. At a microscale, forklift trucks and similar machinery have revolutionized in-factory and in-warehouse handling of materials and goods.

Trucks and Busses

Trucks move small quantities of perishable and valuable goods within and between cities. The spatial flexibility of industrial location made possible within the city depends as heavily on the truck as on the automobile. Technical innovations have been slower in trucking than in private vehicle construction because weight and size reduction is far less at issue. Operating costs have had far more impact on the development of truck technology. Busses use truck technology to move streetcar-sized loads of people on flexible routes within and between cities.

Like automobiles, trucks developed fastest in America, although Europe lagged less in truck than in automobile adoption. In 1921 the United States had one commercial vehicle for every eighty-five persons and one automobile for every twelve persons, or just over seven automobiles per commercial vehicle (tables 5-1, 5-3). In Britain the ratio was near parity, and in France it was just over two to one. Whereas it took until the mid-1950s for Britain to achieve automobile adoption rates comparable to those in the United States in 1921, Britain had bettered the United States's 1921 commercial vehicle adoption rate by 1938.

Trucks were particularly appealing to the new, service-oriented industries. Factories could be built on inexpensive land away from fixed transport lines or even in cities poorly served by fixed transportation. Perishable, high-value or low-bulk products could be efficiently distributed by factory-owned trucks over much greater areas. And trucks could control pilferage or the perishing of goods with short shelf lives much better than railroads could. A truck dispatched in the morning could deliver goods 100 to 160 kilometers away even without freeways and return by the end of the working day to the security of the factory. Routine truck maintenance could be performed in the night hours. Truckers gained status and their unions grew powerful because truckers were away from the oversight function of the factory. It was the trucker who made sure that perishable goods did not spoil and valuable goods were not pilfered and that deliveries were made quickly and efficiently.

Truck distribution not only increased the daily range of the factory but it reduced product inventory. Production could be geared almost directly to demand. This was of greatest significance in the food industry, where perishability is highest. Sales of chocolate, for example, are typically in extremely small amounts, in the form of candy bars, in a huge number of outlets. Chocolate is an exotic semiluxury product dependent upon tropical agriculture for its basic raw materials, cocoa beans and sugar. Heavy capital investment is needed in the machinery to "conch" or grind the cocoa beans. To amortize such large capital investments large-volume production is necessary, but because the product is perishable and requires knowledgeable handling, distribution has to be carefully controlled by the factory. The growth of the large specialist chocolate manufacturers in the 1920s and 1930s, in particular Rowntree and Cadbury in Britain, Menier in France, and Hershey in the United States, is thus also the growth of the trucking system. Such companies kept fleets of trucks and geared production to consumption.

In technical terms commercial vehicles have changed little since their initial appearance. Nearly all are still built on a heavy chassis and have a front-mounted engine driving the rear wheels. Such automobile innovations as unit construction and front-wheel drive have had little impact. Rear-engined busses have become common, however, because they offer the advantage of either a low load floor or underfloor baggage storage. In Europe and Japan busses are still of great importance in urban transport. Low load floors allow elderly and infirm passengers to board easily. In the United States less innovation has occurred in city busses because the major producer, GM, saw them as merely a stop on the road to full automobility (Flink, 1988, 367). Long-distance busses, however, have used the rear engine to advantage for heavy baggage storage by providing baggage compartments at almost curb level.

The most important innovation in commercial vehicles has been the use of the compression ignition engine, popularly named after one of its major inventors and proponents, Rudolf Diesel. Diesel was a brilliant student who received as fine an education as imperial Germany could offer in the late nineteenth century, at the Technische Hochschule in Munich. His engine was conceived from first principles, the direct application of the science of thermodynamics to the heat engine (Cummins, 1976, 304). Development, however, took time and the resources of big industry, eventually involving the huge Krupp combine as well as Maschinenfabrik Augsburg-Nurnberg (MAN). Diesel's patent was granted in 1892, the year in which MAN agreed to build an experimental engine. Krupp signed on in early 1893 and development began in earnest, yet it was 1897 before the first successes were recorded.

Krupp took a major interest because Diesel's engine promised efficiency. Oil was expensive in imperial Germany, not only because the U.S. production cartels set high prices but also because of high tariffs (Cummins, 1976, 318). Diesel's engine certainly delivered. It was well over twice as efficient in converting fuel to work as the spark ignition engine. Cummins quotes an efficiency of 11.2 percent for the Otto and Langen four-stroke stationary engine of 1884, compared to 30.2 percent for the diesel in October 1897 (1976, 169, 324). As developed up to

World War I, Diesel's engine was far too heavy, even for trucks. It was fine for electric power–generating stations or for ships, whether submersible U-boats or merchant ships. Diesel himself also thought of it as a substitute for steam power in railroad locomotives.

For truck use it required radical downsizing, a process laboriously undertaken by MAN. Fuel remained expensive in Germany after World War I, and MAN showed the first diesel-engined truck at the Berlin Motor Show in December 1924. By 1932 MAN produced only diesel-engined commercial vehicles. Despite its cost the diesel reduced fuel costs to at least a half and sometimes as little as a quarter of comparable spark ignition engines. The Nazi party gave diesels high priority because they reduced oil imports. The Nazis also encouraged German companies to design diesels for passenger vehicles, and companies such as Mercedes made some progress (Heinze, 1938). Diesel engines, however, are far better suited to trucks than to private automobiles. The thermal efficiency of a diesel engine is so high that fuel consumption is little different at part and full throttle and is extremely low at idle. Spark ignition engines are very inefficient at idle and full throttle. Automobiles spend much of their working life at part throttle, whereas commercial vehicles on the highway are frequently at full throttle and, in city delivery traffic, spend much time at idle. Diesels suit the work pattern of commercial vehicles better, and their poor power-to-weight ratio is less a problem in a heavy vehicle than in one continually striving for lightness. Diesels have a place in urban taxicabs, which spend much of their working lives idling, but are not suited to private automobiles unless the engine's power-to-weight ratio is much improved. The 1982 Volkswagen Rabbit was available with an excellent light diesel, turning out .76 of a horsepower for every horsepower of its spark ignition equivalent (Automobile Club of Italy, 1982, 316). The engines used by Mercedes automobiles are more typical of normal diesel practice, turning out only .55 horsepower for every horsepower of their spark ignition counterparts (Automobile Club of Italy, 1982, 116–17).

Tractors

The replacement of the horse by the tractor amounted to a major revolution in agriculture. Horses were very expensive, taking four years to reach working age and commonly having only a three- to four-year working life. Routine horse care required, according to U.S. government reports, 27 minutes each day per horse and the horse averaged 3.5 hours of work per day, tiring out if worked for 6 hours. On these bases an average tractor of the teens was "as powerful as 25 horses, as enduring as 100 horses; and about as expensive as ten" (Casson, Hutchinson & Ellis, 1913). The involvement of the United States in World War I drove the adoption of tractors hard. Henry Ford entered the market after the British government placed a huge order for light tractors in 1917 in an attempt to improve labor productivity in British agriculture and free men for the war effort.

As with automobiles and trucks steam tractors were tried, but their high weight told against them and they tended to compact soil or even to break through the upper soil horizons and bog down. Their size was also a disadvantage except in newly settled lands, where large fields allowed large tractors. In densely settled regions steam plowing was done with plows pulled by stationary engines through fields by cable (Whitehead, 1977, 46).

In North America steam tractors were common in the wheat-producing areas of the northern Great Plains and prairies and the Sacramento Valley of California. Agricultural implement manufacturers served the first region from Chicago and Minneapolis–St. Paul and the second from the San Francisco area. The Holt and Best tractor companies of California produced the most bizarre of all steam tractors to tear up the virgin lands of the region. Driven wheels up to 4.5 meters wide and nearly 3 meters in diameter were common (Gray, 1975, 12). Such tractors pulled up to twenty-five plows and needed many acres to turn around, thus were totally useless in small fields.

The early pioneers of the automobile saw the potential for a light, internal combustion–engined tractor almost immediately. Two deserve brief mention, the almost unknown British pioneer, Dan Albone, and Henry Ford. U.S. steam tractor makers quickly substituted spark igni-

Avery 30-horsepower steam tractor, 1916. Such monsters were derived from railroad locomotives and were of use only where fields were huge, usually in the newly opened wheat lands of the continental interiors. In older, more densely settled areas, they were unusable. Ford Museum, by the author.

tion engines in their tractors, but the vehicles remained large and heavy. Dan Albone, who had built several automobiles and motorcycles in the late 1890s, made a successful spark ignition–powered light tractor in 1901, which he named the Ivel after the river Ivel in his hometown. Unfortunately Albone died young in 1906, and the life went out of his company (Williams, 1980, 12–16).

Henry Ford began experiments with automotive tractors almost as early as he did with cars. His first tractor of 1905 was assembled from components of two failed passenger automobiles. It used the four-cylinder Model B engine and much of the running gear from the six-cylinder Model K. Ford concentrated on Model T after 1908 but returned to the tractor in World War I, designing the Fordson in 1917 for mass production. The Fordson contributed little to the war effort. U.S. tractor production doubled between 1916 and 1917 and doubled again in 1918. Only about 7,000 of the 132,000 tractors produced that year were Fordsons (Gray, 1975, 50, 53), but by 1920 Ford claimed annual sales of 100,000 and total dominance of the light tractor market. Part of the Fordson's success was its cast-iron, unit-construction frame, which gave strength with low weight.

Subsequent manufacturers had to provide more than mere plow pull-

Fordson tractor, 1918. Ford was as obsessed with inexpensive tractors as with inexpensive cars. Although the Fordson was compact, it was just a tractor, not an integral part of a farming system. Specialist farm machinery manufacturers such as John Deere, Farmall, and the like were able to respond with systems that better suited the needs of farmers. Ford Museum, by the author.

ing to sell tractors against Ford. The midwestern-based International Harvester Company produced its successful Farmall tractor in 1926 by recognizing that competing against Ford was "a case of realizing a necessity for foreseeing a demand, not a case of developing something to meet a demand" (Coleman & Burnham, 1980, 78). The John Deere Company did essentially the same with their Model D of 1924 (Deere, 1987, 6). The entry of the large agricultural implement manufacturers with a superior product quickly drove Fordson from the U.S. market. Ford moved the Fordson plant to Ireland in 1928, then to Dagenham. Fordson demand in the United States was met by importing units from these plants.

Ford's tractors were successful only because of the work of an Irish engineer, Harry Ferguson. If a plow hit a stump or other immovable object the tractor could rotate around its rear axle, throwing and even crushing the driver if it went all the way over. Ferguson patented a two-point hitch that transfered the load forward causing the front wheels to dig in rather than rear up. Ferguson's hitch was initially designed only for the Fordson. By 1933 Ferguson had produced the improved, three-point hitch, which allowed automatic control of plowing depth for any tractor. Ford licensed this Ferguson system in 1938 and launched the important 9N tractor to regain his American market for light tractors after 1939 (Williams, 1980, 56–60).

Modern tractors have become increasingly sophisticated, but three technologies are still central: unit construction, pioneered by Ford, for lightness and strength; power takeoffs, pioneered by International Harvester and John Deere; and the three-point hitch, pioneered by Harry Ferguson. Diesel engines have been added to improve fuel utilization and four-wheel drive for ease of movement in heavy soils, but the base technology is unchanged.

Outside the midwestern mainstream of development the Holt and Best companies of California both vied to make big-wheeled tractors to operate on soft local soils. By 1904 Holt had perfected a viable alternative in the track-laying tractor, which spread its load over a much wider area of ground. Holt and Best merged in 1925 as Caterpillar, a nickname earned by Holt tractors in World War I (Gray, 1975, 85). The British army had an exclusive agreement to use Holt tractors to pull artillery and supply wagons in the mud of Flanders. When the Royal Navy became interested in this technology it had to turn to other U.S. companies to develop the tracks.

Tanks

The Royal Navy, driven to radical solutions by incessant technological competition with imperial Germany before World War I, conceived the tank as a "land dreadnought." The first British tanks, with tracks running completely around their lozenge-shaped sides, were intended as armored trench crossers to clear a path for infantry. French thought produced a smaller tank designed merely to flatten barbed-wire entanglements, again for infantry (Macksey & Batchelor, 1970, 24, 26). The French were able to use Holt tracks but produced smaller, less capable tanks. German tanks similarly evolved from Holt tractors (Macksey & Batchelor, 1970, 36). The Holt tanks proved too small and thus not powerful enough to carry adequate armament and armor as the proper role of the tank was worked out. The Germans spent the last two years of the war trying to copy British tanks and catch up, but they were unsuccessful. By late 1918 British thinking about the use of tanks had evolved from "the direct art of tactical breaching to the plane of strategic decision" (Macksey & Batchelor, 1970, 48). Plan 1919 was to move out of the trenches to truly mobile warfare in what is now better known as blitzkrieg, with rapid air and ground advances coordinated by wireless telephony. Although the British conceived blitzkrieg, German and Russian forces put it into practice, ironically with the aid of a U.S. designer.

J. Walter Christie was a remarkable engineer. In the early 1900s his greatly improved technique for making battleship turrets was adopted by both the U.S. and British navies (Christie, 1985, 2). He then produced a remarkable series of front-drive automobiles that laid the ground for the Mini. Next Christie turned to a front-drive tractor to be substituted for horses, then to four-wheel-drive trucks for the military. Christie took one of his trucks to Texas in 1915 to aid Pershing in his pursuit of Pancho Villa in and out of Mexico. From this experience grew designs for a mobile gun carriage designed to travel by road on rubber wheels to the battlesite, then to mount tank tracks and take to open country (Christie, 1985, 16–20).

European tank designers had not progressed beyond the infantry support tank in World War I. Had Plan 1919 gone into effect they would presumably have realized that tanks with a top speed around 8 kilometers per hour were not suited to blitzkrieg. Christie made major innovations in suspension that allowed steel-tracked vehicles to move at speeds up to 50 kilometers per hour without shattering tank and crew. After a series of arguments with a skeptical U.S. Army, Christie fell on hard times by 1930. To recover he sold his ideas to Russia (Christie, 1985, 43). Attempts by the Nazis to attract Christie to Germany to supervise tank construction were unsuccessful, despite the offer of a million dollars in 1935 (Grayson, 1976, 272). German engineers examined and photographed enough Christie tanks to remove most of Christie's technology to Germany, however (Christie, 1985, 46, 50).

Christie's excellent designs were compromised by his ideal of speed. Russian redesign added protection, durability, and firepower. The Russian BT-5 was the first modern tank, with carefully sloped armor in the Christie pattern, a 45-millimeter gun, and a 350-horsepower diesel engine (Macksey & Batchelor, 1970, 72–73). Diesel fuel made a great deal of sense in a battle tank because it is hard to ignite. Despite their lead in diesel engines, German tanks used modified aircraft engines fueled with gasoline. This and excessively light armor made German tanks inferior to the Russian descendants of BT-5 in the great tank battles of World War II.

The northern German plain was as good an area for automotive cavalry as it had been for the medieval horse variety. In World War II it seemed initially as if German armor and blitzkrieg would carry all before it, thrusting both west into Europe in 1939 through 1940, then east into European Russia in 1941. Of the many forces that halted the eastward thrust, the Russian T-34 battle tank was by no means least. It was certainly the prime agency by which the Russian return blitzkrieg against Germany was carried out after the German advance ground to a halt. "T-34 was in direct line of succession from the Christies . . . a beautiful balance between hitting power and self-protection, long range, speed, and reliability" (Macksey & Batchelor, 1970, 113).

The T-34 had broad tracks and could often keep going when German tanks bogged down. The Germans finally broke the backs of their tank battalions against the Russian tanks at the battle of Kursk in 1943 (Macksey & Batchelor, 1970, 128). From Kursk on, Red Army tanks easily outnumbered those of Nazi Germany. Although the Germans somewhat recovered technical superiority late in the war, they lacked it when they most needed it, in 1943. Later, vastly greater numbers of T-34s more than made up for the slight technical superiority of German tanks.

Mere possession of Christie's ideas did not guarantee success. The Russians were simply the most effective at improving Christie's technology. Their use of lightweight diesel engines is perhaps most notable. Russian tanks consumed much less fuel and had far better range than German tanks, as well as less risk of explosion once hit. German designers preferred firepower to armor protection, and without a Christie prototype to work from, they failed to understand the need for carefully sloped armor to bounce shells off (Macksey & Batchelor, 1970, 113). The British automobile company, Morris, bought a Christie tank to try to break the monopoly of British tank design by the Vickers armament company as a result of the success of Vickers tanks in World War I (Macksey & Batchelor, 1970, 54–58). The British Crusader, a Christie-style tank deployed in the North African campaign of 1941 and 1942, suffered large numbers of breakdowns from inadequate development and was undergunned (Macksey & Batchelor, 1970, 109, 120).

There is an interesting reversal of national styles of warfare here. British ships at the climactic battle of Jutland in 1916 were overgunned and underarmored. Britain is an island fortress best defended by aggressive control of the seas. German ships were undergunned and overarmored, an essentially defensive posture. In the land battles of World War II the positions were reversed. German tank designs were offensive; British tank designs were defensive because the British army's function in European wars was defensive. Germany has no natural boundaries and took an aggressive posture toward its neighbors on land. Russia was a land-based polity with only the western frontier in need of offensive protection. The eastern extension of the Soviet state, once the Asian population was brought under control, offered merely defense in depth against an aggressor from the west. Warfare reflects the shared values of a culture, thus all its strengths and defects.

The U.S. military, which had totally rejected Christie, finally managed to come up with a viable tank in the Sherman. It was slower, heavier, and less powerful than either German or Russian tanks. But industrial capacity told heavily in World War II. The United States built 49,000 Shermans, Russia perhaps 40,000 T-34s. The Germans built only 5,508 Panthers, the only tank that could take on a T-34 (Macksey & Batchelor, 1970, 113, 129). After World War II Christie's design principles became universal. Field experience in the various Middle Eastern wars, with Israeli use of U.S. tanks and British tanks captured from Iran, suggests that modern Russian tanks are no match for U.S. and European tanks, at least when the Russian tanks are used by Arab countries.

Mobile warfare tactics are still based on blitzkrieg, as modified by German and Russian experience. Tanks are the weapon of choice in such open areas as the northern German plain and the deserts of North Africa and the Middle East. Improved air power, however important, is no substitute for mobile land warfare, although blitzkrieg requires local air superiority.

CONCLUSION

The flexible extension of military power over space has been as important in the Neotechnic as has the flexible extension of economic power. The tank and the airplane, united in blitzkrieg, put land warfare back in the position it occupied when eclipsed by naval power, which had a stunning success from the late 1400s on. Blitzkrieg destabilized the old order of Europe established after the death of Charlemagne and the creation of the feudal frontier. Blitzkrieg has given settled agricultural societies the land mobility previously reserved to pastoralists.

Automobility destroyed the defensive advantage of Russia. The British geostrategist, Sir Halford Mackinder, clearly understood this loss of advantage when he propounded his "heartland" thesis in 1904 (Blouet, 1987, 110). Mackinder believed that an aggressive imperial Germany would push east against a Russia weak from corruption and internal dissension to dominate the vast resources of the Eurasian heartland. But whereas imperial Germany foresaw Mitteleuropa, a customs union from the Pyrenees to the Elbe, the Nazi state saw a super-Germany, a Third Reich from the Pyrenees to the Urals (Fischer, 1967, 247–56; Herwig, 1976, 185–86; Blouet, 1988b). Only automotive transport and warfare made this dream possible, and only superior understanding and adaptation of blitzkrieg allowed Russia to resist and ultimately destroy the Nazi threat. In 1945 Russia had the ability to extend its state to the Atlantic but was restrained by demonstrated U.S. competence at blitzkrieg, achieved more through industrial than technical or tactical superiority.

The industrial supremacy of the United States was also due in no small measure to automobility. Automobiles, trucks, and busses radically lowered the fixed costs of production in societies where the supply of land was limited by the need for access to fixed lines of transportation.

Automobiles and trucks used fixed capital in factories and machine tools efficiently, and low selling prices achieved by Fordist mass production encouraged their mass adoption. Sloanist flexible mass production used the natural cycle of machine-tool wear to institutionalize innovation in design and production. McNeill has pointed out that the military version of the command economy returned in the naval race that began in the late nineteenth century (McNeill, 1982, 269). Yet a command economy is not totally alien to a market economy, as Sloanism demonstrates. A managerial desire to innovate in order to maximize sales (the organizing principle of a market economy) was modified to a production technology demanding innovation to reduce fixed costs (the organizing principle of a command economy). This innovation was willingly supplied by a new cadre of professionalized industrial managers in the countries that made the earliest transition to the Neotechnic.

Chapter Six

Aviation and the First Global System

After 1945 . . . the United States built an empire without annexing any land. Under an aerial nuclear umbrella, the nation established bases in forty-four countries around the globe. These bases Washington maintained and largely manned by air. At the same time American businessmen, traveling the same airways, branched overseas and created multinational corporations that enlarged the American presence on every continent.

Carl Solberg
Conquest of the Skies

Despite all that has been written about the development of a world system after the late fifteenth century, the development of a truly *global* system has been contingent upon the rise of aviation and is thus very recent indeed. Earlier world systems merely extracted bulk agricultural surplus, first from the coastal regions, then from the continental interiors.

Civil aviation moves two things very well: information and people. It began in the United States in the 1920s with the air mail, but had an important precursor as a people carrier in Germany before World War I. Given the cost of movement by air, until recently civil aviation has concentrated on the movement of information, managers, and people with highly specialized job skills. For the most part this has meant people with the skills to repair or maintain pieces of equipment critical to production facilities. Since the late 1960s, however, declining costs have made recreational air travel commonplace. Air freight has come into use to move high-value, low-bulk cargoes.

Military aviation has geostrategic importance, and many advances in civil aviation have military origins. In the 1940s aircraft armed with nuclear weapons represented the ultimate ability to project the power of the state at a distance until the emergence of long-range missiles. Conventional air power remains important in modern warfare, but it has not replaced sea power as the measure of a state's military potential. Nuclear missile–armed submarines are now the dominant forces of the world's strategic arsenals (Modelski & Thompson, 1988, 43). Manned aircraft figure mainly in blitzkrieg, a strategy perfected by Nazi Germany and Soviet Russia in the 1930s, and in what are now referred to as "surgical strikes." These precision attacks on highly defined targets, such as nuclear reactors and terrorist training camps, have become the hallmark of the Israeli Air Force since the 1960s. Gone is the notion of long-range strategic bombing that was important to air power in the premissile era.

Unlike such vehicles as the automobile, and despite some grandiose dreams, civil aviation has never become a tool of individual mobility. Airplanes require three-axis control for the three dimensions they inhabit. Such control requires much skill and training on the part of the pilot. Airplanes require sophisticated electronic communications, blind-flying devices, and automatic pilots if they are to move in anything but good weather and broad daylight. For almost any foreseeable future the notion of individual transportation through flying machines is in the realm of science fiction. Civil aviation thus operates like a bus system: it is flexible as to the routes vehicles can operate over, even avoiding temporary obstructions like bad weather, and sensitive to the changing market.

Railroads were risky investments unless they were agglomerated into very large systems and thus were poorly matched in the long term to a market economy. A railroad built between two cities would suffer terribly if the economy of one of the cities declined. Even spreading the risk over ten to twenty cities was not enough if the regional economy declined. Although early and short-term investment in railroads usually paid handsome dividends, over the long haul the decline of cities and regions made such investments unattractive. In many polities investment in railroads became, of necessity, a function of the state. The state could accept a poor return on investment in exchange for economic and strategic benefits. By the end of World War II only in the United States did the railroad system remain in the hands of private investors, and radical decline set in once U.S. airlines began to take the profitable trade in passengers.

Civil aviation is better suited to a market economy because it localizes immobile fixed capital and thus the penalty for failure not related to transport service. If a city's economy declines the city is penalized, not the airline, which can switch the capital represented by its airplanes to a city that is booming. A true market economy approach to civil aviation has never come into being, even in the United States, because of a sensible preference to spread local risk more evenly by involving the federal government. To begin with city and state government paid for airports: Chicago began such a process in the 1920s (Young & Callahan, 1981, 146–51). State and, eventually, federal funds were added in the 1930s. Private enterprise runs the vehicles as mobile fixed capital. If a city's economy declines, the primary loss to fixed capital is at the local level. Some loss is taken by local taxpayers, but the loss is spread over all taxpayers by state and federal government participation.

Civil aviation has developed in five major, overlapping technical phases: the pioneer phase, 1799 to 1919; lighter-than-air flight, 1852 to 1937; long-range flying boats, 1914 to 1939; the passenger-carrying landplane, 1919 to 1958; and the jet, from 1958 on. The two major contenders for global hegemony in the fifth world leadership cycle, Germany and America, did most of the technology forcing because they sought to integrate the economies of large, continental world systems. Britain attempted to use civil aviation in a maritime-based world system to hang on to hegemony, but failed. Russia was able to use air power to integrate a continental system but did not have the concept of a world system and continued to operate as a world empire.

THE PIONEER PHASE, 1799–1919

All theoretical material necessary to achieve flight was published and a multitude of viable vehicles developed in the pioneer phase of civil aviation. The problems were the same for all pioneers: control in three axes; enough power to lift and propel a vehicle; structures strong enough to resist aerodynamic forces yet light enough to fly and carry reasonable loads of fuel, mail, or passengers.

Most pioneers failed to learn the lessons of three-axis control that were available from birds and had a "chauffeur" mentality rather than that of an airman (Gibbs-Smith, 1970). Up and down for them were only part of forward motion, not integral to lateral control. The Wrights' observations of bird behavior and their careful reading of such previous pioneers as the German glider pilot, Lilienthal, led them to understand that an airplane would be an essentially unstable vehicle under constant control, balancing as it turned.

Much of the Wrights' work lay on the shoulders of "the father of aerial navigation," Sir George Cayley (1773–1857) (Gibbs-Smith, 1970, 21–25). Cayley was an English country gentleman who devoted his life to applied science. In 1799 he isolated the crucial forces governing flight—lift, drag, and thrust—and in 1853 built and had flown a glider with a pilot-operated system of three-axis control. His ideas were not taken seriously by his contemporaries. Powered flight needed lighter structures than were available to Cayley, as well as more powerful engines. Both became possible in the late 1880s. Safety bicycles emphasized light structures and mechanical components. Would-be automobilists began to improve the internal combustion engine. It is no accident that the Wright brothers were both bicycle builders and ornithologists (Crouch, 1986, 11). In 1903 they achieved powered flight using a less than satisfactory engine. By 1905 a better engine and far better mastery of three-axis control gave them an airplane capable of an hour's flight. In 1908 they took this vehicle to France and stunned

European would-be aviators with their complete mastery of the air. The Wrights had become true airmen as the Europeans stumbled along with their chauffeur mentality.

The 1908 Wright demonstrations galvanized Europeans. Discontent with the outcome of the Franco-Prussian War of 1870 had caused France to look to new technologies as a solution to the German threat. Aviation was obviously useful in war. The Germans responded to rapid French progress in aviation after 1908 (Morrow, 1976, 15). The British recognized the threat to their insularity posed by aviation (Gollin, 1984). Only Americans felt themselves secure behind their ocean walls. Bleriot's flight across the English channel in 1909 underscored British concerns. The flight was in response to one of a series of prizes offered for spectacular flights by the British newspaper, the *Daily Mail*. These prizes acted as efficient technology forcers, culminating in 1913 in the offer of £10,000, then $50,000, for the first aircraft to cross the Atlantic (Allen, 1984, 15–24).

The most serious contender seemed to be Curtiss's large, twin-engined *America* flying boat, launched at Hammondsport, New York, in August 1914. Curtiss was helped in his design by B. D. Thomas, an engineer from the British Sopwith company (Knott, 1979, 24), and he chose an international crew. When World War I broke out the British co-pilot, naval lieutenant John Cyril Porte, returned to Britain and persuaded the Royal Navy to buy *America* and a sister ship. The potential of a long-range flying boat for open-ocean patrol work against U-boats and surface cruisers was not lost on the British. Sixty-two more Americas were ordered from Curtiss (Knott, 1979, 31).

Porte modified, upgraded, and eventually totally redesigned the Americas. In the process he laid the foundation for the British flying boat industry between World Wars I and II. Curtiss received regular feedback from Porte and modified his designs as well. In the United States Curtiss technology spread quickly. U.S. military aircraft procurement called for the winner of a design competition to sell to the government not only the design but also the manufacturing rights. Boeing built many Curtiss flying boats in World War I because of this system (Davies, 1982, 6). Appropriately enough it was a descendant of the Curtiss *America,* the four-engined Curtiss NC-4, that completed the first air crossing of the Atlantic in May 1919, albeit via the Azores and Portugal (Allen, 1984, 47–62).

The success of the Curtiss and Curtiss-based flying boats led aviation technology down a blind alley. Curtiss chose a flying boat because underpowered engines meant long takeoff runs with fuel for long flights. On land such long runs were all but impossible. Flying boats could use several miles to take off when necessary. Since air travel initially offered speed advantages over ocean travel more than over railroads, many long flights were expected to be over water. Flying boats also seemed safer in the event of inevitable engine failures. Until the mid-1920s few engines powerful enough for heavier-than-air flight could run reliably for more than a dozen hours or so. Commercial operators quickly saw possibilities in flying boats. As early as 1914 scheduled service for a passenger or two was offered between Tampa and St. Petersburg in Florida (Davies, 1982, 9). The success of the Curtiss *America* produced serious proposals by 1914 for a network of services down the East Coast from New York to Florida and Cuba as well as across New York state to the Great Lakes and hence to Detroit.

The *Daily Mail*'s transatlantic competition also acted as a technology forcer in imperial Germany. Count Zeppelin was initially determined to enter one of his lighter-than-air vehicles. He was dissuaded from doing so. First, consultation with leading German engineers convinced him that his Zeppelins were technological dead ends. And second, he was influenced by the successful four-engined airplane built by Russian designer Igor Sikorsky in 1913 (Gibbs-Smith, 1970, 170). This plane had a fully enclosed cabin and, within a month of its first flight and with eight people aboard, stayed up for almost two hours (Cochrane et al., 1989, 36).

Zeppelin organized two companies. One, at Staaken just outside Berlin, was to build wooden bombers as quickly as possible. The other was to concentrate on metal flying boats and long-term development. He transferred Claudius Dornier from work on an airship with a steel structure intended to win the *Daily Mail* prize to working on flying boats at Lindau on Lake Constance (Haddow & Grosz, 1969, 100–101). This company, though founded and owned by Zeppelin, was always called Dornier. As Zeppelin predicted, the fastest gains came at Staaken. Staaken built the best of the German *Riesenflugzeug* (giant airplanes) of World War I. The Zeppelin-Werke Staaken began with an R-plane project by Hellmuth Hirth, who had intended to enter the *Daily Mail* transatlantic race with a giant, six-engined float plane (Haddow & Grosz, 1969, 209).

Technology forcing is always costly in resources and time, and the R-planes were no exception. Many were built but no series production of a standardized type was ever attempted. Although R-planes were used to bomb Britain in the first Battle of Britain of 1917 and 1918 (Cross, 1987, 41), they were too few to enjoy the success they deserved. R-planes were labor intensive as well as expensive. The six R-planes used to bomb Britain required 750 men to keep them operational, of which only 42 were flight crew. Each plane cost around half a million marks (Haddow & Grosz, 1969, 37, 252, 278).

Yet R-planes were capable of remarkable performance. In May 1918 the treaty of Brest-Litovsk (between Germany and Soviet Russia) established the Ukraine as an autonomous state under German protection. Austro-German troops withdrew from the Ukraine with the armistice of November 1918, and Russian and Polish armies rushed to occupy the region. The Ukrainians wanted to pay for their war for independence with currency being printed in Berlin, but the Germans regarded land delivery through hostile Poland, Czechoslovakia, and Rumania as too risky. Staaken R-planes were chosen to make the currency flights. They made an unknown number of flights in 1919, flying nonstop nearly 800 kilometers from Breslau, now Wroclaw, to Kamenets-Podolskiy in the Ukraine (Haddow & Grosz, 1969, 49–51). R-planes, which were designed so that engines could be repaired in flight, could have easily made a nonstop crossing of the Atlantic in 1919, but the *Daily Mail* specifically excluded German planes and pilots when it reopened the competition after World War I (Allen, 1984, 26).

Dornier forced airplane technology even more spectacularly than Zeppelin-Staaken. In 1915 he built the RS-1 flying boat using airship construction techniques, alloy-steel strip riveted at joints, and much Duralumin for unstressed parts. The RS-1 had nearly twice the wingspan of the Curtiss *America* and was almost three times as long and four times the loaded weight (Casey, 1981, 181, 186; Haddow & Grosz, 1969, 107). Underpowered, it never flew before it was destroyed in a storm. Subsequent Dorniers used much more Duralumin. In early 1918 the RS-III completed an 840-kilometer delivery flight nonstop in seven hours, with two hours of fuel in reserve (Haddow & Grosz, 1969, 120). Dornier's designs became some of the most successful between World Wars I and II and firmly established Germany as the third country seriously interested in flying boats after the United States and Britain.

Adolf Rohrbach was transferred from Dornier to Zeppelin-Staaken in 1916–17 to develop all-metal landplanes. At Staaken Rohrbach brought into being the first modern civil aircraft, thus ending the pioneer period. A first-rate theoretician, Rohrbach pioneered all-Duralumin, stressed-skin construction, where the skin carries part of the structural load so that the internal structure can be lighter. Such a low-drag surface gave high performance (Haddow & Grosz, 1969, 289; Wolko, 1981, 82). The Zeppelin-Staaken E.4/20 of 1919 was a modernistic four-engined, high-wing monoplane. It was scrapped by order of the Inter-Allied War Commission in 1922 because of its obvious military threat. With a range of more than 1,000 kilometers at a cruising speed greater than the top speed of almost any fighter then extant, it could have been a bomber to fear (Haddow & Grosz, 1969, 292–93).

Rohrbach was not alone in his belief that all-metal construction was critical to success. Hugo Junkers was a Berlin-based manufacturer of internal combustion engines, water heaters, and sheet-metal boilers who became interested in flight. His technical education convinced him that thick, internally braced, metal cantilever wings would be superior in drag reduction to the thin, externally braced wooden wings common in the pioneer phase (Schmitt, 1988, 20; Wolko, 1981, 72). Junkers's 1910 patent also laid the basis for the development of all-wing aircraft, again conceived by Junkers to minimize the parasitic drag of fuselage and tail (Wooldridge, 1983, 18).

Junkers relied on the cantilever structure of his wings to carry most of the load. He corrugated his thin Duralumin wing surfaces for strength; they were not load bearing. The Junkers J-9 fighter of 1918 was as remarkable a plane for its time as the E.4/20 and its equal in speed (Schmitt, 1988, 35–37, 180–81). After World War I Junkers evolved the

F-13 single-engined, low-wing cabin monoplane from the J-9. This had seats for four passengers and would not look out of place in the general aviation section of an airport today (fig. 6-1). The Inter-Allied War Commission saw no military threat to the F-13 and allowed Junkers to enter production.

By 1919 in Germany, the essentials of aircraft structure had been worked out and aerodynamic theory had developed scientifically. Pioneering use had been made of airplanes to carry passengers and mail in the United States. The Germans had begun to use the airplane as a strategic bomber, and the British were about to do the same at war's end. In Plan 1919 the British laid down the theory of tactical use of air weapons as part of a coordinated air and ground assault, later known as blitzkrieg. By the end of 1919 one of the world's most dangerous oceans, the Atlantic, had been crossed not once but four times: once by the Curtiss NC-4, once by a British Vickers Vimy bomber to win the *Daily Mail* prize, and twice by the British airship R-34.

Both Germany and Britain saw the potential of airplanes early in this century, and the United States was close behind. It was an American who posed the crucial geopolitical question most clearly. A 1908 article in *Century* magazine commented that, for two centuries, Britain had been "the Princess of the Power of the Sea and by the same token unassailable whether in her insular stronghold, or upon the waves which Britannia has ruled [but if a] Prince of the Power of the Air [arises] then the distinction, the unique advantage of the British Empire van-

German Zeppelin-Staaken E.4/20 all-metal, stressed-skin airliner, 1919. Designed by Adolf Rohrbach, this was the first airplane to stress speed, which was achieved by a high wing loading and low drag, over good takeoff and landing performance. It was more than a decade ahead of its contemporaries. NASM, © Smithsonian Institution. Reprinted with permission.

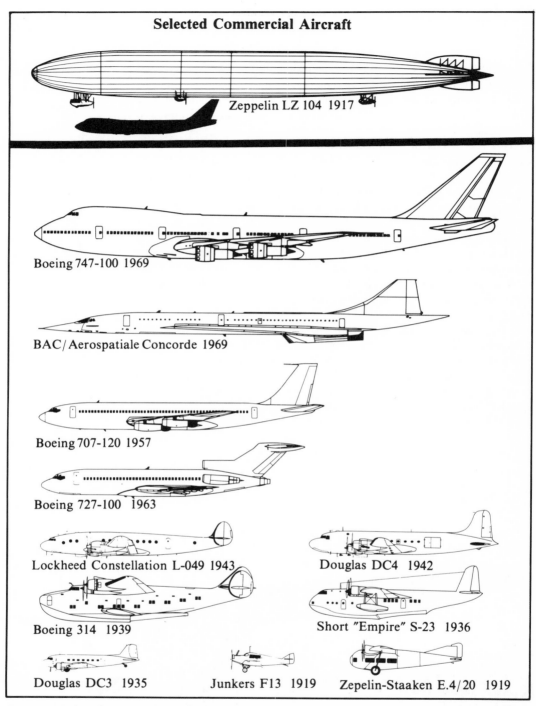

Selected Commercial Aircraft

Zeppelin LZ 104 1917

Boeing 747-100 1969

BAC/Aerospatiale Concorde 1969

Boeing 707-120 1957

Boeing 727-100 1963

Lockheed Constellation L-049 1943

Douglas DC4 1942

Boeing 314 1939

Short "Empire" S-23 1936

Douglas DC3 1935

Junkers F13 1919

Zepelin-Staaken E.4/20 1919

Figure 6-1 Selected commercial aircraft. The only asset of an airline is its seating capacity, and capacity not sold by the time a plane takes off is lost forever. Large aircraft burn less fuel and use fewer highly paid crew members per pound of payload carried. Faster planes fly more passengers per hour of airframe life. Each of the four greatest commercial airliners so far, the Junkers F-13, the Douglas DC-3, and the Boeing 707 and 747, has represented a great breakthrough in speed and in economies of fuel and crew. Designs that represent breakthroughs in one or two of these areas are also shown.

ishes, and Great Britain must take her place on a level with all the other sovereign great powers" (Stedman, 1908, 381–82). In such a case Britain would no longer be able to rely on Admiral Mahan's principle that sea power gave to insular polities remarkable advantages over continental polities. The implications for the United States were unstated, but the author's Anglophilic leanings led him to portray Germany as the potential devil of air power.

Germans saw the possibility of aerial navigation equally clearly, not only for ending Britain's insularity, but also for integrating a Germany unified politically but not in terms of land transport. Figure 6-2 shows the potential radius of the experimental airships under construction by Zeppelin in 1901, "able to reach every principal capital of Europe from the borders of German territory and return." On such a trip even primitive airships could carry a 5-ton payload (Dienstbach & MacMechen, 1909, 430–31). By 1910 transatlantic flight was under discussion, a

Figure 6-2   Potential radius of Zeppelin air transport, 1901. Zeppelin saw his airships as people carriers, not weapons delivery systems. Redrawn from Dienstbach and MacMechen 1909.

"probability that marks the supremacy of Germany's position in the contest for commercial supremacy, for Emperor, press, and people are united on the creation of large air-ships for international traffic" (Mac-Mechen & Dienstbach, 1910, 325–26).

To contemporary Americans this progress seemed both explosive and dreamlike, if clearly threatening an American insularity that was even more finally ended by the events of 1919 (MacMechen & Dienstbach, 1912, 284, 292). For all the pioneering work on airplane control systems by the Wrights, it was apparent as early as 1912 that the United States had fallen signally behind in the race for aviation technology, disastrously so by 1919.

THE LOST CAUSE: AIRSHIPS, 1852–1937

Henri Giffard flew the first airship in France in 1852, but its steam engine was too weak to move it against air currents and was clearly unsafe in a vehicle lifted by hydrogen. Storage batteries made electric airships far too heavy. The internal combustion engine was first used experimentally in two unsuccessful airships in 1897, that of David Schwarz of Austria being important because of its aluminum structure and covering (Beaubois, 1976, 29–31). In 1900 the first Zeppelin, LZ-1, flew in Germany; it was based on Zeppelin's 1895 patent for an airship with a rigid internal frame and on the Schwarz airship. After three short, rather uncontrolled flights, Zeppelin abandoned airships for five years, but then made remarkable progress: LZ-3 flew for 8 hours in 1907; LZ-4 carried passengers on 12-hour flights covering more than 300 kilometers in 1909, and in 1910 Zeppelin formed the Deutsche Luftschiffart Aktien Gesellschaft (DELAG), the German Air Ship Transportation Company (Vissering, 1922, 43–49). Zeppelin envisioned a remarkable network of services between large German cities, many of which invested substantially in ground handling facilities (fig. 6-3) (Vissering, 1922, plate 36). By 1914 DELAG had carried 34,228 passengers for 3,167 hours over 172,535 kilometers on 1,588 flights with no fatalities, a remarkable record (Vissering, 1922, 49).

German military use of airships developed on two lines: as fast, aerial scouts for the imperial German Navy, and as strategic bombers. Airplanes, in particular floatplanes, were to prove far better at scouting than the cumbersome, expensive Zeppelins. Hydrogen lifting gas proved lethally vulnerable against antiaircraft guns and minimal aerial opposition: the life of the Zeppelin bomber was short and fiery. Many raids were aborted simply because Zeppelins were blown off course by winds that would not have affected airplanes. Out of eighty-eight German airships built in World War I more than sixty were lost, twenty-five to enemy action and thirty-four to accidents and bad weather, a loss rate of nearly 70 percent. Just under 200 tons of bombs were dropped on Britain in fifty-one raids. German Gotha twin-engined bombers flew twenty-two raids, dropped 42 tons of bombs, and had a loss rate of 16 percent (Cross, 1987, 80). R-planes flew eleven raids, dropped 13.5 tons of bombs, and suffered 7 percent losses (Haddow & Grosz, 1969, 56).

Figure 6-3 DELAG Zeppelin route network, 1912–1913. DELAG might have made it possible to move administrators and business executives around Mitteleuropa in the absence of an integrated railroad system among the German states. Vissering 1922.

Despite their manifest faults, airships had spectacular range and payload capabilities, far greater than those of any airplane foreseeable in 1918. LZ-104 was despatched in late 1917 to relieve General von Lettow-Vorbeck's troops in German East Africa; it carried 9 tons of machine gun ammunition, 4 tons of medical supplies and 21 tons of fuel (fig. 6-1) (Vissering, 1922, 21). British messages that von Lettow-

Vorbeck had surrendered persuaded LZ-104 to turn back south of Khartoum, returning to base "with its entire cargo after a [nonstop] flight of 6791 kms in 90 hours" (Beaubois, 1976, 121).

At the end of World War I Britain, France, Italy, Belgium, and Japan took German airship technology as war reparations. The United States commissioned LZ-126 as reparation. Germany was refused permission to build new airships for internal use by the Inter-Allied War Commission.

The logical users of airships in the 1920s seemed to be the United States and Britain. Of the U.S. naval airships, however, only LZ-126, named *Los Angeles* and delivered by air from Germany in 1924, had a safe career. *Shenandoah,* copied from one of the later Zeppelins, was launched in 1923 and torn apart by a squall in 1925 (Beaubois, 1976, 137). Similar weather-induced accidents destroyed *Akron* in 1933 and *Macon* in 1935. Like *Shenandoah,* each had a two-year operational life (Beaubois, 1976, 167–68). The much vaunted U.S. use of noninflammable helium to lift airships did not provide much of a safety margin in practice.

Britain's R-34, copied from the same captured German raider as

"DELAG"-Zeppelin Harbor at Leipzig, 1913.
"Sachsen" landing for first time after completion of harbor June 1913.

DELAG Zeppelin facilities at Leipzig, 1913. Each of the cities on DELAG's route network invested in similar facilities. Vissering 1922.

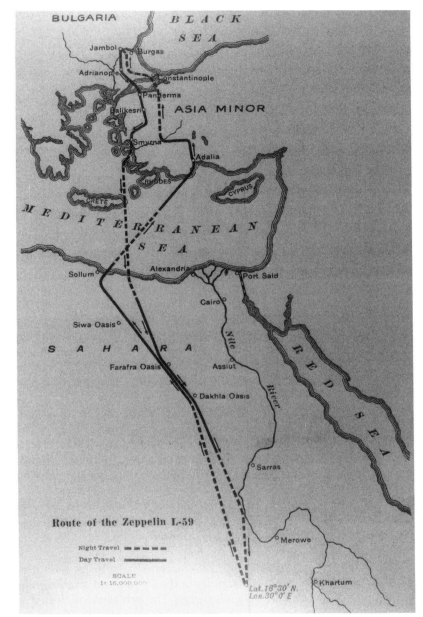

Route of the Zeppelin L-59 (LZ 104) in its nonstop flight of 6,791 kilometers in late 1917, a vain attempt to provide von Lettow-Vorbeck's troops in German East Africa with 13 tons of much needed ammunition and medical supplies. It was turned back by superior British intelligence work and a false radio message that von Lettow-Vorbeck had surrendered. Vissering 1922.

Shenandoah, made an uneventful double crossing of the Atlantic in 1919. The logic of the airship for long-distance air travel seemed to be confirmed, but R-34 was broken up as a postwar austerity measure. In 1924 the Imperial Defence Committee therefore had to begin over, since it was clear that nonstop flights to India could be made only by airship. Heavier-than-air service was technologically possible and had been proposed for converted Handley-Page bombers as early as 1919, but it was impractical as long as the numerous countries along the route refused refueling facilities. The Imperial Defence Committee commissioned two large airships, R-100 and R-101. R-100 was designed by Barnes Wallis using an innovative geodesic structure (Morpurgo, 1972, 129–30). The only successful airship that did not use German technology, it crossed the Atlantic to Canada and returned by August 1930 with forty-two crew and thirteen passengers (Beaubois, 1976, 168–69). The success of R-100 was, however, so overshadowed by the appalling end of R-101 that it was broken up.

In October 1930, at the height of the Imperial Conference held that year in London, and as the delegates were stressing the problems created by the time taken to move information and personnel around the empire, R-101 took off for India on its long-distance proving flight. Of the fifty-four aboard, including the secretary of state for air and the minister of civil aviation, only four survived. Out of control in stormy weather, R-101 slammed into a hillside in France and exploded in flames (Beaubois, 1976, 170). The tragedy guaranteed the supremacy of airplanes in Britain, with particular emphasis thereafter on the flying boat for empire air services.

Only Germany seemed immune to disaster. Germany was allowed to resume Zeppelin production in 1926 because the vulnerability of the bomber airship was recognized (Beaubois, 1976, 165). LZ-127, *Graf Zeppelin,* was launched in 1928, completed a twenty-one-day round-the-world cruise in 1929, and went into regular service in 1931, making 144 Atlantic crossings in six years. LZ-129, *Hindenburg,* proved to be a different story. Coming in to land at Lakehurst, New Jersey, after its thirty-third Atlantic crossing, *Hindenburg* without obvious reason simply exploded into flames in full view of the motion picture camera and in the middle of a live radio broadcast.

Forces other than safety would have doomed airships in any case. Airplane technology advanced rapidly in the 1930s, and airplanes enjoyed a huge speed and crewing advantage. The best airships cruised at only 110 kilometers per hour, about a third the speed of the first commercial airplanes with Atlantic range. A two-day crossing by airship was faster than four days on a crack ocean liner but far slower than fifteen hours in a propeller-driven airliner. *Hindenburg,* specially designed to be an efficient passenger ship, still needed at least one crew member per passenger (Wall, 1980, 100). Propeller-driven transatlantic airliners carried ten passengers per crew member, and jets manage twenty.

LONG-RANGE FLYING BOATS, 1914–1939

The impact of the Curtiss *America* and its British cousins was felt before the end of the pioneer phase because of their fine record on antisubmarine patrol. By early 1917 Porte's Felixstowe was patrolling the North Sea in strength, virtually denying the area to U-boats. U.S. flying boats escorted 6,243 ships through U-boat-infested waters between late July and early November 1918; only three were lost (Knott, 1979, 36–37). Curtiss and Felixstowe boats were ordered in large numbers. Before U.S. entry into World War I a need for 852 Felixstowe F-2A boats was projected for 1918. U.S. entry cut British production requirements to 234 (Bruce, 1957, 242).

Porte continually improved the Felixstowe boats through his last important design, the F-5. Too late for the war, it was selected for production in the United States as the standard U.S. Navy boat in the early 1920s (Bruce, 1957, 254; Knott, 1979, 36). Some 138 were built, two being converted for civilian use by Aeromarine West Indies Airways. The Japanese bought fifteen F-5s for the Imperial Japanese Naval Air Service, and the British Shorts Company, famous for its flying boats in the 1920s and 1930s, pioneered an all-metal hull for the F-5 (Bruce, 1957, 254–55). The flying boat technology of the United States, Britain, and to a large extent Japan can thus be traced to one progenitor, the Curtiss *America*. Only Germany went a different route, first with Dornier's designs and then with those of Rohrbach. Of the three great pioneers of metal airplanes, only Junkers stayed with landplanes after 1919.

Operational experience with the wood-hulled Felixstowe boats showed a life expectancy of about six months, less on tropical duty. They not only rotted, they also became water logged, increasing the boat's takeoff weight and reducing hull strength. Shorts's aluminum-hulled F-5, moored out for two hundred nights, showed no corrosion and weight savings of 25 percent over water-logged wood (PRO/AIR2/750). Vickers claimed in 1919 that an entirely Duralumin airplane would weigh 70 to 80 percent of a wooden one. The same memo pointed out that wet spruce had only a third the strength of the dry wood (NASM/Y4000320). British adoption of metal was also driven by ecological problems. Britain produced no aircraft-grade lumber, relying on the states of Oregon and Washington for high-grade silver spruce. When the United States entered the war it diverted all its wood to U.S. production (Penrose, 1969, 222–23). Although British Columbia was able to fill some of the gap, British airplane constructors began looking to metal. John North commented in the *Journal of the Royal Aeronautical Society* that the British move to metal airplanes was a result "not . . . of the great engineering advantages attending it, but . . . of a world shortage of . . . timber . . . suitable for light structural purposes" (North, 1923).

It was, however, the Germans who most understood the advantages of metal over wood and the necessity for flying boats at this stage of development. Dornier was committed to flying boats from the start of his career in 1915, but Rohrbach's conversion is of interest. After the destruction of his E.4/20, he founded the Rohrbach Metal Airplane Company in Denmark to build his twin-engined *Ro-II* flying boat, a design that set ten world records in 1923. In one respect Rohrbach was returning to his early career as a ships' hull designer at the Blohm und Voss shipyards in Bremen. After the success of the *Ro-II* Rohrbach sold production licenses to Britain and Japan (Mikesh & Abe, 1990, 94–95, 165–67). In 1929 he began a U.S. subsidiary to build the *Romar,* a descendant of the *Ro-II,* but found Americans "unwilling . . . to place orders with his firm" (*New York Times,* obituary, July 10, 1939). Rohrbach was responding to a Pan American World Airways 1929 request to tender for five Romars with U.S. engines. He even offered to "resign all profit" on these machines if PanAm would fly them across the Atlantic (NASM/CO529300). That same year PanAm requested Dornier to tender for eight to sixteen Wal flying boats (NASM/AO187900).

National politics, increasing anti-German sentiment, and the Depression persuaded PanAm to buy American. PanAm began with Sikorsky airplanes, the great Russian designer having emigrated to the United States after the Russian revolution (Cochrane et al., 1989, 101–6). The company later ordered giant flying boats from Consolidated, Martin, and Boeing. PanAm's growth began with the Kelly Foreign Air Mail Act of March 1928, which gave it a monopoly in the Caribbean. It bought out competitors and expanded south down the Atlantic and Pacific coasts of South America with constant encouragement from the U.S. government. Behind this support was fear of Junkers-led German competition in the region (Knott, 1979, 119). By 1932 PanAm had a remarkable network of routes in South America (fig. 6-4).

PanAm next moved into the Pacific. Because Russia refused landing rights along the great circle route to Japan, China, and the Philippines via the Aleutians, PanAm had to fly across the mid-Pacific island chain (Davies, 1982, 249). The longest leg of the trip, from San Francisco to Hawaii, was made possible by the four-engined Martin M-130 flying boat of 1935 (fig. 6-5) (Davies, 1982, 252). Even larger vehicles were needed to carry enough fuel to fly a decent payload across the Atlantic, and this was PanAm's next goal. Under the gentlemen's agreement of the early 1930s PanAm was to share the Atlantic route with Britain's Imperial Airways (Beaty, 1976, 90–101). In return Imperial Airways ceded the trans-Pacific route exclusively to PanAm (PRO/DO35/214/1). Implementation of the Atlantic route was slowed by the need for a vehicle large enough to fly from Europe to America against the prevailing headwinds. By 1939 PanAm had begun to take delivery of the 38-ton Boeing 314, more than half as heavy again as the 23-ton Martins used on the Pacific and with nearly twice as much power (fig. 6-1) (Davies, 1982, 257). In summer this boat was intended for the northern route via Bermuda, the Azores, and Portugal (fig. 6-6). New York's La Guardia

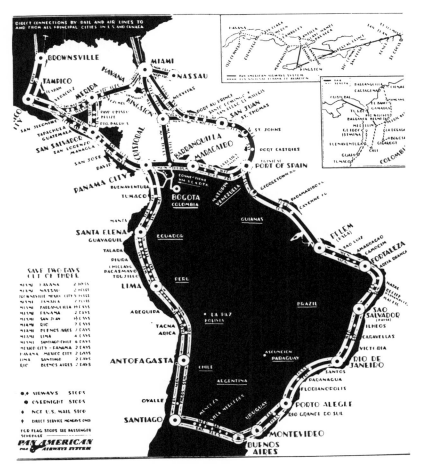

Figure 6-4   Pan American Clipper Latin American air routes, 1932. The flying boat was ideal for the early expansion of air routes around the coast of Latin America, although success was more political than economic. By permission of Pan American World Airways.

airport was built as a dual landplane and flying boat terminal to accommodate the transatlantic service that began in 1939.

Imperial Airways underwent similar development to PanAm, with commercial and political aspirations intersecting. Until the R-101 tragedy the British government saw Imperial Airways as a European airline. Thereafter Imperial Airways became the tool of empire its name suggests. In 1934 the Empire Air Mail scheme began. Imperial Airways was guaranteed a fifteen-year subsidy to operate four-engined flying boats through most of the empire, although the subsidy was to be reduced steadily over time to promote free enterprise. Daily service was offered to Egypt, four flights a week to India, three a week to East Africa and Aden, and two a week to South Africa and Australia. The Foreign Office acclaimed it as "a forward step almost equivalent in importance to the first transoceanic cables" (PRO/AVIA2/1911).

A certain amount of jingoism was evident in the scheme, although a leader in the London *Observer* on Empire Day, 1934, harkened back unknowingly to Stedman's article of 1908.

> The air must be the highway of the future for all who would live great-ly . . . whether individuals or communities . . . Those who are laggards in the race must prepare to resign their dreams of Empire or Commonwealth or anything that spells hegemony . . . Our leadership of the world has rested upon our infinity of world-contacts, on our primacy in travel and commerce, on an elder brotherly responsiveness to all cosmopolitan problems in their emergence. (*Observer*, May 24, 1934).

The buildup to the Empire Air Mail scheme was long and intriguing. The scheme was initially attacked for its supposedly large subsidies. British government data, however, showed that the U.S. subsidy to air-lines through air mail was twelve times as high as the Empire Air Mail subsidy (PRO/AVIA2/1911). The affair emphasizes the different views of aviation in Britain and the United States. In Britain aviation became

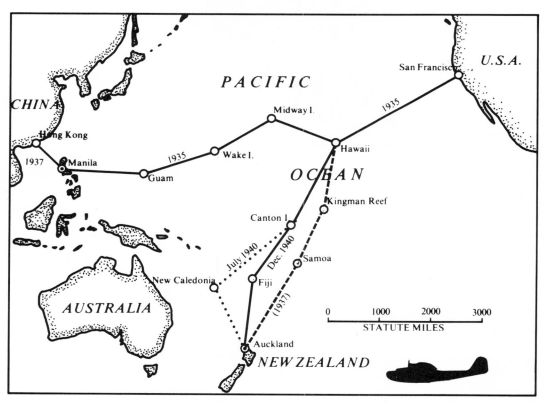

Figure 6-5 PanAm Pacific air routes, 1935. As with contemporary Latin American routes the aim was to politically connect such far-flung American outposts as Hawaii, Midway, Guam, and the Philippines to the North American mainland at a time of rising Japanese expansionism. Redrawn from Davies 1972.

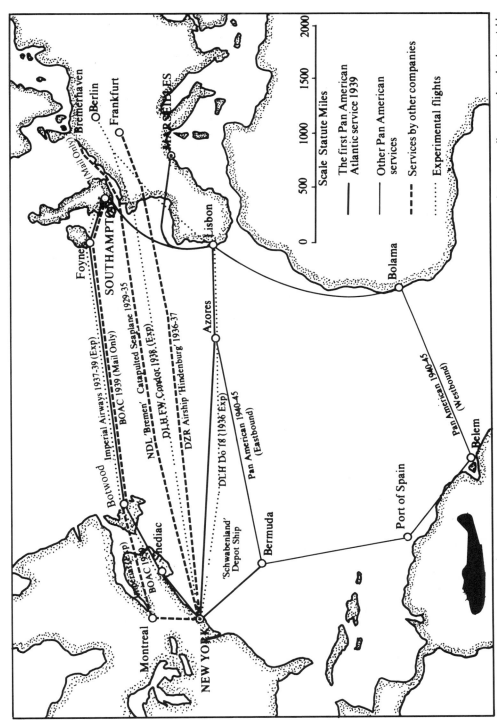

Figure 6-6 PanAm, Imperial Airways, and experimental Atlantic air routes to 1940. The United States, Britain, and Germany all competed to develop viable Atlantic routes in the 1930s. Only PanAm, with its Boeing 314 flying boat, was successful. Redrawn from Davies 1972.

the tool of empire, driven by Conservative party politicians who, however much they were obsessed with low public expenditures, were even more obsessed with a need to hang on to a global empire that was slipping inexorably away. Sir Samuel Hoare, later Viscount Templewood, Conservative secretary of state for air between 1922 and 1929, with a brief period out of power in 1924, makes this quite clear in his memoir of the period, *Empire of the Air* (1957). Hoare was heavily influenced by Lord Trenchard, founder of the Royal Air Force in 1918 and the first real prophet of strategic air power as opposed to blitzkrieg. Trenchard saw that one major advantage of air power was that it provided "control without occupation" (Hoare, 1957, 48). In the Middle East and on the Indian frontier this policy substituted inexpensive, mobile air power, backed by armored cars on the ground, for long trains of infantry backed by camel and horse cavalry (Cross, 1987, 67–69). The Iraqi uprising in 1920 was put down at vast cost, and about $100 million per year was being spent to garrison the country. After 1922 air power reduced that expenditure to $37.5 million a year (Hoare, 1957, 47).

Hoare could thus push air power on the grounds of economy, and the RAF survived the threat of total obliteration during budget cutting after World War I. Hoare's other responsibility was civil aviation. He demonstrated a clear grasp of the geopolitical problem.

> We had grown great on naval supremacy and the insular security that the command of the sea had given us . . . It became apparent . . . that if we were to hold our place in the new world, we must make use of the new element of the air for improving our Imperial communications . . . I knew nothing about the technical problems of air transport services, but as a Conservative who had been brought up in the days of Rudyard Kipling, Joseph Chamberlain and Milner, I saw in the creation of air routes the chance of uniting the scattered countries of the Empire. (Hoare, 1957, 12, 90)

Hoare had imperial vision but no technical education. At school at Harrow he kept to "cricket and the classics" (Hoare, 1957, 246). Although he claims to have sought out technical advice, his autobiography shows him full of the faults of his class and period. He listened to Moore-Brabazon, "my fellow old Harrovian" (Hoare, 1957, 246), but there is no mention of other advisors. In his account of the remarkable proving flight to India with Imperial Airways that he undertook starting in late December 1926, he names his companions in the cabin as well as the pilot and navigator, who were analogous to a ship's captain and first officer, but he merely lists a "wireless expert, an engineer from the Bristol Jupiter [engine] Company, and a photographer" (Hoare, 1957, 125–26).

This lack of integration between the new skilled technical class and the class that ran the empire produced blind spots. Hoare failed to understand the advice he was given in two crucial areas. First, his memoir claims that he had no qualms about airships and that their clear range advantage over heavier-than-air flight seduced him as much as others. "I could no more foretell the future of civil air transport in these early days of aviation than any of my advisors, and I felt it necessary to experiment in every likely direction" (Hoare, 1957, 221). Lord Thomson of Cardington, Hoare's successor as secretary of state for air and from the same social class as Hoare, albeit a member of the Labor party, paid with his life on R-101 for the same lack of qualms. Second, Hoare failed to understand how advances in aerodynamics and engines being made for national prestige could also be used to speed civil aviation. It is therefore ironic that Hoare strongly supported the successful British efforts to win the Schneider trophy for racing seaplanes. As he correctly points out, the technological lessons learned in these races led to the Supermarine Spitfire fighter, the tool that gave RAF pilots a critical edge in the Battle of Britain in 1941. But Hoare saw the Schneider victories only in terms of national prestige leading to military superiority, not in terms of commercial advantage (Hoare, 1957, 205–6). This replacement of a commercial focus, which the British had in large measure in the third and fourth world leadership cycles, with a political one clearly accounts for the failure of Britain in the fifth.

The four-engined, 18-ton Shorts Empire flying boats that entered service in 1936 carried twenty-four passengers and a ton of mail at 250 kilometers per hour over 1,050-kilometer legs (fig. 6-1). They cut the surface time from London to Capetown by two-thirds. As Imperial Airways publicity boasted, it was "Africa in days instead of weeks" (Wall, 1980, 162, 108). Shorts then turned its attention to a transatlantic plane. The Shorts G-class boats were half again as heavy as the Empire boats and, at 33 to 38 tons, not much lighter than Boeing's 314 (Gunston, 1980, 109, 131, 133). They never entered transatlantic service because of the outbreak of World War II. Like the Curtiss *America* in World War I, however, they provided the basis for a valuable, long-range, antisubmarine flying boat, the Shorts *Sunderland*.

By 1939, however, the star of the large flying boats was waning fast. The need to stress the hull to operate in rough seas made their structure heavy relative to carrying capacity. For all the advantages of needing almost no airport, loading and unloading passengers by small boats was an awkward operation, few bases being equipped with pontoon bridges such as those at La Guardia. Flying boats also had difficulty taking off in heavy seas. In winter the faster, northern route across the North Atlantic was unusable for this reason, and the route via Bermuda, the Azores, and Portugal had to be used instead. Finally, few large flying boats were built in any case. Three American companies built only twenty-five long-range boats and one British company built forty-five (Gunston, 1980, 86, 99, 109, 131, 133). Douglas built nearly six-hundred DC-2 and DC-3 landplanes in the same period (Gunston, 1980, 83, 107).

The development of long-range landplanes for strategic bombing and troop transport in World War II ended the flying-boat era. PanAm sold its flying boats in 1945, bought forty-five Douglas C54s (militarized DC-4s) for $90,000 each, and converted them back to commercial specifications (Beaty, 1976, 174). The Boeing 314 took $17\frac{1}{2}$ hours from

Junkers' "Europa Union" Air Routes, 1925

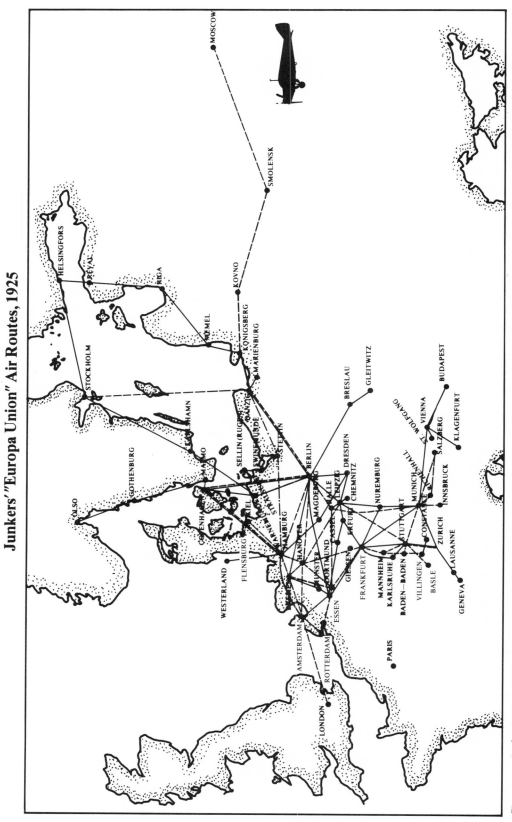

Figure 6-7 Junkers's Europa Union air routes, 1925. German dreams of Mitteleuropa did not die with the loss of World War I, and Junkers's map of the routes flown by his F-13s makes clear the commercial underpinning of such dreams. Redrawn from map in PRO.

San Francisco to Honolulu: by 1946 landplanes took 9¾ hours (All-ward, 1981, 127). Post—World War II air travel revolved increasingly around the twin concerns of speed and economy of operation. Flying boats were too large and cumbersome to offer either.

The Takeoff of the Landplane, 1919–1958

Despite the very real achievements of German landplanes in the late teens, the obsession with flying boats in both the United States and Britain overshadowed landplane development. German technology was not unknown but was ignored in practice. Nationalism ensured that even such vastly superior aircraft as the Junkers F13 could not thrive in an international market. Finally, a vast number of war surplus aircraft went onto the civil market at throwaway prices.

The Development of Civil Aviation in Germany and the U.S.S.R.

The collapse of the German war effort in 1918 did not end dreams of Mitteleuropa, in particular for Junkers. Commercial forces might still achieve the goal that had eluded military ones. Unlike Rohrbach and Dornier, Junkers was a businessman and a visionary as well as an engineer. He clearly understood the commercial implications of aviation better than any contemporary. He conceived a network of air services using Junkers airplanes throughout Europe and integrating local airlines into his proposed Europa Union (fig. 6-7) (PRO/AIR 10/1324).

Junkers's ambitions worried the British. In 1925 the British ambassador to Holland analyzed the Junkers-run World Traffic Company.

> Under the title of the *Europa Union,* the Junkers company, with its dependencies, aims at attaining to the hegemony of Europe in the domain of civil aviation. They believe, not without reason, that they will succeed in monopolizing the European system, and impose the use of their all-metal aeroplanes . . . It is not only for the capture of European traffic that the Junkers company has plans. They have sought, and will seek again, a concession in the Dutch East Indies, from whence they contemplate eventual service to Australia. The company has already some foothold in Persia, and thus they have a further potential link in the Middle and Far East chain. (PRO/FO371/10715, 10394, 10680)

Another British observer commented in 1927 that, "as regards the development of German aviation on international routes . . . Germany's geographical situation gives her a great advantage of which . . . she makes the most" (PRO/AVIA2/265). This advantage was conferred in part by the intersection of private and public funding of the route network. By 1927 Europa Union had been forged, but as the state-owned Deutsche Lufthansa, not as a commercial company owned by Junkers. Junkers's air transport companies had been on shaky financial ground in a Germany still staggering under the terms of the Treaty of Versailles. Before bankruptcy Junkers had paid about 70 percent of his running

Lufthansa, Deruluft, Eurasia "Europaisch-Asiatischer" Air Routes, 1932

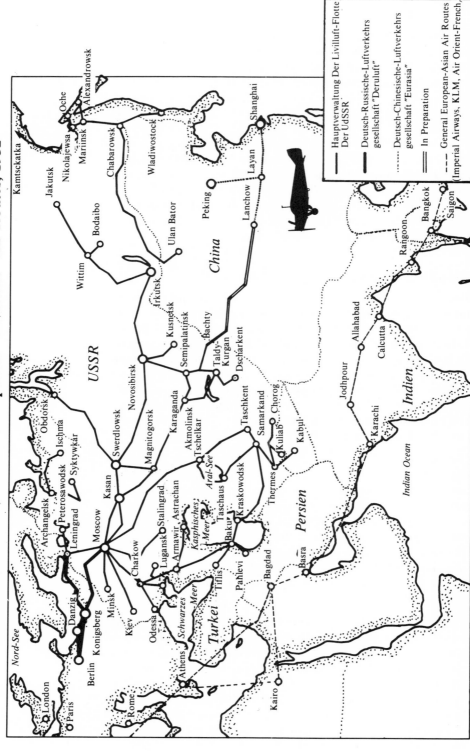

Figure 6-8 Lufthansa, Deruluft, Eurasia Europaisch-Asiatischer air routes, 1932. Junkers F-13s in the hands of these three airlines were capable of linking Europe to Asia across the Eurasian land mass by the early 1930s, although the connection across the Gobi Desert was never finalized. Western European airlines had to fly the much longer routes around the southern edge of that land mass because they could not overfly the Soviet Union. Redrawn from map in NASM.

costs from subsidies from the Reich, its states, and its cities, the last because of "local patriotism and inter-urban rivalry. The fact that few of the larger cities are willing to dispense with the prestige and convenience of a place in the *Lufthansa's* network evidently facilitates . . . extend[ing] that network into every corner of Germany" (PRO/AVIA2/265).

In 1921 Junkers signed a secret agreement to build F-13s in Russia. The production target was three-hundred machines a year, and Russian engineers were to be allowed to study Junkers's production techniques (Boyd, 1977, 9). British intelligence reported to the Air Ministry in 1927 the formation in 1926 of Deruluft (the Deutsche Russische Luftverkehrs gesellschaft), an airline jointly owned by German AeroLloyd and the Russian government. Deruluft monopolized the important Berlin-Moscow flight. Russian air activity pushed out from Europe over Asia. Demonstration flights were made to Beijing and Tokyo as early as 1925, and to Iran and Turkey the following year (PRO/AIR10/1322). Deruluft's Europaisch-Asiatischer route map of 1932 shows a remarkable net, complete with published fares and schedules (fig. 6-8). Deruluft linked at Moscow and Leningrad into a vast but tenuous Russian network and had proposed links to Eurasia, a German-run Chinese company flying scheduled service from Shanghai to Lanchow with a link to Beijing. The rise of the Nazi party in Germany, Stalin's consolidation of power in Russia, and the expansion of imperial Japan into China brought such pioneering ventures to an end.

Yet the vast reaches of Eurasia were clearly amenable to landplane development. Russia struggled mightily with the problem, forging an aircraft industry from scratch. Forging air-mindedness was another matter, however much the Communists pushed. The rest of the world recognized, as did Britain's secretary of state for air, Lord Thomson, that Russia's "wide and distant spaces offer an ideal field for aviation . . . If . . . Russia builds up a national air transport system with an adequate aircraft industry for her needs, so vast are her resources and so great her natural advantages, that in a few years time she may quite possibly be in a position to control the air routes over Asia to the East" (*Observer,* September 11, 1927).

In the short run the failure of Russia to develop an aircraft industry was predictable. As late as 1927 Duralumin could not be produced, most critical engine components and even whole airplanes still had to be imported, and German and Italian experts ran Russian factories (*Izvestia,* April 10, 1927, quoted in PRO/AIR10/1322). Such faults were not tolerated in Stalin's Russia, and the aircraft industry slowly matured, albeit on a much less commercial footing than in the Leninist period. Russian engineers trained at Junkers's Fili (Moscow) plant founded their own design bureaus, most notably Tupolev, Petylakov, Sukhoi, Arkhangelski, and Myasishchyev (Gunston, 1983, 10).

Russians were obsessed with large, heavy, long-range aircraft on the Junkers model through the late 1920s and early 1930s. Tupolev built such obvious copies that he was sued (unsuccessfully) by Junkers in 1926 (Gunston, 1983, 287). By 1931 Tupolev's ANT-6 four-engined bomber was the "first Soviet Aircraft to be ahead of [the] rest of [the] world" and comparable with planes entering service in Britain and the United States a decade later (Gunston, 1983, 290). This line of development ended for political, and not technological, reasons. Stalin purged those Russian officers who believed in strategic air power rather than blitzkrieg and refocused the efforts of Russian designers on short-range attack planes that provided little crossover technology for civil aviation.

In 1919 Junkers negotiated a license with John M. Larsen of New York to build the F-13 in America as the Junkers-Larsen JL-6 (Schmitt, 1988, 57–59). The JL-6 had a high-compression BMW engine that seemed the answer to high-altitude flight on the airmail routes proposed over the Rockies (Leary, 1985, 119). The Duralumin structure of the JL-6 was also believed impervious to fire in the air. Four crashes caused by fires started by fuel leakage, three of them fatal, proved this belief fallacious (Leary, 1985, 122–23, 126, 139). Junkers maintained in 1922 that Americans were using improper fuels, which caused the engine to misfire, the fuel feed to loosen, and the engine to overheat (NASM/AO277700).

The F-13 was the most successful civil airplane before the Douglas DC-2. Between 1919 and 1929 a total of 314 F-13s were built. France,

Junkers W33 "Bremen." The W33 was a slightly enlarged and upgraded F13, Junkers's radical all-metal design of 1919, the first really successful commercial airplane. In 1928 "Bremen" made the first crossing of the Atlantic from Europe to America, a more difficult feat than flying the other way because of the constant winds from west to east. Ford Museum, by the author.

Belgium, Britain, and Japan all received F-13s as war reparations (Schmitt, 1988, 66). Junkers sold 29 F-13s in complete airline packages to several Latin American nations, and it was largely to drive out Junkers that PanAm became the "chosen instrument" of the U.S. government after 1931.

The United States Experiments with Airmail

U.S. interest in long-distance air services began at about the same time as that of Junkers but focused exclusively on airmail and made no attempt to carry passengers. Junkers's prescient mix of private and various levels of public finance was initially wanting because the U.S. airmail was almost all flown by federal subsidy. The availability as war surplus of thousands of U.S.-built British de Haviland DH-4 day bombers, built for shipment to the European front in 1919, drove equipment procurement. The U.S. government simply transfered a hundred DH-4s, their Liberty motors and spares, from the war department to the post office on November 15, 1918, four days after the armistice (Leary, 1985, 66).

At first sight the DH-4 was adequate. Shorn of bomb load, protective machine guns, and observer it had room for extra fuel and 180 kilograms of mail. With stops for fuel every few hundred kilometers a transcontinental service was clearly within its capabilities (Bilstein, 1983, 32). Its inadequacies were more obvious in service. Because war planes in 1918 lasted only hours in actual service, they were not built to last. Reliable high-power engines did not exist, and although the Liberty engine was no worse than its contemporaries, engine failure in a single-engined plane was eventful. The main attraction of airmail was commercial. Banks saw the advantage quickly. Checks and other items in process of collection represented a "float" of idle money that was losing interest as long as it was in transit. The twenty-one leading banks in New York saved just over 0.1 percent of all such transfers by using airmail. Overnight airmail between New York and Chicago saved a full day's interest. The thirty-two-hour service between New York and San Francisco begun in 1924 was a great improvement over the four and a half days taken by crack mail trains (Bilstein, 1983, 33, 35, 41). The price paid in the loss of aircraft and human lives was high: aviation technology in the early 1920s had not confronted the problems of flight at night or in bad weather. In the first six years 3,850 forced landings were made and twenty-nine pilots killed, but just over 255 million letters were carried (Leary, 1985, 249, 253–56). After complaints from banker J. W. Harriman that over $63,000 in checks and drafts had been lost to fire in the airmail's second fatal crash in May 1919, mail planes were fitted with asbestos mail compartments (Leary, 1985, 84). After a crash, mail could simply be forwarded by train: pilots and airplanes were not always so durable. The four bugbears of this pioneering system were lack of reliable airframes, low vehicle speed, and lack of reliable engines and navigational systems. Solutions to the first two had to wait, respectively, until the mid-1920s and early 1930s. In the meantime mail delivery was a reasonable way to pioneer an air system. It created a commercial basis for air traffic as well as the beginnings of an infrastructure of landing fields and a cadre of experienced pilots.

The first problem, that of airframe unreliability, was solved by the metal structures developed in Germany. That it took so long for American designers to adopt them must be put down to cultural conservatism (the Not Invented Here or NIH syndrome), the drastic reduction of government spending after World War I, and the cheapness of war-surplus airplanes. There were also lingering doubts about corrosion in Duralumin. Despite positive British reports based on experience with captured German airplanes (PRO/AIR10/494), the U.S. Navy refused to endorse aluminum for salt water use until the anodizing process for aluminum was discovered in 1926 and the Alclad process was perfected that same year to coat aluminum for airframe use (Howard S. Wolko, personal communication, June 1988). Since the military was about the only customer for airplanes in the 1920s the conservatism of the U.S. Navy affected the entire industry.

The second problem, low vehicle speed, was also solved by German work. Rohrbach pioneered the high wing loadings needed for high-speed flight as well as stressed skin structures (*Flight*, obituary, July 13, 1939). The pioneer British theoretician of flight, Frederick Lanchester, pointed out that "the dead weight of wings, fuselage etcetera amounts to a constantly increasing part of the total . . . and . . . there exists, in consequence, a limit beyond which the wings, etcetera grow so heavy that there remains no useful (available) load" (Rohrbach, NASM/CO523900).

Lanchester's work was best understood in Germany before World War I, where it was refined by Ludwig Prandtl, the "founder of the science of aerodynamics" (Wolko, 1981, 32). Rohrbach recognized that Lanchester's comments referred to airplanes with low wing loadings and that Lanchester's limit could be postponed by the use of high wing loadings. Higher wing loadings would make airplanes "cheaper in initial cost and running work, . . . less sensitive in squalls and winds . . . , superior in general robustness . . . and . . . [able to] carry considerably more useful load with a noticeably higher speed, than "old" type large planes of the same weight" (Rohrbach, NASM/CO523900). Rohrbach's E.4/20 of 1919 was recently calculated to have been as inexpensive to operate as a 1934 DC-2 on trips up to 400 kilometers and only 20 percent higher between 400 and 800 kilometers (Brooks, 1986, 231). High wing loadings, however, require very long takeoff runs. It was for this reason that Rohrbach abandoned landplanes for flying boats after World War I.

Other designers sought a technical solution in the shape of devices to generate higher lift. A focus for the development of these was the Guggenheim International Safe Aircraft competition of 1929. The leading-edge slot required on all British military airplanes by the Air Ministry in 1928 (Gibbs-Smith, 1970, 192) first showed its paces in public at the

Guggenheim competition. A second high-lift device, the flap, came into common use on U.S. landplanes of the early 1930s, notably the DC-3.

The third problem, that of engine reliability, was tackled and solved in the early 1920s by U.S. designers. By 1918 the potential advantages of the air-cooled radial engine were appreciated over the water-cooled in-line and air-cooled rotary engines that powered World War I planes. British designers produced the ABC radial in 1917 but were unable to resolve persistent cooling problems in an otherwise promising design. Bristol eventually produced workable British radials in the mid-20s, but at high cost. In the United States Charles Lawrance found the key to success in aluminum cylinders with steel liners (Taylor, 1971, 41–42). By 1921 the U.S. Navy had stated it would order only aircraft powered by air-cooled radials. Lawrance helped found the Wright Engine Company, which developed his design into the Cyclone, the first really reliable, light, simple aircraft engine. Lindbergh's nonstop New York–Paris flight of 1927 used a Cyclone-powered Ryan airplane. The other great U.S. company to produce radial engines, Pratt & Whitney, was set up by Wright's chief engineer (Taylor, 1971, 45–46). By the onset of World War II every major radial engine in the world, whether in Britain, the United States, France, Germany, Italy, Japan, or Russia, could trace its descent to Wright, Pratt & Whitney, or Bristol engines of the mid-1920s, most through formal license agreements (Smith, 1981; Gunston, 1989).

The fourth and final key to commercial aviation was navigation. Airmail experiments with such ground beacons as searchlights had worked well in clear weather, but the airmail was an all-weather operation. The radio beacon was patented in Germany in 1907, but it took until 1927 for U.S. engineers to work it out in practice (Hallion, 1977, 103–6). It was immediately successful in airmail use.

These four developments set the stage for the remarkable growth of U.S. civil aviation in the 1930s. Since the United States was clearly behind Germany and even Britain in the race for civil aviation in the early 1920s, its success needs accounting for. The slowness with which it adopted metal construction and high-lift devices suggests that the United States suffered more than other countries from the NIH syndrome in this period. Airmail was the key, proving that a commercial civil aviation was feasible even at a low technical level. Such an airmail could have been just as easily developed in Europe, and Europa Union operated a more imposing network in 1925 than the U.S. airmail. Lack of commercial and political integration brought Europa Union to a halt, especially in its dreams of links to Russia and the East. Even so, had a German victory in World War I brought about Mitteleuropa, German aviation would have had commercial and political integration. Certainly Junkers, Dornier, and Rohrbach could have delivered the airplanes. Americans learned, albeit slowly, from German technology.

The key to U.S. success was the establishment of the Morrow commission in mid-September, 1925, as a result of charges by Colonel William ("Billy") Mitchell that the crash of the U.S. navy dirigible *Shenandoah* on September 3, 1925, was caused by gross incompetence in the military establishment. The Morrow commission, chaired by financier Dwight Morrow, dismissed most of Mitchell's charges, and Mitchell was court-martialed. But the commission also recommended far-reaching changes in military aviation as well as the establishment of the Bureau of Civil Aviation in the Department of Commerce (Hallion, 1977, 14). The establishment of the Guggenheim chairs in aviation technology at major U.S. universities helped bring U.S. theoretical understanding in the late 1920s up to that of Germany in the early teens. In 1929 Theodore von Karman, perhaps Prandtl's greatest student, was appointed to such a chair at Caltech in Pasadena and brought American aerodynamics up to date (Hanle, 1982, 120–31). German structural engineering also passed into U.S. practice through a series of technical memoranda translated by the National Advisory Committee for Aeronautics (NACA). Radial engines developed faster and better in the United States than in other polities.

Yet U.S. success was also European failure. Junkers saw the future with amazing prescience in 1921 when he visited the British commercial secretary in Berlin to protest the destruction of the Zeppelin-Staaken E.4/20 (fig. 6-1).

> Professor Junkers . . . showed very great bitterness and resentment at the allies' demands. He is evidently convinced that it is trade rivalry which has produced the present situation even more than any apprehensions as to national security . . . He considers that the Allies and particularly England will have cause to regret the action they are now taking, as it is certain to impede their own aerial development which in a small continent like Europe is entirely dependent upon close co-operation between all countries. Professor Junkers thinks that the situation created in aircraft construction by the Ultimatum will be largely to the benefit of the Americans who have ample space to develop their civil aviation on a large scale and will now be able to do so without serious European competition. (PRO/FO371/5877)

In the light of such prescience it is ironic that the Americans had withdrawn from the Inter-Allied War Commission, leaving the British to do their work for them.

The Success of the American Twin-engined Airliner

The success of the U.S. aircraft industry in the early 1930s, so accurately forecast by Junkers, is a well-known story. It hinged on adoption of the high wing loadings and stressed-skin construction pioneered by Rohrbach and of British technologies for high-lift wings; improvements in radial engines and drag reduction; and development of high-lift flap systems. By 1930 three advanced U.S. airplanes had appeared, all of which were to leave a major mark on civil aviation: Lockheed's *Orion* used an efficient wooden structure with a plywood skin but marked the limit of wood construction; the Boeing 200 *Monomail* was most true to Rohrbach's principles; Northrop's *Alpha* was genuinely innovative. The *Orion, Monomail,* and *Alpha* had a huge influence elsewhere in the world.

Plywood structures like that of the Lockheed *Orion* were inferior to Duralumin only in life expectancy and were in use for combat airplanes in Russia and Britain throughout World War II. The *Orion* entered Swissair service between Zurich, Munich, and Vienna in 1932. Lufthansa demanded from the reborn German aircraft industry a plane with better performance and received it in the metal-structured Heinkel He-70 (Gunston, 1980, 78). The He-70, in particular its efficient and graceful elliptical wing, provided the basis for Heinkel's He-111 twin-engined airliner. The He-111 was then developed as a bomber to become the mainstay of the German bomber force of World War II. Reginald Mitchell's brilliant *Spitfire* used the same wing design to knock innumerable He-111s out of the air and deny the Germans victory in the Battle of Britain (Nowarra, 1980, 18).

Boeing's *Monomail* led to genuine U.S. dominance in air transport in the shape of the Boeing 247 twin-engined airliner of 1933. As with the Heinkel, the line to a major bombing airplane is clear. Lessons learned from the 247 were instrumental in the success of Boeing's great World War II four-engined bomber, the B-17 *Flying Fortress.* Seventy-five 247s were built and a standard 247 came third in the 1934 MacRobertson air race between Britain and Australia. The winner of the MacRobertson was a British de Haviland 88 *Comet,* an all-out racing machine, and second place went to another U.S. airliner, a DC-2 flown by the Dutch airline KLM and carrying fare-paying passengers.

The Northrop *Alpha* used a genuinely innovative multicellular wing structure, which showed its true mettle when Northrop moved to Douglas to work on the DC-1, DC-2, and DC-3. Transcontinental and Western Airlines (TWA), unable to buy Boeing 247s because production for the foreseeable future was earmarked for United Airlines, contacted the other leading and not-so-leading U.S. manufacturers. Douglas, a relatively small company, proposed the Douglas Commercial 1 to carry twelve passengers faster, farther, and more comfortably than Boeing's 247. The DC-1 was stretched slightly to carry fourteen passengers, four more than the Boeing, as the DC-2. Wing flaps made landing and takeoff far easier than in the Boeing, and it had a huge impact. In eight days in 1934 TWA DC-2s "broke the speed record from New York (Newark) to Chicago four times, and virtually chased the Boeings off the route, knocking half an hour off the 247's $5\frac{1}{2}$ hr flight time" (Davies, 1982, 186). By late 1934 TWA was scheduling Chicago nonstop and Los Angeles in only eighteen hours from New York, stopping at Chicago, Kansas City, and Albuquerque.

American Airlines had the longest transcontinental route of the major U.S. domestic airlines of the 1930s (fig. 6-9). It requested a wider version of the DC-2 to accommodate sleeping berths. Yet the DC-3 (fig. 6-1) was more than just an enlarged DC-2. Wind-tunnel testing at Caltech under the direction of von Karman considerably refined its aerodynamics. In fact the Guggenheim chairs paid off handsomely in the development of the DC-1, -2, and -3, because each change was carefully tested at Caltech before being put into production (Hallion, 1977, 194–97; von Karman & Edson, 1967, 168–72). For the first time

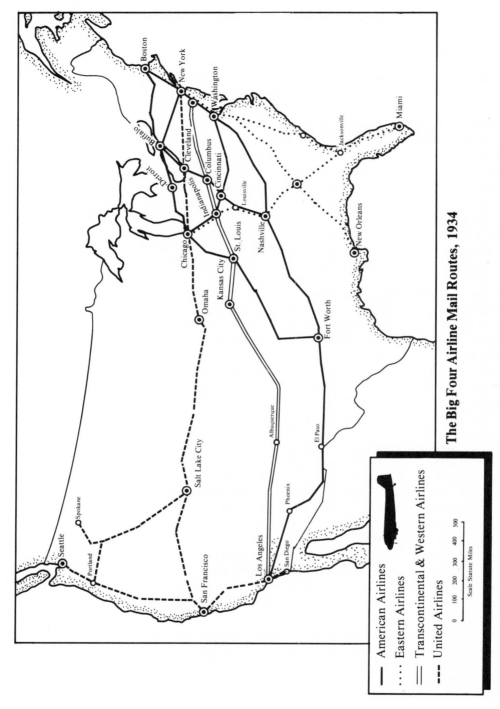

The Big Four Airline Mail Routes, 1934

Figure 6-9 The "Big Four" airline mail routes in the United States, 1934. Improved airliners, the Douglas DC-3 in particular, made possible rapid movement of business executives and the emergence of a truly national U.S. economy in the late 1930s. Redrawn from Davies 1972.

American Airlines
Eastern Airlines
Transcontinental & Western Airlines
United Airlines

Scale Statute Miles

0   100   200   300   400   500

the German model of interaction between universities and large corporations was paying off in the United States.

The DC-3 was little faster than the DC-2. More powerful engines just compensated for greater weight. Its advantage was in productivity: it carried twenty-one passengers with only three crew. Good aerodynamics allowed "a 50 percent increase in payload . . . for only a 10 percent increase in operating costs" (Davies, 1982, 191). The U.S. airmail service demonstrated that numerous problems remained to be solved if loss of life and equipment was to be reduced. Better engines, airplanes, and blind-flying technology were the keys to safe, reliable, all-weather service, and by the end of the 1920s they were in sight. These assets allowed explosive growth of civil aviation after 1930, because more powerful engines in more aerodynamically efficient and stronger airplanes allowed passengers to be carried as well as mail. By 1937 Douglas's DC-3 allowed airlines to "escape from dependence on mail payments to make up the difference between operating costs and passenger revenue" (Davies, 1982, 191). By 1938 DC-3s carried 95 percent of the United States's burgeoning air traffic (Hallion, 1977, 197–98), and in almost complete safety. In a twelve-month period in 1939–40, when some 80 percent of commercial flights were in DC-3s, not a single life was lost in commercial operations (Davies, 1982, 193).

The DC-2 and -3 proved that high speed through aerodynamic refinement was commercially sound. After much huffing and puffing even the British began to concede that speed had advantages, although Imperial Airways and too many British constructors remained convinced that space and luxury were more significant on the long empire runs. KLM's adoption of the DC-2 proved how wrong they were. KLM flew DC-2s over essentially the same routes to Asia as Imperial Airways (figs. 6-8, 6-10). The DC-2's second place in 1934's MacRobertson air race should have made Imperial Airways realize their error, but it did not. By the late 1930s the Association of British Malaya was complaining that Imperial Airways was taking 1.5 days longer to fly mail from Singapore to London than if KLM were used via Batavia and Amsterdam (PRO/ AIR2/1302).

The Development of the Four-Engined Airliner

The remarkable success of the twin-engined planes of the early 1930s brought on their second-generation successors of the late 1930s. The DC-3's shortcomings, in particular the fact that it took sixteen to twenty hours to cross the United States, prompted investment in aircraft to fly farther and higher, thus faster and "over the weather." Without radical increases in engine power, which seemed unlikely in the 1930s except to visionaries of jet power, only four-engined airplanes offered the possibility of getting enough fuel off the ground to cross the American continent nonstop with a paying load of passengers. Boeing's model 307 *Stratoliner* of 1938 offered the first pressurized cabin in a production airliner, allowing it to fly at fourteen thousand feet, over most of the

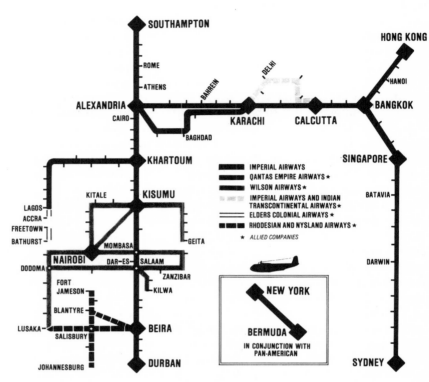

THE LINK OF EMPIRE

Figure 6-10 Imperial Airways route network, c. 1938. Britain attempted to use air transport in the form of such planes as the Shorts Empire flying boats to politically administer the empire forged by sea trade. Redrawn from diagram in NASM.

lower air turbulence (Davies, 1982, 205). It reduced coast-to-coast time to fourteen and a half hours, although stops were still needed. Boeing's *Stratoliner* was eclipsed by the war and the new planes it spawned. In 1946 Lockheed's *Constellation* began eleven-hour service between New York and Los Angeles (fig. 6-1). By 1953 improved Constellations did the run in less than eight hours.

Nonstop service across the American continent also meant the possibility of transatlantic service. The Atlantic gap was finally closed in World War II by the construction of airports at Gander in Newfoundland and Shannon in Ireland. These made it possible to ferry four-engined bombers and transports from the United States to Britain. After the war aircraft such as the Douglas DC-4 used these airports to cross from New York to London in fifteen hours (fig. 6-1). Not until 1957 did a reciprocating-engined plane appear that could fly nonstop from New York east to London but also from London west to New York against substantial head winds. Once again it was a Lockheed *Constellation* that achieved this, the *Starliner* version. London to New York times fell below eleven hours. Air traffic mileage grew almost in direct proportion to speed and convenience (table 6-1). Of the just under two million passengers who crossed the Atlantic in 1957 over a million went by air. For the first time air traffic exceeded sea traffic. A decade later the ocean greyhounds were on their last legs.

The ten years or so after the end of World War II thus saw established an essentially global network of profitable airline services. This greatly accelerated the development of a global economy because mail and high-level personnel could be moved almost anywhere in the world in twenty-four to thirty-six hours, albeit at some cost. Although Ford was a multinational corporation as early as 1913 and other companies, notably General Motors, had quickly followed suit, overseas divisions had considerable autonomy in decision making. Key personnel were rarely exchanged: Ford of England was run as an almost independent company, and Henry Ford treated Percival Perry, who ran Ford of England, as a partner, not an employee. General Motors afforded remarkable latitude to its nominal subsidiaries, Vauxhall in England and Opel in Germany. Four-engined, pressurized aircraft made it possible to change that and control overseas operations directly from a single corporate headquarters. Ford-Detroit took back Dagenham's nominal control of European operations in 1948, "the first step in the resumption of the American company's control over its foreign operations" (Wilkins & Hill, 1964, 366). By the 1950s U.S.-trained engineers and stylists were arriving at the European subsidiaries of Ford and General Motors in large numbers and were having a disproportionate influence on production. By 1961 Ford-Dagenham was entirely controlled by Ford-Detroit. Although some European corporations also became multinationals during the 1950s, the period was one of marked U.S. hegemony, and most multinationals were American.

Table 6-1   Passengers crossing the Atlantic on U.S. flag airlines, 1956–1988

Year	Passengers	Comments
	thousands	
1956	536	
1958	646	First jet service
1960	817	
1962	873	
1964	1,265	
1966	1,799	
1968	2,539	
1970	4,084	First jumbo jet
1972	5,376	
1974	4,820	OPEC oil embargo
1975	4,532	
1980	5,491	
1984	6,618	
1988	9,923	

Source: FAA Statistical Handbook.
U.S. flag airlines carried from 39 to 62 percent of total passengers in years for which data are available (1956–75), averaging some 55 percent.

Aviation also proved its worth in military and quasi-military operations. Troops could be rushed to sensitive areas in a matter of days rather than weeks. The Berlin airlift of 1948–49 demonstrated that large amounts of material could be moved by air when necessary. More than 2 million tons of goods were delivered, some two-thirds of it coal. Aviation also allowed totally artificial environments to be created by commercial companies abroad. Companies like Aramco, the Arab American Oil Company, maintained oases of U.S.-style life in the desert of Arabian Islamic culture. Middle Eastern oil was very much cheaper than U.S. oil, but its extraction needed large numbers of American managers and technicians. Moslems were hostile to such Western cultural attitudes as the consumption of liquor and the relative equality afforded women. Aramco therefore sequestered its personnel on bases in military-style seclusion. Serviced with food and entertainment by air, and with personnel routinely repatriated for rest or the education of children, the oil companies achieved a style of colonialism only dreamed of a few generations previously. Even at the height of the British presence in India, the British had never been totally insulated from Indian culture. Britain was months away by sea and, perforce, Indian foods and living styles had to be at least somewhat adopted. Long-range aircraft removed such a necessity for Americans living in a radically different culture.

As in the early 1930s Douglas led, Boeing and Lockheed fought hard for a place in the market, and the European manufacturers struggled to catch up, with only the Germans keeping close. The first DC-4, the DC-4E (for "experimental") of 1938 was a mistake. It emphasized luxury through voluminous passenger accommodations rather than speed, and its operating economics were bad. The single plane built was sold to Japan in 1939, where it was used as the basis of an equally unsuccessful long-range bomber (Francillon, 1970, 423). The second DC-4 of 1942 had both transcontinental and transatlantic capabilities, and 1,242 were built, 1,163 of them military. A "large number" were "civilianized" in 1945 (Davies, 1982, 325). The DC-4 was not only a success in the United States, but it completed U.S. commercial hegemony of the world civil aviation market by providing a landplane with transatlantic range that was profitable carrying passengers alone. Despite attempts to produce local equivalents of the DC-4, European nations reequipped their airlines after World War II with the DC-4 and its derivatives, the DC-6 and DC-7, of which 875 were sold, or with competing planes from Lockheed and Boeing. No European producer could match U.S. airplanes in the postwar years.

Three competing U.S. companies deserve brief mention—Boeing, Lockheed, and Republic, a new company that had risen to prominence in World War II. Boeing developed its four-engined B-17 bomber into a civil airliner in the model 307 *Stratoliner* of 1938. This plane lacked transcontinental range but pioneered cabin pressurization on large airplanes (Gunston, 1980, 129). Only ten Stratoliners were sold. After World War II Boeing modified its four-engined B-29 bomber to make an airplane with transcontinental and transatlantic range, the 1944 model

377 *Stratocruiser.* Although PanAm was attracted to this airplane by its range and luxurious accommodations, and even canceled an order for DC-7s in Boeing's favor, it was hard to maintain, unprofitable, and ugly. Only fifty-five were sold (Gunston, 1980, 144–45).

Lockheed's L-049 *Constellation* of 1943 was as beautiful as the *Stratocruiser* was ugly but suffered protracted development problems. Many airlines bought the *Constellation* for its pressurized cabin, speed, and range, as well as for its beauty. Eventually Lockheed sold 556 Constellations, Super-Constellations, and Starliners, the last being the most beautiful piston-engined airliner ever built (Gunston, 1980, 171, 192). Beauty was not enough once the much faster jets appeared in 1959. In any case Douglas consistently beat Lockheed to the post with more workmanlike, less expensive, and more profitable planes. Four-engined planes such as the DC-4 were produced in much greater quantities than the *Constellation,* and the improved DC-6 was a much more profitable aircraft than the Lockheed, but it was the *Constellation* that set speed, range, and altitude standards after World War II (Davies, 1982, 325–34). The first Douglas design to match the *Constellation* in speed was the DC-7 of 1953, but not until the DC-7C of 1957 did Douglas catch Lockheed in range (Gunston, 1980, 171, 188).

Republic had no airliner experience but an excellent design team that pushed the performance of piston-engined airliners to the limit. Its elegant, highly streamlined *Rainbow* was designed to reach Los Angeles nonstop in eight hours from New York and London nonstop in nine and a half hours. PanAm and American Airlines placed orders on this basis. Republic intended to defray development costs by selling twenty to the United States Air Force for long-range reconnaissance, but the greater promise of jet aviation and postwar budget cutting ended the contract in 1947. American Airlines decided operating costs would be too high and dropped out, forcing PanAm, which wanted a high-speed plane for the Atlantic, to follow suit (Martin, 1969). Republic's experience is typical of many innovators. The *Rainbow* was a superb airliner, but it represented the maximum extension of a technology. Had jets not been on the horizon, its considerable increase in performance relative to even the *Constellation* would have been sought after. Unlike the DC-4 and the *Constellation,* it also came too late in the war for its development to be paid for by the taxpayer.

A good case can be made that the DC-7C and the *Starliner* had extended piston-engined aircraft technology to, if not slightly beyond, its limits. The *Rainbow* almost certainly had. Although neither of the first two airplanes proved dangerous to their passengers, the Wright turbocompound engines they used to achieve their performance were too unreliable for a commercial airliner, even if the turbocompound did fly 20 percent farther on the same amount of fuel (Setright, 1975, 6). Too much time was spent in servicing, and too many flights were delayed by mechanical problems. The trouble-free performance of jets was welcome.

European challenges to U.S. hegemony of civil aviation faded to almost nothing by the late 1930s. Many European makers emulated the DC-3, and licenses were bought in Russia and Japan to build both the DC-2 and -3. Europeans might have mounted a challenge in long-range, four-engined designs, as shown by the performance of the German Focke-Wulf 200 *Condor*. In 1938 Condors flew from Berlin to Tokyo in just over forty-six hours, and nonstop from Berlin to New York in twenty-five hours (Gunston, 1980, 124). In the event the main success of the *Condor* was as a commerce raider in the battle of the Atlantic, and no other European aircraft came close to U.S. airliner performance levels through the rest of the piston-engined era.

Attempts to develop the very successful British four-engined strategic bombers of World War II into airliners, or to design new planes to compete with U.S. airliners, were dismal failures. Despite government policy that the British Overseas Airways Corporation "buy British," BOAC canceled its order for the problem-ridden Avro *Tudor* in 1947 in favor of proven U.S. designs (Gunston, 1980, 150), pointing up the quite remarkable U.S. superiority by 1945 in airframe design, product development, and engine reliability.

In the 1930s propeller-driven landplanes provided the basis for the thorough integration of continental economies. They did so by greatly speeding air mail and high-status passengers. Airlines in the United States developed as tools of business. Had Mitteleuropa come about, they might also have been important as a tool of centralized, bureaucratic governments. In the United States's market economy this function was less noticeable, if still present. In a British Empire that had lost sight of its commercial origins, the bureaucratic function was uppermost, although the air mail clearly served commercial needs. The British Empire was, in any case, too large for propeller airplanes. The United States was small enough that propeller planes put anywhere in the country within a day's reach of anywhere else. Even the DC-3 allowed a businessman to leave New York in the early morning and do half a day's work that same day in Chicago. A further half day's work and a flight back to New York the next day meant less time away from the office compared to trains, which required sixteen hours each way, thus three days away from the office instead of two to achieve one day's work in the field. Put simply, airplane travel increased the productivity of managerial labor. Even in the 1920s Standard Oil of Indiana reckoned its executive Ford *Trimotor* saved $16,000 a year in executive travel time. The train took thirty-eight hours from Chicago to Standard's Casper, Wyoming, base. The *Trimotor* took fourteen hours. If four executives, each worth at least $100 a day by Standard's calculation, flew to Casper and back, the company saved two working days for each, or $800 (Bilstein, 1983, 71).

Civil aviation also helped unify the U.S. economy. Airplanes carried not only bank deposits by mail but also advertising plates for use by local newspapers and movies for local theaters. Together with radio broadcasting these plates created a national mass market in place of a series of regional markets. Movies began to mold tastes without necessarily intending to, presenting the image of the "good life" in southern

California. Bizarre incidents such as the drop in undershirt sales when Clark Gable revealed he was not wearing one in *It Happened One Night* (1934) only serve to illustrate this persuasive, understated power.

Deliberate manipulation of tastes came through advertising. National radio broadcasting evolved in the United States around the sponsored show. Sponsors used better highways and trucks to deliver their goods nationally, local advertising campaigns managed from national headquarters by telephone, and airmail delivery of newspaper plates. Thus, continental economies of scale emerged in America in the 1930s. No other integrated, continent-sized, market economy was possible at that time. Continental economies of scale, achieved first in America, along with managerial innovations on the production side, do much to explain the United States's achievement of commercial hegemony in the fifth world leadership cycle, which was finally established with transatlantic airplanes after World War II.

THE TRIUMPH OF THE JET, 1958 TO THE PRESENT

The process of global spatial integration begun by four-engined, pressurized aircraft after World War II has been radically accelerated since 1958. That year marked the successful onset of the Jet Age, with the first commercial flights of Boeing's 707 both across the Atlantic and across the North American continent. Passenger traffic growth on the Atlantic accelerated hard after 1958 and did not slow until the oil price shock of 1973–74 (table 6-1). New York–London schedules fell to seven hours, and comfort improved substantially. Transatlantic flight in a reciprocating-engined airliner left the feeling, as my father remarked in 1959, that one had been in a cement mixer for thirteen hours. In jet planes lack of vibration allowed you to balance a penny on edge, although a new form of passenger fatigue entered the language as "jet lag." What the jet offered was a much greater facility for mixing people, ideas, and economies. Juan Trippe, the mastermind behind the growth of Pan American World Airways, once claimed that the postwar world was a race between air travel and the atom bomb (quoted in Beaty, 1976, 200). Jet planes have remarkably improved humanity's chance to win that race.

Inventing the Turbojet

Thirty years before the 707 Frank Whittle had laid out the problem of high-altitude, high-speed flight. Whittle's 1928 thesis, with which he graduated from the British RAF College, stated that "for it to be possible to fly fast and far it would be necessary to fly at very great heights— far higher than was possible with the then conventional piston engine and propeller" (Whittle, 1979, 3). In 1929 Whittle proposed the gas turbine, and in 1932 he patented the idea. Even though he was a serving RAF officer, the patent was not put on the secret list. Progress was accelerated by World War II. For military use the limiting factors of early jets, poor range and poor fuel economy, were far less important than the advantages of speed and altitude performance.

In Germany Hans von Ohain began speculations on alternatives to reciprocating engines based on an "intuition that a continuous aerothermodynamic propulsion process could be inherently more powerful, smoother, lighter and more compatible with the aerovehicle than a propeller-piston engine" (von Ohain, 1979, 29). Von Ohain knew nothing of Whittle's work, but had a solid academic grounding at the University of Göttingen, where Prandtl had developed aerodynamic theory. Despite his later start, von Ohain obtained industrial backing more quickly than Whittle, in particular from Ernst Heinkel. Heinkel wanted to diversify from airplanes into engine manufacture, but the tooling costs for reciprocating engines were high. Von Ohain's engine, simple and built mostly from sheet steel, seemed ideal (Constant, 1980, 199). The German Air Ministry preferred to let jet-engine contracts to more established engine builders such as Junkers and BMW, although Daimler-Benz refused them. Ernst Heinkel's notorious independence dissuaded them from pursuing his engines.

Ed Constant, in his fine book, *The Origins of the Turbojet Revolution* (1980), points out that the idea of the turbojet developed faster in Germany for many reasons. Most notable was the strong theoretical base of German aerodynamic science, which led three people rather than one, as in Britain, to the "turbojet conclusion." Von Ohain may have been first, but Herbert Wagner of Junkers and Helmut Schelp of the German Air Ministry were not far behind. Wagner was educated at the Technische Hochschule in Berlin and worked for Rohrbach; he greatly improved Rohrbach's stressed-skin structures before being appointed professor of aeronautics at Danzig from 1927 to 1930 and at Berlin from 1930 to 1938. He joined Junkers in 1935 and became its head of development in 1937, producing a viable jet engine design by 1939. Schelp was German but educated in the United States. He returned to Germany in 1936 to work first for Daimler-Benz, then for the Deutsche Versuchsanstalt fur Luftfortforschung (DVL), the official German aerodynamic research institute. DVL had come to the conclusion by 1937 that the limiting factor on speed was now engine design, airframes like Heinkel's He-70 and Messerschmitt's new Bf-109 having stunning performance on unexceptional engines.

Whittle, von Ohain, Wagner, and Schelp all reached the same basic conclusion independently "on the basis of [a] presumptive anomaly deduced from theoretical aerodynamics" (Constant, 1980, 207). The conclusion was not that propeller-driven airplanes were unsatisfactory, just that they could be presumed to become unsatisfactory in the near future, airframe aerodynamics having improved so much that high-speed, long-range flight needed more powerful engines. German advantages were twofold: German engineers had the best technical education in the world; and Germany lacked good reciprocating engines because of the fifteen-year hiatus in engine development enforced by the Treaty of Versailles of 1919 (Constant, 1980, 198).

The United States was left behind in the drive for the jet because of a primary interest in civil aviation. Britain, Germany, and to a lesser extent Italy concentrated on speed and national prestige (Constant, 1980, 151). Yet Constant somewhat underestimates the importance of U.S. and British obsessions with the range of civil airplanes in leading toward the turbojet conclusion. Americans investigated high-altitude flight thoroughly in the 1930s, both theoretically and empirically. Where Americans failed was in not deducing from first principles that continued gains in cruising altitude and speed could only come from higher propulsive efficiencies than propellers could produce.

Britain's obsession with range is clear from Whittle's work. His first company, Power Jets of 1935, was to build a power plant for a fast, high-altitude, transatlantic mail plane (Constant, 1980, 188). Such a mail plane had obvious uses in the rest of the world, in particular throughout Britain's global empire. The concern with transatlantic range only emphasizes the extent to which Whittle was concerned first and foremost with flight at high altitudes, where air was thin and drag was low, rather than with speed in itself.

Diffusion of the Turbojet

Once the turbojet revolution was established, the normal processes of diffusion got under way. Under the pressure of impending war General Electric accepted a U.S. Army contract to develop Whittle's engine in the fall of 1941 (Constant, 1980, 218). General Electric was a sensible choice with considerable experience in the design of closely related technologies and no vested interests in reciprocating engines. Just as British technology was given to the United States as part of the war effort, German technology was given to Japan. Only one Japanese jet flew in World War II, a smaller version of the German Messerschmitt Me-262 jet fighter that entered operational service in 1944. Despite lack of factory drawings, the Japanese were able to develop a working turbojet from photographs of the German BMW 003 (Francillon, 1970, 443).

After World War II diffusion was even more rapid. The Allies removed large volumes of German scientific information, much of it about aeronautics. German engineers were also removed to Russia, the United States, and France. In the United States engine development was satisfactorily established in the early war years, first with British technology, then with rapidly increasing local input. U.S. gains from captured German data and engineers thus came more in the realm of high-speed aerodynamics. U.S. planes were the first to use the swept and delta-shaped wings suggested by German research.

Russia acquired considerable amounts of German technology and experts, but German engines proved inadequate for Russian needs. The Russian jet engine industry was thus greatly helped by its acquisition of British Rolls-Royce Nene engines in 1946 (Gunston, 1983, 13, 174). Britain's action, strange in retrospect, can only be put down to the Labor government's attitude of close cooperation with Russia. It is perhaps understandable only in the light of concern that the United States would, as in 1919, retreat into isolationism. In 1946 neither the Marshall Plan nor the Berlin airlift had demonstrated continuing U.S. support of Europe. Long-term commitment was lacking. Britain's action might thus be regarded as a security-seeking gesture in a world where British power had collapsed utterly but Pax Americana was not yet demonstrated fact.

The most significant U.S. development of the immediate postwar period was Boeing's radical B-47 jet bomber of 1947. Boeing had tremendous experience building durable bombers and unparalleled experience of high-altitude flight at the limits of conventional propeller airplane technology, achieved with the B-29 Superfortress. The B-47 merged the best engine practice of the period, the U.S. development of both British and German technologies, with the best aerodynamic practice, that of Germany. The B-47 had German-style swept-back wings with engines podded in nacelles under the wings (Steiner, 1979, 142–44). Swept wings were first proposed in a German research paper of 1935, then reinvented by NACA in the United States. Boeing's vice president of corporate production development, John Steiner, has noted that it was German wartime research rather than that of NACA that convinced Boeing to use swept wings on the B-47 (Steiner, 1979, 142).

Lessons learned on the B-47 and its big brother, the B-52, convinced Boeing by early 1952 that a civil jet transport was within its grasp (Steiner, 1979, 150). The initial success of Britain's de Haviland *Comet* showed that there was a market for jet airliners. When the *Comet* suffered a series of disastrous accidents, airlines turned their attention to turboprop rather than turbojet propulsion for the next generation of airliners. Boeing persevered with the jet and made its private-venture prototype a success, first as a tanker for the USAF to refuel its B-47 and B-52 bombers, then as the Boeing 707 airliner.

The prototype 707 demonstrated its potential in the fall of 1955 by flying from Seattle to Maryland in just under four hours and back in four hours, twenty-one minutes (Steiner, 1979, 152). Until then Douglas had enjoyed unbroken supremacy in commercial airliners since the success of the DC-2 in 1934. Douglas quickly proposed its jet DC-8 with longer range and more passenger space than the 707 (Steiner, 1979, 155). However good its bombers were, Boeing had failed for more than twenty years to break Douglas's hold on civil aviation. PanAm agreed to order the 707 only if Boeing would guarantee both transatlantic capacity and earlier delivery than Douglas. Only hard work and continued development of the 707 enabled Boeing to sell against Douglas's superior customer base (Steiner, 1979, 156).

The 707 was not the first jet airliner. That honor went to the British de Haviland *Comet,* an advanced and beautiful plane that entered service with BOAC on the London–Johannesburg route in 1952. During World War II the British government set up a committee chaired by Lord Brabazon to investigate British entry into the world airliner market. One recommendation was reminiscent of the power jets proposal: it was for a jet transatlantic mail plane. In the event de Haviland developed a shorter-range aircraft to suit the old British overland routes to the empire and promised long-term development of a transatlantic passenger carrier.

Despite its short range the *Comet I* even attracted a small order from PanAm (Daley, 1980, 400). This apparent success in the critical U.S. market and the very real success of another product of Brabazon committee thinking, the Vickers *Viscount* short-to-medium-range turboprop airliner, seemed to confirm the soundness of Brabazon's policy of government financial backing and technological lead (Bright, 1978, 85–86). The *Comet*'s disastrous failure, explosive decompression at high altitudes caused by the fatigue of constant repressurization of the passenger cabin for high-altitude flight, was almost unpredictable given the technology of the time. Boeing, however, believed that such a problem might occur even before the *Comet* failed. The 707 was stronger, if heavier, and used oval rather than square windows to reduce fatigue cracking at window corners.

Boeing was also successful for reasons other than its aggressive sales campaign for the 707 and the technological edge given by experience with the B-47 and B-52. Boeing gambled 25 percent of its net worth on the 707 but could afford to because the company suffered an 82 percent effective tax rate based on the excess-profits tax (Steiner, 1979, 150). Taxes that should have been paid on the Korean War–induced boom in bombers were diverted to 707 development. In 1962 the tax court ruled against such a diversion, but by then Boeing could afford the taxes. Finally, Boeing's competitors believed that another generation of piston-engined airplanes could be sold to the airlines, and they overrated the possibilities of turboprop-engined planes (Bright, 1978, 89).

The 707 ushered in the jet age for certain when PanAm opened its transatlantic service with the 707-120 in late October 1958. BOAC had begun operation of the *Comet IV* three weeks earlier. Both airplanes had to refuel in Ireland or Newfoundland eastbound, and in both places westbound if headwinds were bad. In August 1959 PanAm introduced the 707-320 with nonstop range from London or Paris to New York, even in the worst headwinds. British failure to develop the *Comet* in like manner greatly intensified U.S. commercial hegemony of world airline markets, and leadership in sales moved from Douglas to Boeing (Davies, 1982, 511–12).

The 707 was not without its flaws. Its early engines were noisy and smoked badly, and the plane was underpowered. The latter was a serious fault at many older airports, and runway extensions were quickly called for. Because the 707 climbed slowly, noise problems worsened for houses and businesses under the approach paths. Ground handling had to change drastically. Passengers had to be loaded and unloaded more quickly because of the short flight times and lack of maintenance requirements, which alone amounted to almost a revolution. Jet engines were vastly more reliable than such complex beasts as the last generation of reciprocating engines. Where reciprocating-engined planes with ten-hour flight times could make one one-way transatlantic crossing a day, jets could make two, three if pressed. The old habit of loading from steps in the open was slow, and the blast from jet engines was far more dangerous than the prop wash from propeller airplanes. Airports had to be rebuilt around not only longer runways but also the now ubiquitous jetways for passenger loading and unloading.

Whatever their flaws the 707 and its competitors were immediate successes. Maintenance costs were extremely low, jets burned kerosene at half the price of high-octane gasoline, and the public loved the speed, smoothness, and quiet of jets at cruising speed. The high altitude at which jets fly allowed weather problems and airsickness to become almost a thing of the past. The result was that airplane companies had a hard time filling the airliners' demand for jets.

After the success of the 707 Boeing turned its attention to the shorter-haul market. Boeing's 727 was the first jet able to take off and land from short runways and older airports. It was designed specifically to fly out of the 1,426-meter instrument runway 4-22 of New York's La Guardia airport for Eastern Airlines's nonstop service to Miami. A very complex wing was required, "the highest non-powered lift system ever used on a swept-wing airplane" (Steiner, 1979, 159). The 727's T-tail also allowed the designers to use wing spoilers to get high descent rates without vibration, which meant excellent landing performance. Boeing also made the 727 much easier to fly than the 707 and thereby enlarged the pool of available or potential pilots.

The very success of the 727 posed problems (Constant, 1988). Jets could now fly into airports with short runways, most of which were in older, more developed parts of cities. This meant more flights, more noise over residential areas, more complaints, thus political pressure to reduce noise. One way to reduce noise was technological, through improved engine design, and that route has been assiduously followed. The other was to restrict the number of flights, in particular those after about 10 P.M. Political pressure tended to be higher in more densely settled areas, simply because more people were affected. Reducing the number of flights was one of the major aims of the first really big "jumbo" jet to enter airline service, the Boeing 747 (fig. 6-1).

Europe, where most cities are old and have high population densities, has adapted more slowly to jet travel than the United States. Before the EEC, Europe also suffered heavily from nationalism. State-run airlines had monopolies of national markets. In the United States the market was not fully open until the deregulation begun at the very end of the Carter administration, but it was far less closed than in Europe. Since deregulation airlines have been able to bid more freely for profitable routes and have had, in theory, access to the entire North American continental market. A similar situation in the EEC would enable airline growth in Europe approaching that in the United States if infrastructure problems are resolved, runways extended, and noise controlled technologically.

The Soviet Union is well suited to jet aviation in theory, but the market has not been developed. In spite of being underpopulated and having plenty of room for a generous infrastructure, a need for rapid long-haul transport, and a perfectly adequate industry, the Soviet Union has never exploited the commercial jet properly. Perhaps one key is lack of capital for infrastructure investment, as well as a different attitude. One aviation historian provides an illustrative anecdote about the Soviet abandonment of a very advanced propeller transport of 1946, the Tupolev Tu-70. The official reason was that the Tu-70 was not suited to existing infrastructure. Gunston comments that

> in 1957 a colleague asked a US airport administrator how he intended to extend a runway at what was then known as Idlewild Airport, New York, because the waters of Jamaica Bay stretched to the horizon. The instant answer "We'll fill in the lousy bay." Cherishing this retort I found myself a few weeks later at Paris talking for the first time with the great Russian airplane designer Tupolev. He side-stepped the suggestion that maybe the airfields could have been improved [for the Tu-70] by pointing out this was no concern of an aircraft designer who, he said, "does what he is told." (Gunston, 1983, 13)

The Soviet Union is still second only to the United States in the number of jet transports built. It has given new life to the early 1930s' obsession with very large, long-range airplanes. The Tupolev Tu-114 is a huge turboprop plane that nearly matches big-jet performance, cruising at only 100 kilometers per hour less than a Boeing 707. Slow development, typical of Soviet civil airplanes, obscured the relative importance of this design, which originated in 1953 but entered service only in 1961 because of "severe operating problems with . . . airfields and equipment" (Gunston, 1983, 339). The Tu-114 simply could not be handled by existing Soviet airport infrastructure, despite its nonstop Moscow–New York range and its ability to fly from Moscow to Havana, Cuba, with only one stop and back with none.

The reason for the Soviet lag compared to the United States is not simply differences in ideology. European adoption of commercial jet aviation also lags, if not as badly, in a society with far fewer ideological differences. But where in Europe it is the persistence of small-scale national markets that bedevils the growth of airline systems serving the EEC as a whole, in the Soviet Union it is the lack of a competent system to finance the infrastructure. In the United States and Europe airports are financed at least in part by local tax dollars. Cities and states compete in a national market—a large one in the United States, a still small one in Europe. In the United States similar competition between cities goes back to that of the early republic, as the misplaced colonial cities of the eastern seaboard sought connections to the trans-Appalachian West with first canals, then railroads. In Europe such competition has been marked since the medieval city-states.

The Geographic Impact of the Commercial Jet

Jet airplanes have remade the transportation geography of the United States and are in the process of remaking that of the globe. Whereas

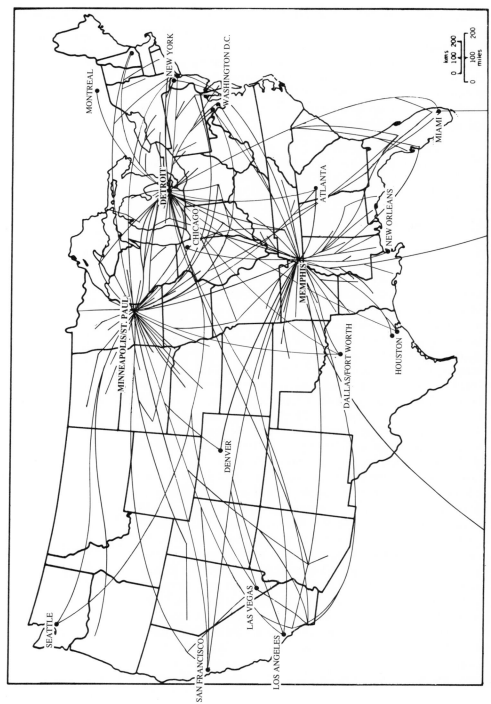

Figure 6-11   The modern U.S. airline hub system as represented by Republic Airlines and made possible by such planes as Boeing's 727. Republic disappeared in the airline mergers of the 1980s. Redrawn from Republic Airlines route map.

propeller landplanes offered the beginnings of economic integration of regions up to about 5,000 kilometers in diameter, jet planes offer even better integration at the continental range, with the promise of global range to come. At least part of the reason for the recent growth of cities in the midsection of the United States is that they offer corporations headquartered in them equal access to both coasts. The four major midsection hubs of Chicago, Atlanta, Denver, and Dallas/Fort Worth, as well as lesser ones such as Houston, Kansas City, Memphis, Minneapolis, or St. Louis, allow business executives to board a plane by 7 A.M. and fly to either coast by late morning. A half-day's work can still be accomplished at the destination. This more efficient use of executive time was well recognized in the initial growth of passenger traffic in the United States in the late 1920s, at least by major corporations. Since the 1920s the percentage of the work force in the executive group has risen substantially, as has the percentage of those in white-collar jobs. At the same time the real cost of air transport has been substantially reduced. The hub system has also allowed considerable local feeder traffic using small turboprop planes to develop from surrounding small cities. It is thus possible to reach almost anywhere within continental North America within a working day, usually within a half day. Airlines have developed as hub specific, and they have networks of feeder lines to smaller cities in their hub regions. Figure 6-11 shows a typical hub system with its feeders, that of the now defunct Republic Airlines. If midsection cities provide better geographic locations of U.S. corporate headquarters, New York is ideally situated to serve transatlantic corporations. European or West Coast cities are only a working day's travel away. U.S. companies with ties to Japan are better located in Los Angeles, although Pacific flight times are still substantial.

The development of hemispheric-range jet planes is now at hand. They had their genesis in 1941, when the United States initiated the design of the propeller-driven Convair B-36 strategic bomber to attack Germany from U.S. bases if Britain collapsed (Steiner, 1979, 146). After World War II the USAF asked for a jet bomber to carry 10,000 pounds (45,000 kilograms) 10,000 miles (16,000 kilometers). The design of Boeing's B-52 began in 1948 to achieve this 10/10 objective (Steiner, 1979, 146). Boeing also built the first commercial jet to come close to B-52-range levels, the 747 *Special Performance*. The genesis of the 747 SP was in PanAm's requirement for an airplane with "the ability to fly with a full payload between New York and Tokyo, non-stop" in thirteen to fourteen hours (Davies, 1987, 80). The 11,000-kilometer range 747 SP entered service in 1976, but could carry only 233 passengers. Lessons learned on the SP have allowed Boeing to create the new generation 747-400 with extremely fuel-efficient engines and lightened structure using composite materials. The 747-400 now entering service carries 380 passengers over a range of 13,500 kilometers. It can fly nonstop along existing routes from London to Tokyo, Los Angeles, or Johannesburg, or from Dallas or Los Angeles to most places in the Pacific Rim or Western Europe all within sixteen hours. The growth of the Asian

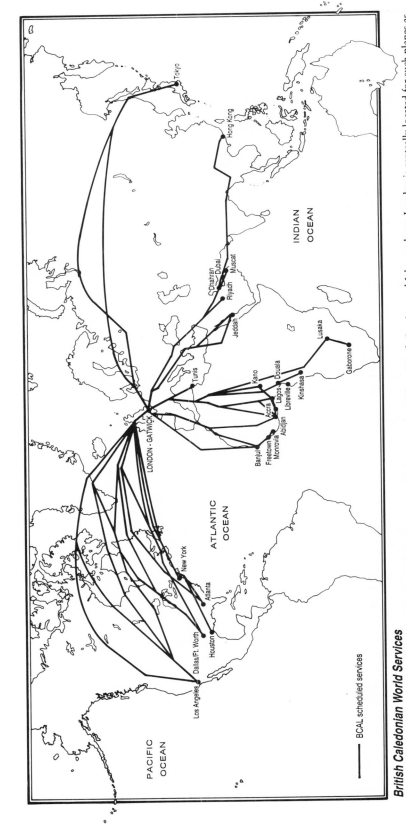

British Caledonian World Services

Figure 6-12 London as a global hub for long-range subsonic jets. Relative to Los Angeles, Tokyo, Rio de Janeiro, and Johannesburg, London is centrally located for such planes as Boeing's 707 and 747. Redrawn from British Caledonian route map.

PACIFIC
OCEAN

ATLANTIC
OCEAN

INDIAN
OCEAN

—— BCAL scheduled services

Los Angeles
Dallas/Ft. Worth
Houston
Atlanta
New York

LONDON · GATWICK

Tunis

Banjul
Freetown
Monrovia
Abidjan
Accra
Lagos
Kano
Douala
Libreville
Kinshasa

Lusaka
Gaborone

Jeddah

Riyadh
Dhahran
Dubai
Muscat

Hong Kong

Tokyo

economies within the world system almost depends on such a plane. Without it, managerial decisions will have to be made locally. With it, the level of integration achieved in the Atlantic with the four-engined, propeller-driven landplanes of the 1940s and 1950s will be possible. London's recent remarkable growth is clearly contingent on its global hub position, well centered for the shortest possible great circle routes around the curvature of our planet, except to South America (fig. 6-12). New York is not so well positioned for the Tokyo flight, although Los Angeles is. Both North American cities are better situated than London for South America. It is possible to project four global hubs—London, Los Angeles, Tokyo, and New York—using current or expectable long-range subsonic airplanes.

The Soviet Union's position in this global system is important. Flights from London to the Far East are more direct if they cross Soviet territory. British Caledonian, now merged into British Airways, was the first major carrier to secure rights to overfly the Soviet Union without stopping at Moscow, which is south of the optimal great circle route to Tokyo. This is essentially the routing proposed by Lufthansa, Deruluft, and Eurasia in 1932 (fig. 6-8). Currently flights to Japan must circumnavigate the world island of Eurasia to avoid the Soviet Union: through the Middle East to India, thence to Southeast Asia and Japan. A map projection centered on the North Pole shows the essential relationships between Tokyo, London, and New York (fig. 6-13). They are at the points of a triangle, but the two direct routes that reach Tokyo are over both the polar ice cap and the Russian confederation. The 1972 map of air traffic in figure 6-13 shows how little developed these critical direct routes are. Currently improving relationships with the Russian confederation and overflight agreements are thus as critical to the emergence of a truly global economy as are long-range aircraft.

The developers of the Anglo-French *Concorde* supersonic airliner have suggested, quite correctly, that for truly global corporations London is an ideal center (fig. 6-14). If supersonic flight had been allowed over populated areas much of the world would have been accessible from London by *Concorde* in eight hours or less, even including fuel stops, Tokyo in nine hours, and the entire globe in twelve hours (Costello & Hughes, 1976, 293; Barfield, 1975, 93). In retrospect *Concorde* was the wrong technology. Fuel consumption was too high for the plane to operate economically after 1973 when the OPEC oil embargo caused oil prices to spiral upward. And noise pollution levels were clearly unacceptable, however attractive increased speeds were. Yet the geographic logic is still sound. A nonpolluting, reasonably economical vehicle that could travel at twice the speed of sound would put London at the center of a twelve-hour world. However, a suborbital space plane cruising out of the atmosphere, where noise pollution is unhearable, would achieve even more, perhaps reducing even hemispheric distances to four hours and rendering London's centrality irrelevant. Corporations could use a suborbital service to locate corporate headquarters near any launch site and enjoy centralized control. Global economic integration beyond the current level offered by subsonic long-range jets thus requires new technologies of suborbital flight in airplanes that would be more rocket than jet. The German theoretical physicist Eugene Sanger proposed such a technology at the end of World War II: a vehicle intended to bomb New York, then skip back to Germany around the globe on the upper reaches of the atmosphere like a stone skimmed across water (Wolko, 1981, 109). The British designer Barnes Wallis proposed a six-hour trip to Sydney using suborbital planes in the 1950s (fig. 6-15). Whether such vehicles would be profitable in commercial operation is questionable. They certainly could not be developed without a huge investment of public monies, perhaps equivalent to that put into the much less commercially sensible space shuttle.

CONCLUSIONS

Air travel has greatly altered the operating efficiency of the world system. In the 1920s airmail service helped create the communication basis

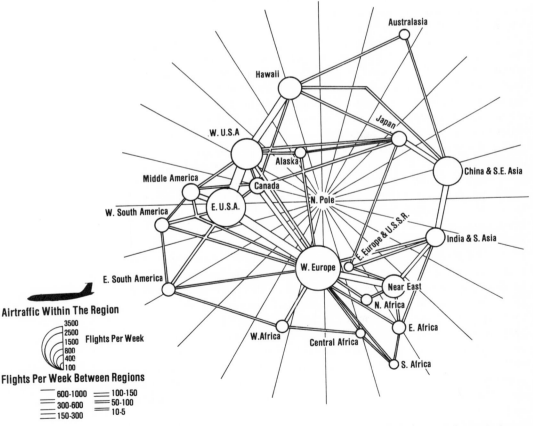

Figure 6-13 World air traffic, 1972. This polar projection shows the new world brought into being by Boeing's 707 and solidified by the 747. Redrawn from Westermann 1978.

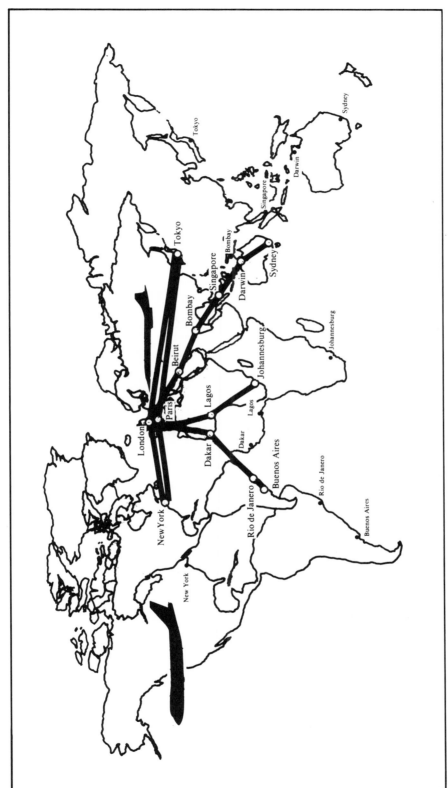

Figure 6-14 Concorde's Mach 2 world compared to that of subsonic air transport: a potentially great idea spoiled by technical failure. Concorde uses too much fuel, is too noisy, and carries too few people to deliver the goods. Redrawn from Barfield 1975.

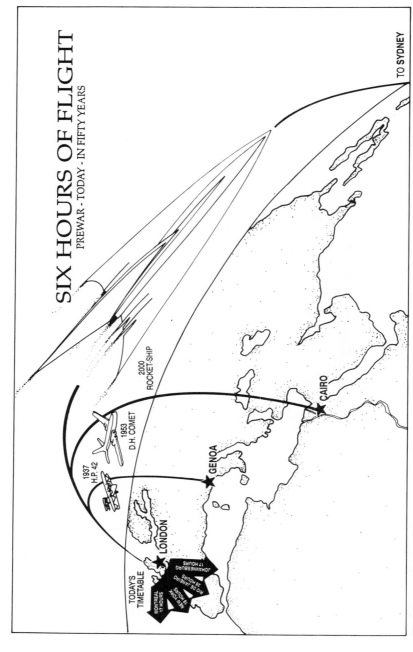

Figure 6-15   Geographic impact of increased air transport speeds. In the 1930s six hours of flight would take a person from London to Genoa, in the 1950s from London to Cairo. This 1950s British prediction of a space plane suggests that six hours will eventually take a person around the globe. Redrawn from Wall 1980.

for a national economy in the United States and could have done so in Mitteleuropa had Germany achieved its aims in World War I. Despite significant technology forcing in Germany before, during, and after World War I and Hugo Junkers's pioneering development of the passenger-carrying F-13 in the 1920s, it was in the United States that successful commercial aviation developed after 1933. Airplanes such as Boeing's 247 and Douglas's DC-2 and -3 provided much more thorough integration of the regional economy of the U.S. manufacturing core, allowing more efficient use of managerial time. Large flying boats extended airmail systems across the Pacific from North America and throughout the British Empire.

A comparison of U.S. success with German and British failure suggests that only in the United States was the optimal set of linkages reached among business, technology, and politics. German aeronautical technology was the best in the world through the early 1930s, but technical advantage alone could not be translated into success. Junkers had a clear vision of the future, but the Inter-Allied War Commission strangled German technical development. Despite Junkers's persistence, the transfer of his technology to the Soviet Union, and such grand schemes as Europa Union, Deruluft, and Eurasia, lack of political integration prevented realization of long-distance flying over the Eurasian land mass.

British failure was more social. British politicians thought globally and saw air communication as vital to retaining the global empire that was challenged by Britain's commercial decline and the stunning success in World War I of German U-boat commerce raiders. The RAF propounded a theory of blitzkrieg in 1918 and went on to develop in both theory and practice the notion of strategic air power to reduce the cost of patrol and punitive expeditions in the empire. The British showed that they could compete and win technologically intense competitions such as the Schneider trophy, and they adopted metal airplane technology far more readily than did the United States. In the crucial decade of the 1920s, however, the British ruling classes (of whatever political ilk) seemed unable to think in technical as well as geopolitical terms. This was first and most markedly a fault in their education, secondly a fault created by a social structure in which the absence of upward social mobility at home drove people abroad, where social structures were less rigid and rewards were based more on ability than origins.

U.S. success, then, was as much a product of a social structure that reduced class division as of the appropriate technology. Entrepreneurs and politicians listened to technologists and took steps in the late 1920s to remedy the marked technical gap between the United States and Europe. The Guggenheim-funded chairs in aeronautical engineering were the result. Americans possessed not only a vision of uniting a nation by air mail, a scheme put into effect in 1919 with remarkable success in view of the technology available, but a vision of profitable, commercial, passenger-carrying airliners that reached reality in the early 1930s. Nor were they, in the agency of Juan Trippe and Pan American World Airways, lacking in global vision. But that vision was firmly based on commerce, not force. In the U.S. prestige and military power followed commerce instead of preceding it as they did in Britain after World War I.

Neotechnic improvements in aviation have had two results. They have allowed truly national, centrally directed corporations to develop even in the largest states, or powerful national governments to emerge where the principle of the command economy predominates, or both. They have allowed transatlantic corporations to develop with headquarters either in the Boston-Washington or the London-Paris-Bonn megalopolises. They promise the development of truly global corporations. If supersonic flight does not transcend current limitations, global corporations could be located almost anywhere within the triangle formed by the current world's three megalopolises: Boston-Washington, London-Paris-Berlin, and Tokyo-Osaka. New megalopolises may well emerge because of the appalling congestion and expense of living in these three areas. New megalopolises are most likely to be in North America because of the need for very large amounts of undeveloped land on which to build housing, highways, businesses, and jetports. They are also likely to be in climatic regions that make year-round movement easy and offer high levels of amenities to residents. The San Diego–San Francisco area is an obvious candidate, as is the central Texas urban triangle formed by Dallas/Fort Worth–Houston–San Antonio. Both areas have the advantage of being in a single state, thus avoiding the problems of split jurisdictions that bedevil the Boston-Washington megalopolis when it comes to raising tax monies for infrastructural development. In both areas particular cities have also taken an aggressive stance with regard to the development of major airports offering global service: Los Angeles and Dallas/Fort Worth. The success of all this depends, of course, on the opening of Russian airspace to commercial overflight. In early 1992 this development seems virtually assured by the collapse of the Soviet Union and the emergence of a Russian confederation on good trading terms with the rest of the Northern hemisphere.

The new map of the world's most powerful economic region which is now emerging will thus no longer be the familiar old Atlantic-centered Mercator projection, which makes Japan and the American West seem to be on the outer fringes of a world centered on London and New York. It will be a projection centered on the North Pole and showing San Diego–San Francisco, Tokyo-Osaka, and London-Paris-Berlin as almost the three points of an equilateral triangle. It will also be a projection that emphasizes the difference between the rich North of the planet, where the major trading areas are all relatively close together geographically, and the poor South, bedevilled by difficult linkages between its constituent parts. This new map will persist as long as we are content with subsonic aircraft. Supersonic planes flying at two or three times the speed of current aircraft would not have a huge effect on it, but the development of hypersonic space planes that could reach anywhere on the planet within four to six hours would clearly rewrite the geographical rules of the game. Such technology will allow global corporations truly global choices of location. Currently underdeveloped and attractive locations such as the western coast of Australia and the Atlantic coasts of South America and southern Africa would almost certainly boom under such conditions, assuming political systems that are not inimical to global capitalism.

The reappearance of the command economy (McNeill, 1982, 269) or of the megamachine (Mumford, 1970, 243) has thus been a function of Neotechnic transportation and communication. On the other hand, the increasingly complex web of air transport linkages has given corporations more freedom of location. This process has been facilitated by the tremendous growth and reduction in cost of long-distance telephone service and data transmission. In this sense Neotechnics is a two-edged sword. It has radically increased the flexibility with which our societies operate in space and time and at the same time has increased the potential for limiting that flexibility through centralized control of operations.

Jet aviation completed the integration of the U.S. national economy by allowing companies to manage all their continental operations from one headquarters. Europe has been restricted from such a level of integration by political barriers due to be obliterated in 1992. At present jet aviation lags in Europe as a result of the political barriers. Better high-speed ground transportation may also restrain aviation, but Europe is moving toward the same advantageous mechanism of infrastructure finance as North America enjoys. Cities and states effectively compete for airline service by offering to build airport facilities, thus reducing airline investment in immobile fixed capital. Airlines are free to invest in new technology, as well as to shift their vehicles as the profitability of routes rises and falls. The hemispheric-range jets now entering service offer at global level the sort of integration that was achieved in the North Atlantic immediately after World War II, thus management of the global economy from a small number of well-located hubs.

At a subsidiary level, jet freight has produced marked increases in marketing efficiencies for low-bulk, high-value goods. The Japanese dominance of consumer electronics is, in part, a result of the ability of jet freighters to carry larger numbers of, say, VCRs, from Tokyo to New York. Jet freight has advantages in security and reduction of inventory and spoilage of an item in which technological change is rapid. The same advantages accrued, in the example given earlier, to trucking in the distribution of high-value, easily spoiled chocolate. The Boeing 747 was, in fact, originally designed not as a passenger carrier but as a freighter with dimensions to accommodate containers that would be transferred to trucks (Steiner, 1979, 167). The width of the plane accommodates, side by side, two containers of the maximum width allowed on U.S. highways, eight feet (2.44 meters).

When the 747 was designed Boeing believed that the passenger plane of the future would be supersonic. It still may be, but not with existing technology. The supersonic Anglo-French *Concorde* cannot sustain speeds of more than Mach 1.7, that is, 1.7 times the speed of sound. The plane is capable of Mach 2, but if held at that speed its aluminum structure could heat to the point of deformity (Howard S. Wolko, personal communication, June 1988). American proposals to build a Mach 3 airplane out of titanium seemed promising in the late 1960s. As with the first jets, development costs were defrayed by military investment in a bomber. Information from the XB-70 would have helped develop a titanium-structured Mach 3 commercial airliner, and NASA took over one of the two XB-70s to collect such data (Jones, 1980, 235–38). Unfortunately one piece of data collected was that, at Mach 3, engines could suffer a problem that became known as an "unstart" if the airflow into the engine was disturbed. Although the flight crew became used to "unstarts" and the airplane continued to fly safely, an unstart was likened to hitting a brick wall at speed. Full-harness safety belts had to be worn continually. When it became clear that a high-flying Mach 3 bomber would be vulnerable to Soviet missiles, the USAF decided against continuation of the XB-70. The focus of bomber development shifted to planes able to fly close to the ground (the B-1) or less visible to radar (the B-2).

As reciprocating engines reached the limit of their usefulness with turbocompound engines, jet engines have reached their effective commercial limit just below Mach 1. Beyond that the noise of the supersonic shock wave annoys people on the ground, fuel consumption becomes extremely high per person carried, and airframe heat is a problem. By Mach 3 the air-breathing turbojet seems unsafe for commercial flight. Without massive government spending on research and development such as was made possible by World War II and the cold war, major technological advances are unlikely.

Chapter Seven

World System Theory and Geographic Reality

While it seems that industrial activity and commerce will tend to become decentralised, it will become more and more important that there should be a single clearing house . . . It does not follow that there should be, along with decentralisation, an actual fall of [industrial] activity in our islands; but it appears to be inevitable that there should be a relative fall. But the world's clearing house tends, from its very nature, to remain in the single position, and that clearing house will always be where there is the greatest ownership of capital. This gives the real key to the struggle between our free trade policy and the protection of other countries—we are essentially the people with capital, and those who have capital always share in the activity of brains and muscles of other countries.

Halford J. Mackinder
"The Great Trade Routes," Journal of the Institute of [British] Bankers

Over the past five hundred years the world system has been obsessed with increases in production, some of it from increased productivity, but most of it from simple expansion. Two frontiers have obsessed the world system: one related to geography, the other to energy. A third frontier, the temporal, has been important nearly as long as the energy frontier. A fourth frontier related to the structure of the labor force has moved in and out of importance.

Geography has been both the most spectacular and the oldest of the frontiers, extending through the whole history of the world system. Energy has been second in importance, a frontier now some three hundred years old. Most energy has been mined, in the form of coal or oil; the exploitation of mineral hydrocarbon resources. A slightly more recent frontier is the temporal. Expansion beyond the confines of the organic cycles of day and night, summer and winter, began in the early nineteenth century but became really important only in this century. The last frontier has been the structure of the labor force. Technology, demography, and the need for particular skill levels have driven children and women in and out of the work force. Social and ethnic groups defined as unneeded in one period have become vital in another. The incorporation of previously untapped labor power is as effective a way to increase production as expansion on any other frontier.

Any would-be hegemon has had to find the right combination of biological makeup, social and demographic structure, economic, and political system, and technology in order to optimize production in each specific world leadership cycle in the history of the world system. Of these variables technological hardware has diffused most easily and therefore generally seems most important at the hinge points between Kondriatiev cycles. Software, usually managerial technology, has diffused more slowly, not least because it is related to social structure. Economic and political systems diffuse even more slowly, and demographic structure perhaps most slowly of all. Since this process began in Europe, it is hardly surprising that in the formative five hundred years of the capitalist world system all the hegemons or would-be hegemons have been European or European-derived cultures, or have consciously copied European systems.

The recent closure of the geographic frontier has slowed the simple production gains that can be bought by expansion and refocused attention on productivity increases. Yet some people have always argued for increased productivity alongside increased production. Adam Smith's argument for laissez-faire economics was not merely to remove the "dead hand" of government. It was to remove that dead hand in order to increase productivity in a system with seemingly finite boundaries. At the level of technical, social, economic, and political organization prevalent in the late eighteenth century, Smith's argument was logical, but a change in any of these components brings it into at least short-term question. The technical forces of production changed drastically with the onset of steam power. The factory system that had seemed to Smith a relatively benign change in the social relations of production turned into a monstrous exploitation of human energies. The specialization and division of labor coupled with steam power, laissez-faire doctrine expanded to the status of ideology, and a Calvinist moral theory caused near disaster in the early Paleotechnic.

Marx analyzed this new conjunction of forces brilliantly, but his work, for all its claims to predictive power, is as historically contingent as that of Smith. Marx's followers, however, have been just as willing to elevate his hypotheses to laws as have the followers of Smith. Neither Smith nor Marx adequately understood the expansionist and dynamic nature of the world system. Braudel and Wallerstein have clarified the nature of that expansionism, although they have concentrated on its geographic basis. Kondratiev and his interpreters have done much to codify the nature of the dynamism.

GEOGRAPHIC EXPANSION

The three phases of geographic expansion have each been driven by transportation technology. In the Eotechnic transportation was largely seaborne, although small ships could penetrate inland as far as the tide would carry them. Inland transportation, which was based on animal power, was very restricted. High-value, low-bulk goods could be carried long distances, but the movement of agricultural surplus was highly restricted. It was uneconomic to move bulk goods more than 50 kilometers or so to or from the head of navigation. Geographic expansion was therefore restricted to peninsulas, archipelagos, and the continental littorals.

Biological problems also restricted early Eotechnic success. Europeans succeeded between roughly 1490 and 1900 in expanding into areas where their biology helped rather than hindered them. In comparison with the other great cultures of Eurasia, in particular China, India, and Islam, Europeans were demographically weak. Except in very special circumstances, such as imposing a new ruling elite on a politically fragmented Indian subcontinent in the later 1700s, Europeans could not go head-to-head with the other Eurasian cultures because of numerical inferiority. Lack of resistance to environmentally vectored diseases kept Europeans out of Africa. The Eurasian disease pool, however, in which Europeans were full partners, wrought havoc among the inhabitants of the Americas and Australasia. As a result the possibility of settling vast new acreages arose, though Eotechnic transportation technology was inadequate to its realization.

The European political fragmentation produced by feudalism greatly helped Eotechnic expansion because it allowed different polities within the overall cultural context of Europe to compete for and to pursue different paths toward hegemony. Monolithic polities such as China lacked this advantage. Feudalism produced a predilection to increase national wealth by easily taxed commerce. It maximized Europe's productivity by allowing each region to specialize in the production of whatever best suited its soils and climate.

Within the European states system the maritime polities worked out a social structure better suited to geographic expansion than those where wealth made from the ownership of land remained more highly valued than wealth generated by sea trading. Within the maritime polities productivity gains came from increasing first the money supply and then its velocity. Radical Protestantism reversed the traditional thrust of Christianity by making it morally wrong to hoard scarce capital or to fail to invest it in the most productive possible way.

Within the Eotechnic hegemony fell first to Portugal, then Holland, then England. Portugal unlocked the door to the Atlantic islands and coastlines with improved ships and navigational technologies. Eurasian diseases slaughtered the indigenous inhabitants of this Atlantic world. On the island groups of the eastern Atlantic, then on the Caribbean archipelago, then on the coastal littorals of South and North America, Europeans achieved massive windfall gains in agricultural production in the wake of European pathogens.

In this organic economy hegemony shifted from Portugal to Holland in the late sixteenth century as the Dutch, under the influence of Calvinism, built less costly, more efficient ships and increased the productivity of scarce capital. Dutch socioeconomic structure proved more appropriate to a long-distance, sea trading economy. Yet the Dutch persisted too

World System
Theory and
Geographic
Reality

long with a Baltic economy and allowed themselves to be driven from the northern Atlantic by the English with the fall of the New Netherlands colony in 1664 at the start of the second Dutch War. Dutch military costs were also higher than those of England, secure behind the "wooden walls" of its navy alone. Holland had to maintain a standing army to fight off its continental neighbors as well as a navy to maintain its profits from seaborne trade.

By the late 1600s England was able to exploit its geographic position astride the Atlantic approaches to Europe to dominate the building Atlantic economy. England's social structure, like that of Holland, accepted wealth derived from trade. In Puritanism England adopted radical Protestanism. Yet long-term success for England came from monopolistic domination of Atlantic trade through the Navigation Acts and a network of both settler and plantation colonies in the northern Atlantic basin which assured profitable, regular cargoes for British ships. The Caribbean sugar plantations fed inexpensive sugar to the British market, and the settler colonies of the North American littoral provided tobacco, fish, naval stores, iron, and ships as well as a market for British manufactures.

By the late 1700s the end of easy production gains from geographic expansion seemed in sight. The islands and littorals of the Atlantic were settled. Cook's voyages were describing new possibilities for seaborne expansion into the Pacific archipelagos and the Australian littoral, but only at much higher transportation cost. By the end of the eighteenth century the integrity of the British Atlantic was under threat, first from the French, then from successful revolution in British North America.

Until this point production gains in the world system had come almost solely from geographic expansion. By the late 1700s Smith, Ricardo, Malthus, and others were aware of the seeming limits to geographic expansion. They were therefore concerned with increasing productivity by removing bureaucratic checks to production, emphasizing regional specialization in production or control of increased consumption by controlling population growth. Such solutions were rendered moot by renewed geographic expansion, driven now by hydrocarbons. Steam-powered transportation, first riverboats, then with railroads, gave access to at least some of the continental interiors.

In the case of riverboats access was controlled by topography and climate. Fast moving water in a river that sloped steeply to the sea meant that underpowered craft could not move upstream. Falls and rapids posed similar limits, although if they were few in number they could sometimes be circumvented by canal or land portage. Rivers also had to flow pretty much year round. Intermittent streams in arid regions were rarely usable. The Mississippi and its tributaries alone provided in 1815 access to a region of at least equal size to the exploited littorals and archipelagos of the Americas. A river accessed by riverboat is itself a littoral, although one with two fronts rather than the single one of an oceanic littoral. Draft animals could haul agricultural surplus to either bank from the newly opened lands. Development of the American Great

Lakes provided coastal access to a region almost as vast again as that opened up by the Mississippi.

Railroads opened lands in the same way as lakes and river systems, but with fewer restrictions from topography and climate. Production of agricultural surplus for export was not economic much away from the rail line. But railroads could penetrate mountain barriers, cross arid regions, and make accessible hitherto useless regions. This, plus the speed with which they moved perishable goods, made their higher cost worthwhile.

The geographic expansion of the Paleotechnic was further aided by the development of steam-powered oceangoing ships, which greatly speeded up oceanic transportation of people and perishable goods. They also lowered losses from shipwrecks and increased the labor productivity of ships' crews. Iron- and later steel-hulled ships could be made larger than wooden ones, and steam required less crew than sail; thus more goods could be carried. Many of these developments were improvements in productivity rather than production, but efficient steamships provided links between the agricultural periphery and the metropolitan core that were crucial to the continued growth of the world system.

Despite the considerable increase in geographic access wrought by the Paleotechnic, even the best Paleotechnic transport system resembled a spider's web. Rivers and railroads, however generously provided by nature and capital, could never completely penetrate a continental interior. Steam railroads were expensive to build, and their infrastructure was privately owned. Since profit was the primary motive for their construction, they were built only in regions that seemed likely to generate profitable traffic. The few that were built elsewhere required considerable government subsidy. The technical nature of steam power encouraged large, infrequently run trains, which were a poor way to amortize the large amount of capital tied up in the lines themselves. Where potential traffic was light or government subsidies were not forthcoming, the railroad never appeared.

British success in the Paleotechnic came more from increasing production by using mineral hydrocarbons than from geographic expansion. Nevertheless, British capital played a crucial role in opening up most of the great continental interiors. British-funded railroads were common in North and South America as well as in India and such British settler colonies as South Africa and Australia. British hegemony in the fourth world leadership cycle was founded on the relatively free movement of capital, raw materials, and agricultural surplus. The fruits of geographic expansion were thus readily available to the British through both the railroads funded by British capital and the British-owned and -operated steamships that moved agricultural surpluses across the world's oceans.

Geographic expansion did not end with the Paleotechnic. It has continued powerfully in the Neotechnic as the bicycle, the truck, the bus, and the automobile have made accessible lands distant from the spider's web of rivers and railroads. Trucks in particular can move small quantities of goods economically to railheads. Railroads early in this century saw the truck as a natural complement to their services, not a competitor. Trucks filled the gaps in the spider's web and, in the process, continued geographic expansion.

With the success of the internal combustion—engined vehicle in the first decade of this century, the problem became one of infrastructure rather than vehicle technology. Trucks, bicycles, automobiles, and busses are no less fixed in their routes than trains and steamboats. Despite considerable experimentation with four-wheel-drive trucks for use where there were no roads, trucks run best on hard-surfaced roads, which cost as much per mile as railroads. Internal combustion—engined vehicles are, however, economical conveyers of small loads, flexible responders to schedule demands and route availability, and representatives of mobile fixed capital. Nevertheless, they still need good roads to run upon. The continuation of geographic expansion into the Neotechnic has thus hinged upon finding acceptable ways to raise capital for road construction. Gasoline taxes raised vast amounts equitably by taxing users directly. Road systems grew fastest in countries such as the United States, where such taxes were applied only to road construction. Where they are applied to general revenues, their impact has been muted. Countries such as Britain, Germany, France, and Japan levy high gasoline taxes but apply little of the money raised to road construction. Only when the role of the automobile in increasing productivity became recognized did polities such as Britain invest in national as well as local road construction.

Production gains from geographic expansion continued in Neotechnic polities such as the United States, a land-based economy that still had a poor land transportation system. Other polities tried to copy U.S. expansionism but failed, usually for nontechnical reasons. Imperial Germany dreamed of the political integration of much of Europe in a Mitteleuropa served by Zeppelins and metal airplanes, and Hitler's Third Reich began railroad, autobahn, and air networks to unite an even large polity. But Germany's neighbors resisted German expansionism, which was aimed at finding markets for German goods, not simply expanding agricultural production.

Imperial Russia and, later, the Soviet Union had the potential for geographic expansion in the Neotechnic, but its underpopulated eastern section between the Urals and the Pacific is climatically difficult for agriculture. The tier of republics to the south is much better suited to agriculture, in particular to cotton, but is strongly Moslem and has a much different demography with a far higher dependency ratio. The Russian-dominated and relatively Europeanized Soviet polity was never able to assimilate these republics, and they have been quick to assert their independence with the collapse of Soviet authority. No such problems confronted the United States in its expansion west and south. The western climate of North America though dry has more moisture than the eastern Soviet Union, as well as far higher winter temperatures. And the humidity of the subtropical southern states has been tempered by technical innovation in the form of air conditioning. The indigenous American population died rapidly of Eurasian diseases, whereas that of the eastern margins of the Soviet Union was part of the Eurasian disease pool.

For the future we can expect no more easy production gains from geographic expansion. The interiors of the great continents are now open, not only along such limited routes as are served by rivers and railroads, but in the interstices of the spider webs created by such routes. The world system of capitalist agriculture has reached the limit of its geographic frontiers on planet Earth.

ENERGY EXPANSION

In the Eotechnic the energy frontier usually was advanced only through geographic expansion or demographic collapse. The former allowed new regions to come into wood production if their product could be economically shipped to market. The latter changed the food-fiber-fuel balance by reducing short-term demand for food. Because the growth period for wood was very much longer than those for food and fiber, wood was rarely popular among landowners, most of whom preferred to maximize short-run profits. Fuel was always the poor relation of the three Fs in an advanced organic economy.

Real energy expansion thus began with the mining of hydrocarbons, which added a vertical dimension to the energy frontier. As the demand for iron cannon for ships increased in the mid-1500s, much wood was diverted to charcoal. Coal production for domestic use skyrocketed, Newcastle's exports going from just under 33,000 tons to nearly 530,000 tons between 1563 and 1658 (Armytage, 1976, 70). By 1661 John Evelyn was complaining bitterly of the sulfurous air pollution of London (Wise, 1968, 12). Abraham Darby's coking technology eventually resolved the resource crisis for the iron industry by transforming coal to coke somewhat as wood was converted to charcoal. Darby's technology spread steadily after 1709, in spite of the lack of an inland transport system to unite iron ore with coking coal. Canals solved much of England's internal transport problem at the same time as Newcomen's simple steam engine drained mines to expand the vertical frontier downward.

On the basis of Newcomen's work Watt in the 1770s produced the first steam engine efficient enough to be used off the coalfield. This, plus Arkwright's contemporaneous software innovation of the factory system, began the Paleotechnic phase of the industrial revolution. These innovations, all British in origin, were more than enough to ensure British success in the fourth world leadership cycle at a time when the British commercial and advanced organic polity seemed in decline in the face of France and revolutionary America. The transformation of commercial England to industrial Britain is a clear example of technologically increased production, thus increased economic strength, the mechanism suggested by Mensch (1979) to drive world leadership cycles. The three world leadership cycles of the Eotechnic were driven more by increased production brought on by geographic expansion, although in the case of both the Portuguese and the Dutch expansions technical forces in ship construction and navigation were at work.

Production skyrocketed in the Paleotechnic for four main reasons: the energy released in the combustion of coal; the managerial improvement of the factory system; the expansion of the temporal frontier into both the winter season, when water power was irregular, and the hitherto unlit night; and the expansion of the labor force to include children and women. Also important was the fact that Paleotechnic factories could be built any place where coal could be delivered economically. Eotechnic factories could be built only where an adequate, exploitable fall of water was available.

At least to begin with in the Paleotechnic wages for factory workers were high relative to agricultural wages. In Britain the age of marriage dropped and fertility jumped. Britain left agricultural self-sufficiency behind and began to import ever larger amounts of first fiber, then food. Increased demand for fiber and food was felt most keenly in the periphery of the world system. The success of British cotton textile mills pushed the cotton production region out of the Caribbean archipelago onto the North American mainland. Crossbred cotton that could tolerate the climate of the southern part of the continental interior, steam-powered riverboats, gins, and slave labor allowed the rapid spread inland of the cotton economy after 1815 or so. Until the 1840s the British Isles were still relatively self-sufficient in food, but the collapse of Irish wheat production after the 1845 failure of the potato crop on which Irish farm laborers depended for food caused Britain to look to the Atlantic world for staples as well as the luxury carbohydrate, sugar. The Atlantic had long provided fish protein. By the late 1800s railroads had driven the wheat frontier far into interior North America.

Without wood- and coal-fired transport capitalist agriculture in the periphery would have been nearly impossible. Without coal-fired engines most of the fiber factories of the core would have ceased to spin. Britain dominated world coal production through 1875, producing almost four times as much as Germany in 1870 and about three times as much as the United States. Thereafter British dominance fell rapidly. By 1910 the United States was producing almost 70 percent more than Britain, although a significant amount of U.S. coal came from British-owned mines and was sold in Britain. By 1910 German production was around 75 percent of Britain's. Such figures spelt the end of British hegemony as long as the world system depended upon coal for most of its production and access to geographic frontiers.

Expansion of the energy frontier to include oil was even more disastrous to British hegemony. With the miniscule exception of some local oil shale production, all Britain's oil had to be imported until the North Sea oil fields opened in the 1970s. Oil importation greatly increased Britain's geostrategic vulnerability, as was made evident in World War I when German U-boats interrupted the flow of food and fiber into Britain. In World War II German U-boats disrupted the flow of all three Fs.

The oil economy benefited most those polities, such as the United States, that still had room for geographic expansion in the interstices of the Paleotechnic transport web and that had geostrategically secure supplies of oil. Polities without oil had to import it and risk geostrategic insecurity, the fate of Britain and, in World War II, Japan. Alternatively they had to seek geostrategically secure fuels or productivity increases in the use of oil. Germany followed a combination of these, converting domestic coal to oil and developing such fuel-efficient internal combustion engines as the diesel.

Russia enjoyed the same theoretical advantages as the United States, but Neotechnic technology fell on stonier ground there than in the West. Although the revolution wrought improvements in social structure over czarist Russia, gains under Lenin were inadequate and were mostly lost in the retreat to a relatively inflexible social system under Stalin. In economic structure the same problems occurred: Stalin recentralized an economy that Lenin had tentatively begun to decentralize. Only in military applications of Neotechnic technology, which enabled the Soviet Union to resist Nazi Germany in 1941, did the Soviet polity reach rough parity with the West.

In production terms the single largest contribution of the oil economy has been to allow continued expansion of the geographic frontier in the newly settled lands of the Americas and Australasia. Its second largest contribution came when German scientists, technologists, and industrialists, shut off from geographic expansion by the policies of the United States, Britain, and France around the turn of this century, sought substitutes for such tropical and subtropical organic products as cotton, rubber, sugar, and the like. They learned to synthesize new products, mostly from hydrocarbons. Oil's third contribution came in lowering the fixed costs of production by making less expensive land available on the outskirts of Eo- and Paleotechnic cities and in making new cities possible. A fourth contribution, air travel, which makes managerial labor more efficient by reducing time in transit, is more a productivity increase than a production increase. Even so, efficient diffusion of the best managerial technology has materially increased production.

Of these four contributions only the first can be regarded as relatively exclusive to specific polities. Only the newly settled lands had the potential for continued production gains through geographic expansion. The technologies to produce synthetic alternatives to such natural crops as sugar, opium, cotton, and rubber diffused rapidly through the industrial world. Beet sugar could be grown in temperate climates and became an alternative to tropical sugar cane as early as the Napoleonic Wars. By early this century, Germany had a substantial acreage in sugar beet. Saccharin, synthesized by an American chemist in 1879, offered an alternative that could free up a considerable amount of land and was enthusiastically adopted in Germany. Until the refinement and marketing of acetylated salicylic acid under the trade name, Aspirin, by the

German company, Bayer, in 1899 the only viable over-the-counter painkillers were various derivatives from subtropical opium poppies. Synthetic alternatives to subtropical cotton and tropical rubber came later. The German firm Agfa patented the first synthetic fiber, chlorinated polyvinyl chloride, in 1933, but du Pont in America patented a superior synthetic fiber, nylon, only five years later. German synthetic rubber technology was even sold to the United States in 1938, and U.S. pilot plants were in production by 1939 (Hugill, 1988b, 123, 124).

The impact of the automobile and truck in lowering the fixed costs of production by improving access to low-cost land was powerful everywhere. Indeed, adoption of the truck proceeded almost as fast in polities such as Britain and Germany as it did in the United States. The U.S. worker enjoyed automobile access to truck-served plants almost from the inception of automobiles in the second decade of this century. Elsewhere in the industrial world workers had to make do with other Neotechnic devices to get to work—the bicycle and the motorcycle. In newly industrializing polities, they still do.

Oil energy expansion clearly benefited all the industrial world, but its advantages converged most spectacularly on the United States, at least as long as it was a major producer of inexpensive oil. Once its inexpensive oil was burned, the United States's advantages were less noticeable. For a time the United States was able to secure inexpensive oil from the Middle East, but this supply is subject to geostrategic disruption, as has been demonstrated by the OPEC cartel formed in 1973, the Iran-Iraq war of the 1980s, and the Iraqi invasion of Kuwait in 1990.

Continued production expansion is now possible only if oil can be *securely* shipped around the planet. In the Pax Americana, global shipping was secure in the same way as it was in the Pax Britannica of the 1800s. And it will remain secure as long as the U.S. Navy dominates the seas in the same way that the Royal Navy did, but the U.S. economy is still subject to the threat of a *guerre de course* fought by submarines just as Britain was threatened by Germany in two world wars and Japan was destroyed by the United States in World War II. In one way the United States is even more vulnerable because the only substance it must import in bulk and by sea is oil, most of which has to pass through the Strait of Hormuz. The Soviet Union could have pursued such a war against the United States but was held back by the threat of escalation to a full nuclear exchange. Such a threat may not hold back a less rational opponent if it saw a terrorist threat to interdict oil traffic as worth the risk. The risk could as easily be taken on ideological grounds (in the form of a Moslem "Holy War" against Western influence) as economic ones. As the Gulf War of 1990 demonstrated the oil-dependent powers will act in unison to secure the flow of oil from the region and ensure the regional polities do not act in a terrorist manner.

Two longer-term problems for the oil economy are the local danger of air pollution from burning hydrocarbons and the less obvious but increasingly discussed problem of global warming. Oil consumption could be decreased if necessary because, with current technology, oil fuel is an absolute necessity only in the mobile applications of Neo-technic transportation. In the United States, for example, some one-third of oil used is burned by transportation, another third by industry, and another third by domestic use. Most of the oil used in industry and the household is burned for heating and cooling. For such fixed applications, electricity would serve as well as oil.

The longest-term threat to continued oil expansion is depletion of resources. Ultimately these resources are finite, and well before depletion is reached, the law of supply and demand will force disastrous price rises. A return to coal fuel would increase the available hydrocarbon energy, but at some cost. Coal is a much "dirtier" fuel than oil or natural gas and would accelerate such real and potential environmental problems as acid rain, carbon dioxide pollution, and global warming. Strip mining of coal can be damaging to the environment, and deep mining tends to take a high toll in human lives in mining accidents. Coal can and will be used, however. By current estimates coal could maintain our industrial society for several hundred years longer than oil could, and coal emissions can be cleaned up. The problem of global warming is less easily solved, assuming that it is a problem. A new generation of earth-monitoring satellites and supercomputers that can model the world's climate in its entirety will confirm whether industrialization is warming the globe or not.

Beyond the few hundred years that coal will last us, continued energy expansion depends on some new technology that is either not yet easily visible or not yet economic. Fusion and cold fusion seem uneconomic ways of generating electricity as yet. Fission has demonstrated dangers of its own and is as subject to resource depletion as fossil fuels. Some parts of the world, such as the Canadian west coast province of British Columbia or South Island, New Zealand, could generate vast quantities of environmentally safe hydroelectric power but are long distances from markets for it. The improved distribution of electricity offered by super-conductors might make it possible to tap such areas, but superconductive transmission of large amounts of electricity is not yet possible.

Even so, the continued prosperity of the capitalist world system probably depends on some such technology for energy expansion. The easiest production gains now that geographic expansion has ended will have to come in this area.

TEMPORAL EXPANSION

Expansion of the temporal frontier has in part been described by Murray Melbin as "the colonization of time" (1978). This colonization began in earnest when the gas released from coal in the coking process was used to artificially light the interior of factories in the early Paleotechnic. Without artificial light production was necessarily restricted to daylight, which is long in summer but short in winter in the high latitudes of northwestern Europe. Expanding the workday into the hours of darkness markedly increased production.

Artificial light also lowered some of the fixed costs of production. Buildings no longer had to catch whatever natural light was available. The need for natural light usually restricted factories to one story and required many small, expensive panes of glass. Artificial light made multistory buildings with little glass a practical proposition.

Gas light had its problems, however. Until the invention of the gas mantle in the 1880s (Armytage, 1976, 208), dangerous and inefficient open gas flames were used. Electric light appeared almost contemporaneously with gas mantles, so that the decision to use gas or electricity after 1880 was usually regional in character as well as technological. In Britain, for example, because most of the population lived close to the coal-producing regions and thus to the coke ovens, most lighting could be by gas. In the United States, the main coalfields were west of the main population concentration on the northeast seaboard and across a mountain range. No gas pipeline was built through the Appalachians until the 1940s. Thus the eastern seaboard invested early in electric light.

Electric light was safe and efficient, even in incandescent form. In the fluorescent form first installed in the 1930s, it was also inexpensive. Interiors could be lit evenly almost to daylight levels, thus completing the invasion of the night begun more than one hundred years before with the use of open gas flames.

The Paleotechnic also produced a second temporal expansion. Eotechnic factories powered by falling water or wind rarely worked all year. Their labor forces were idled at least some of the time by lack of water or of wind. Steam power allowed the expansion of work across all seasons and weather conditions so that production could go on year round.

LABOR FORCE EXPANSION

The most variable of the productive forces has been labor. At the onset of the factory system women and children were taken into the labor force in large numbers. Men tended to remain longer in the agrarian work force of an advanced organic economy than women and children, who moved to new employment opportunities in the factories. But women and children also had some physical advantages over men. Both were smaller than men and could therefore be used in tight spaces, such as mines. Both, but especially children, had more delicate hands than men and could be used to do such tasks as tying broken cotton threads in the textile mills.

As machinery improved, the need for repairs diminished. Male muscle power became necessary instead to control machinery, which grew ever heavier. Social pressure was exerted to protect female and child labor from occupational hazards. In the 1830s technical change and social forces in Britain conjoined to induce Parliament to legislate the restriction of women and children to nonhazardous occupations. As agricultural production grew in an expanding periphery, more men were released from domestic agriculture for the industrial work force.

A second impact on the labor force was the development of industrialized warfare fought by citizen armies, which was achieved first in World War I. The recruitment or conscription of millions of men, coupled with the need to increase production, resulted in a large-scale return to female industrial labor. Depending on the number of men killed, women either stayed in the industrial work force or left again once the war was over. World War I killed vast numbers both directly and indirectly from infection. In France, Germany, and Britain, women therefore tended to stay in the industrial labor force in relatively large numbers. World War II killed more selectively. In the advanced industrial polities mechanized warfare and excellent control of secondary infections among the wounded produced a high survival rate. U.S. and British troops returned in large numbers from that war and, in the United States, the temporary turn to female labor in the factories was reversed. In Britain, where the trend had begun a generation earlier, reversal was harder. In Germany, Japan, and Russia men were lost in much larger numbers, so that women who had entered the industrial labor force stayed.

In the past twenty years the industrial societies have undergone another massive technological shift that has caused a further explosion of the labor force, most notably in the United States, where the shift began. Both the civil rights movement and affirmative action have helped. The displacement of blue-collar by white-collar jobs calls for a more literate, educated work force. The levels of literacy demanded are beyond the capabilities of many people educated in a system designed merely to produce a relatively literate and efficient blue-collar work force. Until the educational system is adjusted, the conjunction of social and technical forces encourages expansion into the potential work force represented by well-educated but underemployed middle-class women. At the same time a reservoir of hitherto overlooked educable ethnic minorities is being tapped.

The cost of this recent expansion, at least in the United States, has been the return to a two-tier society. Educable haves will be increasingly separated from hard-to-educate have-nots. Given the success of Japanese education in producing much higher levels of the new type of literacy, it should be possible to reverse this trend. The forces working against reversal are the resistance of the educational establishment to fundamental change and the resistance of the political establishment both to forcing fundamental change and to paying for it. This is an area where a society's mix of market and nonmarket philosophies is critical. Societies that insist on market solutions to problems are unlikely to prosper if one of the major stumbling blocks to increased production is a high-quality work force that can only be created by such nonmarket forces as government spending. The rewards of investing in education are so long term that market solutions only work where such investments are evaluated in cultural rather than economic terms because most families prefer immediate consumption to long-term gain.

Another potential means of expanding the labor force lies in increased reproduction. This is a dangerous area. The most successful polities in the world system have been lightly populated enough that excessive consumption did not interfere with their savings rate. Also, the evidence of the past century is that more educated people, who are most likely to pass on a belief in education to their children, have markedly fewer children. Children from poorer families are harder to educate because their home environment usually supports the educational process poorly. Encouraging increased reproduction without specifying which social classes you wish most to reproduce is simply increasing the costs of the educational system, reducing its chances of success, and increasing the size of the lower tier in a two-tier society.

PRODUCTION EXPANSION: A SUMMARY

These four types of production expansion explain much of the success of the world system and its past five-hundred-year history. Polities that have increased production most economically and reliably have usually done best. Yet the era of easy production expansion seems to be drawing to a close either because the last frontier has been attained or because the costs of expansion have risen too high. This is most true of geographical expansion.

Robotization has allowed the possibility of a return to a twenty-four-hour workday in industrial production, thus renewed temporal expansion. But most new jobs in highly industrial nations are white-collar, information-processing jobs subject to temporal expansion only if and when artificial intelligences can be affordably produced.

Labor force expansion is also possible, but the risks and costs are high. More people mean more consumption and, if the additional people are reproduced by the lower socioeconomic classes, the risk of a higher dependency ratio. Given the trend of the workplace toward a need for more educated workers, the lag factor is also crucial. A person is well into the third decade of life before any return on a capital investment in education is realized.

Only in energy expansion does there seem to be much hope of continued production gains. There the picture is rendered problematic by the nature of the available technologies. Hydrocarbons and fissionable materials are in finite supply, and both pose serious long-term environmental hazards, either from global warming or from waste and plant disposal. Apart from fissionable waste, a fission plant will itself be radioactive for centuries after its shielding crumbles, thus requiring constant new shielding to be installed after the plant can no longer produce power. Fusion power, of any sort, is not a proven energy producer but has the attraction of an almost inexhaustible energy source in seawater.

More efficient distribution of electricity, plausible if superconductivity can be made to work at high temperatures and for high currents, would help energy expansion. The northern Pacific coast of America has an immense hydroelectric power–generating potential. Two of the world's largest plants generate electricity to produce nothing but aluminum from the flow of rivers such as the Skeena, too far from any population center for the economical delivery of power. More dams could be built inland to tap the high rainfall on the western slope of the coast range. Existing generators are near the ocean only to facilitate shipping raw aluminum in and finished metal out. British Columbian hydroelectricity could make an appreciable dent in North America's energy requirements if superconductive transmissions systems allowed it to be shipped out of the province to the population centers of southern California, the Great Lakes manufacturing core, and the megalopolis of the eastern seaboard.

Other parts of the world also possess the combination of high rainfall and rugged topography enjoyed by the northern Pacific coast. Tropical Mesoamerica, the southwestern Asian archipelago, New Zealand, and Japan enjoy relatively year-round distribution of high rainfall. Monsoon mainland Southeast Asia has a dry winter that would prevent year-round energy production. The Americas are unusually well favored in energy potential, if superconductive transmission becomes possible. A further major attraction of superconductivity is that it would allow maximum use of environmentally benign power sources. Global warming and disposal of dangerous wastes would not pose problems. The energy gain would, however, be limited because hydroelectric production is possible in only a limited number of regions.

Efficient photoelectric cells might eventually convert sunlight directly to energy on a local basis. This process would make at least small amounts of energy available in any sunny area, but only during daylight hours without the use of battery storage, a technique only suited to relatively low power outputs. Solar technology would clearly favor areas of high sun. Suggestions have also been made for a solar-powered hydrogen fuel economy. Photoelectric converters, if necessary augmented by huge mirrors in earth orbit, could crack hydrogen from seawater. Hydrogen fuel could be shipped through existing natural gas pipelines and used for any purpose to which natural gas is now put. The Gulf of Mexico, with its high sunlight and proximity to an optimal earth orbit for spacecraft launched from Florida and to the hub of the North American natural gas pipeline net at Houston, Texas, is a logical candidate for such energy expansion. Hydrogen burns very cleanly, producing only water vapor as a byproduct, and thus avoids the atmospheric contamination producing global warming. Its disadvantages include ease of leakage from the natural gas system, which was designed for a gas with a much larger molecule, the cost of launching the necessary mirrors into orbit, and the current inefficiency of photoelectric converters.

The development of any of these technologies is possible only with relatively massive public and private investment, first in basic research, then in the more costly necessary infrastructure. None of these solutions is likely to be developed before early in the next century. Until such time the efficient use of current energy production systems is critical, as is efficient use of the existing labor force. Marked changes in the educational system or in demography need at least a full generation to produce a result. Just as the transition to the oil economy from the coal economy took from the late 1890s to the 1920s in the United States, where conditions were most suited to it, the best-case transition from the oil economy to something else will take a human generation. Where much is invested in present technology and social structure, as was the case in Britain and is now the case in the United States, the transition will take longer, a generation and a half to two generations.

EXPANSION IN PRODUCTIVITY

Throughout most of the history of the world system the easiest and cheapest way to increase production has been simple expansion on the various frontiers. Productivity increases were not ignored but were generally found to be more expensive. As the frontiers have closed, productivity has once more become important.

From a geographic point of view the most important productivity gain has come from better information flow. Improved transportation has always meant improved information flow, but by the 1600s technologies were being developed to move information even more rapidly than people or things (Wilson, 1976). First came the mechanical telegraph, read at a distance with a telescope. It used much labor and worked only in daylight and good weather. Such telegraphs found particular favor with the military. More rapid and cost-effective progress in information flow came with the electric telegraph in the 1830s. Many people suggested that the flow of electricity along a wire could carry coded information; the problem was how. After many false starts and the development of some complex hardware by people such as Werner Siemens in Germany, Samuel Finley Breese Morse came independently to the conclusion that the main problem in implementation was in software and that a series of long and short interruptions of an electric circuit would produce a workable code (Atherton, 1984, 80–89). Morse was a Renaissance man of the early U.S. republic, a brilliant artist (Hugill, 1986) as well as a scientist. He turned his attention to telegraphy when returning to the United States from Europe in 1832, convinced that his artistic career could go no further after a series of setbacks.

The real success of Morse's code came when Ezra Cornell, of Ithaca, New York, who later left a great fortune to the university that bears his name, merged his telegraph company with that of Hiram Sibley of Rochester, New York, to found Western Union in 1856 (Ellis et al., 1967, 254). Western Union's fortune was made serving the railroads from New York to Buffalo, then on west into Michigan. The huge distances that had to be covered by U.S. railroads in a country short of investment capital meant much single tracking. If trains kept to published schedules they could meet at known times at the few passing points provided. If they got off schedule they were lost in the system, and collisions, often fatal, occurred with depressing frequency. Western Union greatly increased the efficiency of railroad operation because it allowed the position of any given train in the system to be fixed with considerable accuracy. It thus allowed U.S. railroads to avoid the high cost of double tracking. Western Union found that much other information could be carried by telegraph lines, commercial information about prices for agricultural goods at various warehousing and port facilities, information about share prices, and the like.

Morse pioneered a simple system of stringing telegraph wire from wooden poles that worked well for a land-based system: telegraphs proliferated alongside the railroads they served. Maritime empires like Britain looked to the telegraph for better information flow but had first to resolve the problem of laying submarine cables. The British physicist, Michael Farraday, suggested to William Siemens, whose brother had contracts to lay underground telegraph cables for the Prussian army, that rubber might be used as a flexible, waterproofing insulator (Haigh, 1968, 26). Cables also had to be armored to resist attack by sea creatures. The submarine cable first showed its possibilities when one was put into service between Britain and France in 1851 (Haigh, 1968, 34). The British admiralty quickly expressed an interest in submarine cables and had the facilities to lay them: most early experiments used warships to lay the cables. The formation of the Telegraph Construction and Maintenance Company in 1864 brought order to the world of the submarine cable akin to that imposed on U.S. overland cables by Western Union in 1855. The first great achievement of the TC&MC was to charter the steamship *Great Eastern* to lay a transatlantic cable in 1866. *Great Eastern* not only laid the cable in one trip but also repaired the 1865 cable, which had required two smaller ships meeting in mid-Atlantic. The approximate cost to transmit one letter across the Atlantic in 1868 was 3.15 cents. Rates fell to 40 cents a word by 1885 and 25 cents a word by 1888 (Atherton, 1984, 95). Submarine cables were expensive and lasted only a few years in service, but by the mid-1860s they were reliable enough for the TC&MC to begin linking all parts of the world to London (Haigh, 1968, 41). *Great Eastern* went on to lay the first submarine cable to India. Previous attempts overland and through the Black or Red Sea had failed, and the overland route was, in any case, subject to foreign interruption.

The second step forward in information flow came with the telephone in the 1870s. It was again in the United States that the innovation took hold most quickly, and again because it occupied a large continental area. Telephony was a more difficult technology than telegraphy. As Bell conceived it in 1876, the telephone was a short-distance technology powered entirely by the human voice (Atherton, 1984, 105). It was subject to interference from human-generated and natural electricity. The former became common as cities were wired for domestic light, streetcars, and industry in the late 1880s and 1890s (Brooks, 1976, 85–86). Although it was possible to phone 292 miles from New York to Boston as early as 1884 using microphone-modulated current from a battery, the call cost $2 by day and $1 by night. True long-distance telephony required amplification of the flow of electrons to be successful.

Amplification of the electron flow was made possible in 1912 by Lee

DeForest's triode vacuum tube. The purchase and improvement of De-Forest's patents by the Bell system allowed a line to be opened between New York and San Francisco in 1915, but the technology had been pushed to its limits (Brooks, 1976, 137–39). DeForest's use of vacuum tubes to amplify the electron flow was not a technology that suited long-distance submarine telephony, although a transatlantic line was finally laid in 1956 using fifty-one submerged tube amplifiers between Scotland and Newfoundland (Atherton, 1984, 112). Lee DeForest's triode made possible an alternative form of long-distance telephony, by radio. Real-ization that wireless telegraphy and telephony were possible came from the theoretical work of Oliver Heaviside and the theoretical and practical work of Guglielmo Marconi, who first sent a wireless telegraphic signal across the Atlantic in 1901 (Atherton, 1984, 109–10, 186–89). Wireless telegraphy was of obvious worth to ship owners, whose ships were "lost" once they sailed. The rescue of at least some of the passengers and crew of the *Titanic* in 1912 was ample demonstration of its worth. The military took to wireless telegraphy with an enthusiasm not always tempered with common sense. In the judgment of the major historian of *The Great War at Sea,* "instead of providing information for the commander on the spot, [wireless telegraphy] was used . . . to give direct operational orders, sometimes in absurd detail" (Hough, 1983, 84).

Until the development of pulse code modulation (PCM), which digitizes the flow of information and was patented by Alec Harley Reeves in 1938, all long-distance radio telephony, whether in the long- or short-wave bands, was amplitude modulated (AM) and suffered severe variations in signal strength. Fading in and out of voice transmission was a severe problem. Frequency-modulated (FM) transmission did not suffer from fading but worked only in a direct line. AM signals bounce off the Heaviside layer in the upper atmosphere to travel around the earth's curvature, and FM signals require satellites in geosynchronous orbit to pass signals from earth station to earth station. PCM radio messages are digitized, transmitted in AM or FM form, then reassembled. The first PCM system, despite being bulky and cumbersome, was installed on the Atlantic in 1943 for use by Churchill and Roosevelt for war purposes (Williamson, 1988, 46–47). PCM came into its own with the replacement of the triode vacuum tube by the transistor, which when first applied to British telephone circuits, raised capacity twelvefold.

In early telecommunications, by wire or wireless, each circuit carried only a single message transmitted in real time. Ways were worked out to carry two messages (duplex transmission) and, eventually, many messages (multiplexing), but each message still traveled in real time. PCM allows large amounts of information to be compressed by a computer for transmission in a short burst and then reassembled by another computer into a real-time message. Together with extensive multiplexing this has made possible economical global telephony and transmission of commercial and other information through computer networks, fax machines, and the like. Time compression also makes it financially possible to use such high-capital items as earth-orbit satellites. So much data can be transmitted in digitally multiplexed, time-compressed form that high costs become affordable.

All this has led to what we have come to call the information economy. At the level of the individual it is still restricted by the necessity for wires to carry the information, although even that barrier is beginning to fall to digitized, wireless, cellular telephones. High cost and limited geographic availability restrict cellular phones to a relatively small group of users at present. Statewide networks are, however, already in place even in such states as Texas, and it is only the want of computer systems to link short-range cell to short-range cell that prevents long-term expansion. Smaller and less expensive cellular technology will probably make individual portable phones usable from anywhere on the planet to anywhere else early in the next century.

Better telecommunications have increased productivity through better information flow, which has increased efficiency in decision making. The groups that have most benefited from this have been those that produce and manage the flow of information: academics, communicators, and managers. In combination with the greatly accelerated transport ushered in by jet aviation and more efficient ground transportation, a relatively small number of such folk have seen their productivity greatly expanded because their services can be used much farther from their home bases than was possible before. Research and management can now occur anywhere in the telecommunications network—not necessarily where goods are produced—with occasional visits to production facilities by managers traveling by jet.

This is a radical shift from the sort of productivity increase produced by the managerial revolution that began in the United States early in this century. In that case the aim was to increase the productivity of labor, and thus of capital, by submitting the work flow to highly centralized management. Ford, Taylor, and Gilbreth demonstrated how remarkable a jump in productivity was possible. After World War II long-distance airplanes made the multinational corporation a reality by allowing U.S. companies to carry the new managerial technology abroad. Middle managers were, however, still required at each production facility to implement the decisions of central management. Large numbers of site-specific middle managers are no longer necessary for two reasons: information flow has been greatly improved, and production has been largely robotized. What site-specific management is necessary is provided by much smaller cadres of middle managers, or possibly even by the work force itself, as in the Japanese model. Otherwise decisions are made at a distance with occasional site visits by managers located elsewhere.

Improvements in managerial productivity are, of course, more cost efficient than those in labor productivity because management is more expensive per hour of work. It proved relatively easy to raise labor productivity by substituting capital (mechanization, now robotization) and raising the level of managerial skill. Whether it will prove as easy to raise managerial productivity is a more difficult question. Capital can be substituted only in some cases. Computers have so far tended to replace only lower-paid white-collar workers. Word processors clearly reduce the need for secretarial workers but not for managers. Such gains also clearly require more rather than less skill at the managerial level, which tends to raise rather than lower the per hour cost of managerial work, even if that work is used more efficiently by being applied over a larger geographic area.

Other sorts of productivity increase have also occurred. Highly specialized ships have developed over the past twenty years and air freighters have radically reduced inventory costs for businesses. Better transport has greatly facilitated the full-scale adoption of the "just-in-time" system. Robotized factories can produce large volumes of goods relatively quickly and thereby reduce factory inventory. Goods can be shipped almost immediately to market, reducing warehouse inventory at the sales end of the pipeline.

PREDICTIONS

One of the perils of writing in the social science tradition is that it requires prediction rather than conclusion, and in prediction one is open to the judgment of the future. Historians suffer no such constraints. Much of this book has concerned the development of the world system and its cyclical nature. I have taken Wallerstein's evaluation of the importance of production at its obvious value, and I have expanded the notion from the almost solely organic production of Eotechnics through the combined organic and inorganic production of Paleotechnics to the heavily inorganic production of Neotechnics.

I have been concerned with Ricardian notions of agrarian productivity, in particular the Ricardian notion that free trade is an absolute good in promoting productivity, and have extended them into the industrial sphere. The capitalist world system is a product of both the Ricardian desire to maximize productivity and the steady expansion of the major production frontiers. Open systems are, however, fundamentally dynamic. The most productive polity in a given pair of Kondratievs dominates the world system. We are currently leaving the fifth world leadership cycle. If the United States is no longer the hegemonous power, who will succeed it? The temptation to anyone with an historical imagination is to look back to the last changeover. As Britain lost power many contemporaries expressed their concern. To a modern audience writers such as H. G. Wells and H. H. Munro (Saki) were writing what we now call science fiction. In *When William Came*, Saki postulated an imperial German assault upon Britain in the second decade of this century. The British navy is overwhelmed by German aircraft (Munro, 1914, 705), a common theme in the popular press of the time, which viewed "Zeppelins" with particular concern (Gollin, 1984, 315–52). Such accounts can no longer be read as predictive because they clearly failed to predict accurately. Yet that body of literature accomplished something: by suggesting a possible future of imperial German hegemony, it helped to produce the willingness to resist it. A predictive exercise that seems to fail in the judgment of history is not a total failure if it reduces the chance that an unwelcome future will occur.

I believe the data suggest two futures are possible at this juncture. Future 1 is a sixth world leadership cycle and a new hegemon, with or without a period of war. Future 2 is the emergence of a truly global system with no dominant polity.

Future 1 would likely be marked by a transitional war, though not necessarily war in the traditional, military sense. Such a war could be economic. If military, the war could be nuclear or conventional. If nuclear, it could be total or limited. If total, it would be disastrous for humankind. Because so many recognize its potential for disaster, total nuclear war is the least likely possible future; nevertheless it is possible. Limited nuclear war would destroy the productive capacity of some polity or set of polities in a way that conventional war cannot. Limited nuclear war would render the affected area uninhabitable beyond a dangerous subsistence level. Conventional war would have long-term consequences of a different sort, all previous experience having shown that war tends to beggar the economies involved over the short term and that recovery promotes long-term excessive military spending to protect short-term gains.

In favor of economic war is the fear of escalation from conventional to limited nuclear to total nuclear war. The major economies may well retreat behind tariff walls in the way Britain did early in this century. Even in the most market-oriented economies forces urging retreat are strong. Economic war would be fought and won over a generation by the power that best developed its productive forces. Unfortunately, those productive forces are no longer easy to predict. In the Eotechnic land was dominant, in the Paleotechnic land plus coal was critical, and in the Neotechnic land plus oil plus managerial and technical skill is the key. The next phase will require at least the addition of research skill and a better-educated white-collar work force.

In varying ways the forces for this economic war are now gathered around three actors or groups of actors: North America, Germany, and Japan. North America is the powerhouse economy of the two American continents, with Canada as a huge source of largely untapped raw materials and Latin America as a huge source of inexpensive labor, either to be imported into North America or to play a semiperipheral role *in situ,* as Mexico already does. North America has the major advantage of having long enjoyed a huge, coherent, wealthy, domestic market and is extending that market north and south. The recent free trade agreements with Canada and Mexico are necessary preconditions for a North American customs union, itself a precondition for a larger system covering both American continents.

Germany has, since unification in 1871, been the powerhouse economy of Europe. In 1992 Germany will dominate the European Economic Community (EEC), a system that bears a remarkable resemblance to an enlarged Mitteleuropa extending from the Atlantic to the Oder. Events in Eastern Europe promise the possibility of expansion east. The EEC will provide a market at least as large as the Americas, although less coherent in terms of language, culture, and economic traditions. Eastern Europe will give the EEC the sort of raw material and labor resources that North America has in Canada and Latin America.

Japan is obviously the powerhouse economy of eastern Asia, though it is heavily dependent upon markets in North America and Europe. Of the potential hegemons, Japan has the smallest domestic market, although one rendered even more coherent than that of North America by lack of ethnic differentiation. Japan also lacks natural resources. Even so, a Japan at the head of a resurgent greater East Asian coprosperity sphere is not a total impossibility. The Asian economies might be willing to make common cause with Japan in the next century if they are shut out from access to the North American and European markets by tariffs. China would be Japan's logical source of raw materials and inexpensive labor.

Future 1 has three clear subsets. In 1(a) North America would achieve hegemony, succeeding the United States rather as Britain succeeded England; in 1(e) the EEC would achieve it; and in 1(j), Japan. The conditions for each gaining hegemony would be supremacy in production, thus in the conditions promoting production. These conditions no longer depend upon land but increasingly upon skill. Each would-be hegemon has, in reality or potentially, adequate resources in the agricultural sphere at current levels of technology. Each has access to adequate levels of investment capital because of the international nature of the global money market. Access to affordable land for infrastructure (transportation systems, factories, and housing) varies somewhat. All three polities have severely congested areas, but Japan's costs for land are clearly very much higher overall than those of the EEC or North America. Strengths in industrial production also vary markedly, with each power having advantages in potential for production or in quality of product. Raw material resources could be a critical control on potential if tariff walls are erected. Canadian resources would make tariff walls particularly advantageous to North America. As long as the world system is driven by oil, access to the great reserves of inexpensive oil in the Middle East is vital. If those supplies are cut off, North America would be the least affected area: Texas, the Gulf of Mexico, Mexico, and western Canada all have substantial reserves, though they might be expensive to produce.

In terms of the social forces of production there seem to be two current keys to productivity increases. Japan has innovated in areas such as the "just-in-time" system and "worker responsibility" to make its factory labor more productive. Such managerial technologies have, however, diffused, relatively easily in recent years, in particular to North America. It is not clear, therefore, that they will provide a long-run advantage to Japan in the way that managerial technology proved advantageous to the United States early in the Neotechnic, when other countries found the U.S. mix of managerial technology and social mobility hard to copy.

The second key is to improve the efficiency of the more expensive segment of the labor force, management. Management efficiency has been developed most extensively in America, but at least one key component of the EEC, Britain, has been developing along similar lines, with France moving toward an equally highly computerized society on a national rather than a corporate level. Inexpensive jet transport to move managerial labor around is well developed in the United States, but Europe is not far behind. In the rush to increase managerial efficiency there is no clear advantage to North America or Europe, only a disadvantage to a Japan that has not developed multinational corporate management as thoroughly. Language problems also handicap Japan away from Asia because of the massive trend to English as the lingua franca of the business world.

In terms of the social forces of production, there is the question of what is to be produced in the future. The so-called information economy is, in many cases, merely the expression of improved productivity of managerial labor. A true information economy would be production of research knowledge and training of skilled managers. This requires an increasingly complex educational system. Here Japan is disadvantaged because its universities are not based on research in the same sense as North American or European ones. Long-term basic research is likely to prove critical to long-term success, assuming that basic research eventually leads to useful applications. Avoiding both esoterica and excessive concern with application will be the key to success. Types of research that do not seem promising include biotechnology and robotics. As it has developed so far, biotechnology has focused on the needs of agriculture and improving the productivity of crops and domesticated animals through higher yields, greater resistance to disease, and the like. These gains can be useful, but agriculture no longer employs more than a small percentage of the population and is no longer the most important source of raw materials for industry. The current level of robotization is reducing the percentage of the work force employed in blue-collar jobs, at least in the industrialized polities. Only where the cost of relatively unskilled labor is low can humans still compete cost-effectively with robots. Research that leads to lowered energy costs, safer energy production, and higher managerial productivity is clearly to be favored. Because it is extremely expensive, research on prolonging human working lives falls somewhere between these extremes. If biotechnology has a major contribution to make it is likely to be in the field of prolongation of human working lives. The more educated the work force becomes, the more important it is that its working life be extended; otherwise the increasing investment in education will not realize an adequate return. Long lives and rapid shifts in technology also suggest that such high investments will have to be protected by regular reeducation of the work force, which further implies that the more we know about how people learn, the better.

World System
Theory and
Geographic
Reality

The set of polities which comes up with the best set of decisions with regard to what sort of research knowledge to produce will clearly do best. Excessive military research, beyond that needed for national security, is likely to be counterproductive, diverting resources from basic scientific and useful applied research.

In this sense the indicators suggest that North America is doing best at overall research volume, but its research is directed into military applications at a far greater rate than is that of Japan or Europe. Japan clearly leads in patents, which measure, at least to some extent, commercial application of research.

Probably none of the three contenders would emerge with ease as the victor of economic war. Because information now diffuses with ease and rapidity, no polity has a monopoly on even the most innovatory change. Social structure no longer provides much advantage, all of the industrial economies having learned the necessity of a broad, socially and geographically mobile middle class. Minor differences in social structure, such as the tendency of North American society to promote consumption and the tendency of European and Japanese society to promote saving, are probably unimportant. Local capital accumulation through local savings is less relevant to economic success in a period when capital is immensely mobile. Strong economies can borrow all the capital they need without generating it internally.

Just as information and capital now flow freely around the world so do resources. National possession of raw materials, which was critical in the Paleotechnic and early Neotechnic, no longer seems necessary for economic hegemony. No developed economy controls the world's major hydrocarbon resource, the oil fields of the Arab world. The Pax Americana has ensured a free flow of resources since 1945, with only a brief hiatus caused by the creation of OPEC in the early 1970s. Even that proved of minor importance, since the industrial economies found it relatively easy to slash consumption and switch to alternative fuels. The Gulf War of 1990 hardly even affected the price of oil.

The only advantage to national control of raw materials, the ability to fight a protracted military war, has disappeared. The threat of escalation to all-out nuclear war has, at least since 1945, removed the likelihood of protracted campaigns: nuclear war would be fought quickly and with an existing stock of weapons.

I therefore suggest that future 2 is marginally more likely than future 1. Military war is too dangerous and the conditions for economic war do not encourage the emergence of a hegemon. The global economy is already so integrated that it will be hard to sunder. Economic groups such as G7, the group of the seven most highly industrialized nations, are already cooperating at transnational levels in terms of facilitating global trade. Recent recognition that pollution is not a national problem but a global one is also significant. Only global action can resolve the potential consequences of destruction of the ozone layer, global warming, and the like.

Such globalism was proposed at the end of the fourth world leadership cycle as a solution to the problems attendant upon that transition.

International capitalism, socialism, and communism were proposed, but nationalism prevailed. King Camp Gillette, for example, proposed that all shareholders in all other corporations trade in their shares for shares in his World Corporation, which would then manage the global economy (Gillette, 1910). Gillette envisioned vast headquarters in a huge city overlying Buffalo, New York, powered by electricity from the falls of Niagara. No one took Gillette seriously, and he later made his fortune in an entirely different way with the invention of the safety razor.

An international labor union was also proposed, in the form of the United States–based International Workers of the World (Renshaw, 1968). Such a union fell far short of the global unification proposed by Gillette, its aim being merely to advance the cause of one social class. Although tiny, persecuted, and ultimately a grandiose failure, the IWW had an impact out of proportion to its size. It ran counter to the strains of possessive individualism and nationalism that have dominated U.S. and, indeed, global capitalist society for the past two hundred years.

The labor union movement in general believed that international labor would rise in unison against war among the capitalist classes of Britain, France, and Germany (Williams, 1972). It was not entirely wrong: all three countries suffered industrial unrest during World War I, but not enough to prevent production. The internationalist elements in the unions were insufficiently aware of the nationalist fervor of the working classes, which was skillfully developed by such populist conservative politicians as Britain's Disraeli in the late 1800s and Churchill in the early 1900s. The unions also failed to account for the ability of the capitalist system to mobilize female labor at a time of crisis.

Most serious of the three moves toward internationalism was that represented by communism under the able leadership of Lenin. Marx's writings were an avowedly universalistic, deterministic model of historical development. Lenin naturally adopted an international perspective, claiming that communism was predestined to dominate the world. In the aftermath of World War I such a claim seemed much less silly than it does today. Compounded by the economic crisis of capitalism begun by the Great Depression in 1929, the claim for communism looked far more powerful than it ever came to be. From the vantage point of a capitalism resurgent after 1945, communism now looks like a failed experimental phase of the long-term westernization of Russia. Internationalism never properly caught hold in the Soviet Union in any case because much of the revolution and the development of the communist state were driven by nationalist forces.

Internationalism generally failed to take hold at the transition into the fifth world leadership cycle because nationalism was powerful everywhere, because capitalism was still organized on a nation-state basis, and because the costs of war were still low. These conditions are no longer true. The costs of war are likely to be prohibitive, capitalism has been reorganized on a transnational basis, and even nationalism is showing signs of fading in some parts of the world, notably Western Europe.

Most significant of all these forces is the transnational reorganization of capitalism, which first became apparent with the creation of the first large-scale, multinational corporation in the form of the Ford Motor Company (Wilkins & Hill, 1964). By 1913 Ford had demonstrated that transnational capitalism would work, although its success was hampered by the reimposition of tariffs on imported goods, first in England, then in the rest of the world. Ford did not move to England to avoid tariffs; there were no tariffs when Ford began production there. Ford assembled vehicles away from Detroit simply to reduce transport costs, it being less expensive to ship less bulky kits than completed vehicles. Where Ford led, others—notably General Motors—followed, and by the late 1920s U.S. multinationals dominated the world's automobile industry (Hugill, 1988a; Maxcy, 1981).

After World War II the transnational reorganization of capitalism accelerated hard. U.S. success in the production war underlying World War II was convincing proof that U.S. corporations were better organized and managed than those of the rest of the capitalist world. U.S. production technology and organization were assiduously copied. The need for a broadly educated, consumption-oriented middle class was also recognized. The major industrial nations retooled their educational systems to place greater emphasis on white-collar work and higher education. This revised educational system began to stress transnational values rather than the national values emphasized in the earlier version of mass education that took form in the last decades of the nineteenth century. The generation produced by these postwar educational changes is now moving into political power. It predictably will prefer transnational approaches to national ones, given that national approaches emphasize conflict between national states rather than consensus.

We are thus at an interesting juncture of the world system. The rather hesitant attempts to move to a truly global system early in this century failed because nationalism was strong. Yet nationalism is only a relatively recent force in the development of the world system, and it has rarely been a powerful force in the polities that achieved hegemony in the different world leadership cycles. Nationalism became possible only for a brief period with the better land transport of the Paleotechnic. Eotechnic hegemony rested on global trade across national boundaries. Neotechnic success now shows signs of requiring similar transnationalism. No one polity can be self-sufficient in all the resources needed in a Neotechnic economy, in particular as long as hydrocarbon energy is in use. All polities must begin to act in concert if the effects of pollution are to be minimized because it is becoming abundantly clear that global industrialization is producing environmental problems of global proportions. All polities must act in concert also to reduce the risk of a totally destructive nuclear exchange. All-out nuclear war no longer even requires that opponents hit targets in each other's territory. It matters little where the Soviet or U.S. nuclear arsenals are exploded if the result is nuclear winter and the extinction of most or all human life on the planet.

In effect we are returning to the conditions of the late eighteenth century so forcefully analyzed by Smith, Ricardo, and Malthus. There are no more easy production gains to be won by geographic, temporal, or labor force expansion. Dreams of getting off the planet into space, in effect renewing the geographic frontier, are likely to founder on the energy expense of escaping the earth's gravity well and on the time expense of voyaging to even our nearest, probably unusable planetary neighbors. Only a technology that radically reduces the time of travel to usable planets and that radically lowers the energy cost of such trips would allow renewed geographic expansion. Such is the stuff of science fiction: thinkable but unlikely as long as Einstein's theory of the way the universe runs continues to be valid.

If we cannot expand we must win production gains by productivity increases, and we must return to Smith, Ricardo, and Malthus. We must hold down consumption by avoiding excessive population growth relative to production growth (Malthus). We must maximize production by encouraging each region of the world to specialize in producing what it produces best because of local conditions of resource availability, capital accumulation, and socioeconomic structure (Ricardo). And we must encourage the global exchange of regional surplus through the most efficient mechanism we know of, the price mechanism freed from the narrow national interests represented by the government of nation-states (Ricardo and Smith).

A successful shift to future 2 would not be without problems, but the problems would be less pressing and onerous than the problems of any version of future 1. Future 1 would reduce surplus accumulation through the inefficiencies attendant on attempts at regional self-sufficiency. Selfish behavior at a regional level would reduce the chance of resolving issues such as global warming. A polity or set of polities might win the production war but so pollute the planet that one hundred years from now sea-level rise from global warming would be a severe threat to its major cities.

Transnational production with global trade and a global consensus on rules about pollution, resource depletion, and possibly consumption would prevent the more expensive of such mistakes and slow their onset. It is true that we could protect our major cities against the prospective rise in sea level from global warming, but the capital cost of such protection would be as horrendous as the capital loss if we had to abandon the cities.

Transnationalism would also not reduce the need to move away from technologies that use nonrenewable resources. By reducing waste it might, however, give us more time to make the transition to technologies that are less heavily dependent upon nonrenewable resources. The transition depends, however, upon the continuation, perhaps the acceleration, of basic research. In this sense it matters little whether we reach future 1 or future 2. The long-term survival of humans as the dominant species on this planet depends on matching production (and production expectancies) to consumption. We expect to consume so much in the way of food and product during our lifetimes. Expectancies are very high in the developed, industrialized nations and relatively low elsewhere. More people mean more consumption, especially if the population increase is in developed nations. More development also means more consumption, as consumption expectancies increase with industrialization. We could not feed, house, and clothe our current planetary population with its current expectancies using Eotechnic or even Paleotechnic technology. Even with reduced expectancies we might be close to planetary capacity. We are dependent upon Neotechnic technology at the minimum, and that technology has finite limits imposed by the geological supply of hydrocarbons and fissionables. However efficient we become, the supply will eventually peter out.

A transnational global system will prosper in the long haul only if we use our hydrocarbons and fissionables with wisdom to fund the basic research for renewable energy sources and renewable raw materials. In this vital sense the means of production for the next cycle of the world economy, be it national or transnational, will be the research laboratories and universities rather than simply land, labor, energy, and industry as in previous cycles. If we fail to make the transition to a renewable high-energy source, we will slip slowly back into the Eotechnic, with a probable return to petty regional interests, including the use of war, as a declining production base is spread more and more thinly.

Unpublished Materials and Research Collections

Commonwealth Society, London.
Ford Archives: Henry Ford Museum, Dearborn, Mich.
PRO, Public Records Office, Kew, London. AIR and AVIA files.
NASM: National Air and Space Museum, Washington, D.C.

General Reference Materials

Bank of Japan. 1966. *Hundred Year Statistics of the Japanese Economy.* Tokyo: Bank of Japan.

Central Statistical Office. Annually. *CSO Annual Abstract of Statistics.* London: Her Majesty's Stationery Office [H.M.S.O.].

The Europa World Year Book (*The Europa Year Book* before 1989). Annually. London: Europa Publications.

Eurostat. Basic Statistics of the Community. Annually. Luxembourg: Office for Official Publications of the European Communities.

Eurostat Review, 1972–1981. 1983. Luxembourg: Office for Official Publications of the European Communities.

FAA Statistical Handbook of Aviation. Annually. Washington, D.C.: Federal Aviation Administration.

Mitchell, B. R. 1980. *European Historical Statistics, 1750–1975,* 2d ed. New York: Facts on File.

Scherer, John L., ed. 1984. *China Facts and Figures Annual,* Vol. 7, *1984.* New York: Academic Press.

The Statesman's Year-book. Annually. London: Macmillan.

Statistical Abstract of the United States. Annually. Washington, D.C.: U.S. Department of Commerce.

Statistical Handbook of Japan. Annually. Tokyo: Statistics Bureau Management and Coordination Agency.

Urquhart, M. C., ed. 1965. *Historical Statistics of Canada.* Toronto: Macmillan.

U.S. Bureau of the Census. 1960. *Historical Statistics of the United States, Colonial Times to 1957.* Washington, D.C.: Bureau of the Census.

———. 1975. *Historical Statistics of the United States, Colonial Times to 1970.* Washington, D.C.: Bureau of the Census.

ATLASES

Curtin, Philip D. 1969. *The Atlantic Slave Trade: A Census.* Madison: University of Wisconsin Press.

Espenshade, Edward B., ed. 1986. *Goode's World Atlas,* 17th ed. Chicago: Rand McNally.

Herrmann, Albert. 1966. *An Historical Atlas of China,* 2d ed. Chicago: Aldine.

International Association of Agricultural Economists. 1969. *The World Atlas of Agriculture.* Vol. 1, *Europe, U.S.S.R., Asia Minor.* Italy: Instituto Geografico de Agostini-Novarro.

Langton, John, & R. J. Morris, eds. 1986. *Atlas of Industrializing Britain, 1780–1914.* New York: Methuen.

Miller, Theodore R. 1969. *Miller's Geographic History of the Americas, Prehistory to the Present.* New York: Wiley.

Modelski, M. 1987. *Railroad Maps of North America. The First Hundred Years.* New York: Bonanza Books.

Palmer, R. R., ed. 1965. *Atlas of World History.* Chicago: Rand McNally.

Pemsel, Helmut. 1977. *A History of War at Sea. An Atlas and Chronology of Conflict at Sea from Earliest Times to the Present.* Annapolis, Md.: Naval Institute.

Westermann, 1978. *Grösser Atlas Zür Weltgeschichte.* Braunschweig: Westermann.

CATALOGS, DIRECTORIES, AND ENCYCLOPEDIAS

Angelucci, Enzo, ed. 1982. *World Encyclopedia of Civil Aircraft from Leonardo Da Vinci to the Present.* New York: Crown.

Automobile Club of Italy. 1982. *World Cars 1982.* Pelham, N.Y.: Herald.

Bruce, J. M. 1957. *British Aeroplanes 1914–1918.* New York: Funk & Wagnalls.

Cars International 1989. A Comprehensive Guide to Current Model Specifications. London: PRS.

Conway's All the World's Fighting Ships, 1860–1905. 1979. New York: Mayflower.

Culshaw, David, & Peter Horrobin. 1974. *The Complete Catalogue of British Cars.* New York: Morrow.

Georgano, G. N., ed. 1982. *The New Encyclopedia of Motorcars, 1885 to the Present,* 3d ed. New York: Dutton.

Gibbons, Tony. 1983. *The Complete Encyclopedia of Battleships and Battlecruisers. A Technical Directory of All the World's Capital Ships from 1860 to the Present Day.* London: Salamander.

Gunston, Bill, ed. 1980. *The Illustrated Encyclopedia of Propeller Airliners.* New York: Exeter.

———. 1983. *Aircraft of the Soviet Union.* London: Osprey.

———. 1989. *World Encyclopedia of Aero Engines,* 2d ed. Wellingborough, Northants: Patrick Stephens.

Hollingsworth, Brian. 1982. *The Illustrated Encyclopedia of the World's Steam Passenger Locomotives. A Technical Directory of Major International Express Train Engines from the 1820s to the Present Day.* London: Salamander.

———. 1984. *The Illustrated Encyclopedia of North American Locomotives. A Historical Directory of America's Greatest Locomotives from 1830 to the Present Day.* New York: Crescent.

Hollingsworth, Brian, & Arthur Cook. 1983. *The Illustrated Encyclopedia of the World's Modern Locomotives. A Technical Directory of Major International Diesel, Electric, and Gas-Turbine Locomotives from 1879 to the Present Day.* New York: Crescent.

Jane, Fred T. 1903. *All the World's Fighting Ships.* London: Sampson Low, Marston.

Jones, Lloyd S. 1980. *U.S. Bombers, 1928 to 1980s,* 3d ed. Fallbrook, Calif.: Aero Publishers.

Olyslager Auto Library. 1973. *British Cars of the Late Thirties.* London: Warne.

Oswald, Werner. 1977. *Deutsche Autos 1920–1945.* Stuttgart: Motorbuch Verlag.

Silverstone, Paul H. 1984. *Directory of the World's Capital Ships.* New York: Hippocrene.

World Almanac and Encyclopedia. Annually. New York: The New York World.

BOOKS AND ARTICLES

AAR (Association of American Railroads). *Research Report, 1984–1985.*
———. *Research Report, 1986–1987.*

Agnew, John. 1987. *The United States in the World-Economy.* New York: Cambridge University Press.

Ahlstrom, Sydney E. 1972. *A Religious History of the American People.* New Haven: Yale University Press.

Albion, Robert Greenhalgh. 1926. *Forests and Sea Power. The Timber Problem of the Royal Navy, 1652–1862.* Cambridge: Harvard University Press.

Alden, John D. 1979. *The Fleet Submarine in the U.S. Navy. A Design and Construction History.* London: Arms & Armour.

Allen, G. C. 1981. *A Short Economic History of Modern Japan,* 4th ed. London: Macmillan.

Allen, G. Freeman. 1978. *The Fastest Trains in the World.* New York: Scribner's.

Allen, Michael. 1985. *British Family Cars of the Fifties.* Newbury Park, Calif.: Haynes.

Allen, Peter. 1984. *The Ninety-one before Lindbergh.* Shrewsbury: Airlife.

Allward, Maurice. 1981. *An Illustrated History of Seaplanes and Flying-boats.* Ashbourne, Derbyshire: Moorland.

Altshuler, Alan, et al. 1984. *The Future of the Automobile: The Report of MIT's International Automobile Program.* Cambridge: MIT Press.

Andrews, Allen. 1964. *The Mad Motorists. The Great Peking–Paris Race of '07.* London: Harrap.

Armytage, W. H. G. 1976. *A Social History of Engineering,* 4th ed. London: Faber & Faber.

Atherton, W. A. 1984. *From Compass to Computer. A History of Electrical and Electronics Engineering.* San Francisco: San Francisco Press.

Bagnall, William R. 1893. *The Textile Industries of the United States.* Cambridge, Mass.: Riverside.

Baker, Alan R. H. 1973. "Changes in the Later Middle Ages." In Darby, 1973a, 186–247.

Bardou, Jean-Pierre, et al. 1982. *The Automobile Revolution. The Impact of an Industry.* Chapel Hill: University of North Carolina Press.

Barfield, Norman. 1975. "Aerospatiale/BAC Concorde." In *Aircraft in Profile* 14: 73–112. Windsor: Profile Publications.

Bartleet, H. W. 1931. *Bartleet's Bicycle Book.* London: Burrow.

Bass, George F. 1987. "Oldest Known Shipwreck Reveals Splendors of the Bronze Age." *National Geographic* 172 (6):693–734.

References   Beaty, David. 1976. *The Water Jump, The Story of Transatlantic Flight.* New York: Harper & Row.

Beaubois, Henry. 1976. *Airships: An Illustrated History.* New York: Two Continents.

Bendix, Reinhard. 1978. *Kings or People. Power and the Mandate to Rule.* Berkeley and Los Angeles: University of California Press.

Bennett, H. S. 1937. *Life on the English Manor. A Study of Peasant Conditions, 1150–1400.* Cambridge: Cambridge University Press.

Bent, Silas. 1938. *Slaves by the Billion: The Story of Mechanical Progress in the Home.* New York: Longmans, Green.

Berry, Brian J. L. 1991. *Long-Wave Rhythms in Economic Development and Political Behavior.* Baltimore: Johns Hopkins University Press.

Berry, Brian J. L., Edgar C. Conklin, & D. Michael Ray. 1987. *Economic Geography: Resource Use, Locational Choices, and Regional Specialization in the Global Economy.* Englewood Cliffs, N.J.: Prentice-Hall.

Bilstein, Roger E. 1983. *Flight Patterns. Trends of Aeronautical Development in the United States, 1918–1929.* Athens: University of Georgia Press.

Bloch, Marc. 1961. *Feudal Society.* Vol. 1, *The Growth of Ties of Dependence.* Vol. 2, *Social Classes and Political Organization.* Chicago: University of Chicago Press.

———. 1966. *French Rural History. An Essay on Its Basic Characteristics.* Berkeley and Los Angeles: University of California Press.

Blouet, Brian W. 1987. *Halford Mackinder. A Biography.* College Station: Texas A&M University Press.

———. 1988a. "American Geopolitics, 1900–1942." Association of American Geographers, Annual *Program Abstracts,* 17.

———. 1988b. "The Geostrategic Engine of History." Paper presented at conference, "What Is the Engine of History?" Texas A&M University.

Bobrick, Benson. 1981. *Labyrinths of Iron. A History of the World's Subways.* New York: Newsweek.

Borchert, John R. 1987. *America's Northern Heartland: An Economic and Social Geography of the Upper Midwest.* Minneapolis: University of Minnesota Press.

Borgeson, Griffith. 1980. "In the Name of the People. Origins of the VW Beetle." *Automobile Quarterly* 18:340–61.

———. 1984. "Edward Rumpler: Icarus Bound." *Automobile Quarterly* 22:264–81.

Botting, Douglas. 1979. *The U-Boats.* Chicago: Time-Life.

Boyd, Alexander. 1977. *The Soviet Air Force since 1918.* New York: Stein & Day.

Boyne, Walter J., & Donald S. Lopez, eds. 1979. *The Jet Age. Forty Years of Jet Aviation.* Washington, D.C.: National Air and Space Museum/Smithsonian.

Braudel, Fernand. 1972–73. *The Mediterranean and the Mediterranean World in the Age of Philip II,* 2d ed. 2 vols. New York: Harper & Row.

———. 1973. *Capitalism and Material Life, 1400–1800.* New York: Harper & Row.

———. 1981. *Civilization & Capitalism, 15th–18th Century.* Vol. 1, *The Structures of Everyday Life.* New York: Harper & Row.

———. 1982. *Civilization & Capitalism, 15th–18th Century.* Vol. 2, *The Wheels of Commerce.* New York: Harper & Row.

———. 1984. *Civilization & Capitalism, 15th–18th Century.* Vol. 3, *The Perspective of the World.* New York: Harper & Row.

Bridbury, A. R. 1973. "The Black Death." *Economic History Review* 26:577–92.

Bridenbaugh, Carl. 1968. *Vexed and Troubled Englishmen, 1590–1642*. New York: Oxford University Press.

Briggs, Asa. 1984. *A Social History of England*. New York: Viking.

Bright, Charles D. 1978. *The Jet Makers. The Aerospace Industry from 1945 to 1972*. Lawrence: University Press of Kansas.

Brinnin, John Malcolm. 1971. *The Sway of the Grand Saloon: A Social History of the North Atlantic*. New York: Delacorte.

Brodie, Bernard. 1943. *Sea Power in the Machine Age*. Princeton: Princeton University Press.

Brooks, John. 1976. *Telephone: The First Hundred Years*. New York: Harper & Row.

Brooks, Peter. 1986. "Zeppelin-Staaken E.4/20." *Air Pictorial* (June), 226–33.

Brown, F. A. S. 1975. *Nigel Gresley: Locomotive Engineer*. London: Ian Allan.

Brown, Ralph H. 1948. *Historical Geography of the United States*. New York: Harcourt, Brace & World.

Brunel, Isambard. [1870] 1972. *The Life of Isambard Kingdom Brunel, Civil Engineer*. Rutherford, N.J.: Fairleigh Dickinson University Press.

Buenger, Walter L., & Joseph A. Pratt. 1986. *But Also Good Business. Texas Commerce Banks and the Financing of Houston and Texas, 1886–1986*. College Station: Texas A&M University Press.

Buley, R. Carlyle. 1950. *The Old Northwest. Pioneer Period, 1815–1840*. 2 vols. Bloomington: Indiana University Press.

Burke, James. 1978. *Connections*. Boston: Little, Brown.

Burton, Walter E. 1954. *The Story of Tire Beads and Tires*. New York: McGraw-Hill.

Butzer, Karl. 1988. "Diffusion, Adaption, and Evolution of the Spanish Agrosystem." In Hugill & Dickson, 91–109.

Byatt, I. C. R. 1979. *The British Electrical Industry, 1875–1914. The Economic Returns of a New Technology*. Oxford: Clarendon.

Calif, Ruth. 1983. *The World on Wheels. An Illustrated History of the Bicycle and Its Relatives*. East Brunswick, N.J.: Rosemont.

Carlstein, Tommy. 1982. *Time Resources, Society and Ecology*. Vol. 1, *Preindustrial Societies*. London: Allen & Unwin.

Carlstein, Tommy, Don Parkes, & Nigel Thrift, eds. 1978. *Timing Space and Spacing Time*. Vol. 1, *Making Sense of Time*. Vol. 2, *Human Activity and Time Geography*. London: Edward Arnold.

Carter, George F. 1977. "A Hypothesis Suggesting a Single Origin of Agriculture." In *Origins of Agriculture,* edited by Charles Reed, 89–133. The Hague: Mouton.

Cary, M., & E. H. Warmington. 1963. *The Ancient Explorers*. Baltimore: Pelican.

Casey, Louis S. 1981. *Curtiss. The Hammondsport Era, 1907–1915*. New York: Crown.

Casson, Herbert N., R. W. Hutchinson, & L. W. Ellis. 1913. *Horse, Truck, & Tractor: The Coming of Cheaper Power for City and Farm*. Chicago: Browne.

Caunter, C. F. 1970. *The Light Car. A Technical History of Cars with Engines of Less than 1600 c.c. Capacity*. London: Science Museum, H.M.S.O.

Chamberlin, Everett. 1874. *Chicago and Its Suburbs*. Chicago: Hungerford.

Chanaron, Jean-Jacques. 1982. "The Universal Automobile." In Bardou et al., 171–207.

References

Childs, William R. 1985. *Trucking and The Public Interest. The Emergence of Federal Regulation, 1914–1940.* Knoxville: University of Tennessee Press.

Chisholm, Michael. 1962. *Rural Settlement and Land Use. An Essay in Location.* London: Hutchinson.

Christie, J. Edward. 1985. *Steel Steeds Christie.* Manhattan, Kans.: Sunflower.

Church, Roy. 1979. *Herbert Austin. The British Motor Car Industry to 1941.* London: Europa.

Cipolla, Carlo M. 1965. *Guns, Sails and Empires: Technological Innovation and the Early Phases of European Expansion, 1400–1700.* New York: Minerva.

———. ed. 1972. *The Fontana Economic History of Europe.* Vol. 1, *The Middle Ages.* London: Fontana.

———. 1978a. *Clocks and Culture, 1300–1700.* New York: Norton.

———. 1978b. *The Economic History of World Population.* New York: Penguin.

———. 1980. *Before the Industrial Revolution. European Society and Economy, 1000–1700,* 2d ed. New York: Norton.

Cipolla, Carlo M., & Derek Birdsall. 1979. *The Technology of Man: A Visual History.* New York: Holt, Rinehart & Winston.

Clapham, Sir John. 1949. *A Concise Economic History of Britain from the Earliest Times to 1750.* Cambridge: Cambridge University Press.

Cochrane, Dorothy, Von Hardesty, & Russell Lee. 1989. *The Aviation Careers of Igor Sikorsky.* Seattle: University of Washington Press.

Coedés, George. 1968. *The Indianized States of Southeast Asia.* Honolulu: East-West Center.

Coleman, Richard N., & Keith W. Burnham. 1980. "Milestones in the Application of Power to Agricultural Machines." In *An Historical Perspective of Farm Machinery.* Warrendale, Pa.: Society of Automotive Engineers, SP-470: 75–83.

Compton-Hall, Richard. 1984. *Submarine Boats. The Beginnings of Underwater Warfare.* New York: Arco.

Connell, K. H. [1950] 1965. "Land and Population in Ireland, 1780–1845." In Glass & Eversley, 423–33.

Constant, Edward W., II. 1980. *The Origins of the Turbojet Revolution.* Baltimore: Johns Hopkins University Press.

———. 1988. "Cause, Context, and Consequence: Technology as an Engine of Social Change." Paper presented at the conference, "What Is the Engine of History?" Texas A&M University.

Conzen, Michael P. 1975. "Capital Flows and the Developing Urban Hierarchy: State Bank Capital in Wisconsin, 1854–1895." *Economic Geography* 51:321–38.

Cook, Earl F. 1976. *Man, Energy, Society.* San Francisco: Freeman.

Corlett, Ewan. 1981. *The Ship.* Vol. 10, *The Revolution in Merchant Shipping, 1950–1980.* London: National Maritime Museum, H.M.S.O.

Costello, John, & Terry Hughes. 1976. *The Concorde Conspiracy. The International Race for the SST.* New York: Scribner's.

Craig, Robin. 1980. *The Ship.* Vol. 5, *Steam Tramps and Cargo Liners, 1850–1950.* London: National Maritime Museum, H.M.S.O.

Crosby, Alfred W. 1986. *Ecological Imperialism.* New York: Cambridge University Press.

Cross, Robin. 1987. *The Bombers. The Illustrated Story of Offensive Strategy and Tactics in the Twentieth Century.* New York: Macmillan.

Cross, Whitney R. [1950] 1965. *The Burned-Over District. The Social and Intellectual History of Enthusiastic Religion in Western New York, 1800–1850.* New York: Harper & Row.

Crouch, Tom D. 1986. "How the Bicycle Took Wing." *American Heritage of Invention and Technology* 2:11–16.

Cummins, C. Lyle, Jr. 1976. *Internal Fire: The Internal Combustion Engine, 1673–1900.* Lake Oswego, Oreg.: Carnot.

Curwen, E. Cecil, & Gudmund Hatt. 1953. *Plough and Pasture. The Early History of Farming.* New York: Schuman.

Daley, Robert. 1980. *An American Saga. Juan Trippe and His PanAm Empire.* New York: Random House.

Darby, H. Clifford, ed. 1936. *An Historical Geography of England before 1800.* Cambridge: Cambridge University Press.

———. 1973a. *A New Historical Geography of England.* Cambridge: Cambridge University Press.

———. 1973b. "The Anglo-Scandinavian Foundations." In Darby, 1973a, 1–38.

Davies, R. E. G. 1982. *Airlines of the United States since 1914,* 2d ed. Washington, D.C.: Smithsonian.

———. 1987. *Pan Am: An Airline and Its Aircraft.* New York: Hamlyn.

Davis, Ralph H. C. 1962. *The Rise of the English Shipping Industry in the Seventeenth and Eighteenth Centuries.* New York: St. Martin's Press.

Dawson, Christopher. 1932. *The Making of Europe. An Introduction to the History of European Unity.* New York: Meridian.

Day, John. 1976. *The Bosch Book of the Motor Car. Its Evolution and Engineering Development.* New York: St. Martin's Press.

———. 1980. *Engines. The Search for Power.* New York: St. Martin's Press.

Deere, John. 1987. *John Deere Tractors, 1918–1967.* St. Joseph, Mich.: American Society of Agricultural Engineers.

Dienstbach, Carl, & T. R. MacMechen. [1909] 1987. "The Aerial Battleship." In Oppel, 422–34.

Diffie, Bailey W., & George D. Winius. 1977. *Foundations of the Portuguese Empire, 1415–1580.* Minneapolis: University of Minnesota Press.

Diggins, John P. 1986. *The Lost Soul of American Politics. Virtue, Self-Interest, and the Foundations of Liberalism.* Chicago: University of Chicago Press.

Dodgson, Robert A. 1977. "A Spatial Perspective [on Wallerstein's World System]." *Peasant Studies* 6:8–19.

Dohan, Mary Helen. 1981. *Mr. Roosevelt's Steamboat.* New York: Dodd, Mead.

Dollinger, Philippe. 1970. *The German Hansa.* Stanford: Stanford University Press.

Dorsey, Florence L. 1941. *Master of the Mississippi: Henry Shreve and the Conquest of the Mississippi.* Boston: Houghton Mifflin.

Doyle, Michael W. 1986. *Empires.* Ithaca: Cornell University Press.

Dunn, Ross E. 1986. *The Adventures of Ibn Battuta, A Muslim Traveler of the 14th Century.* Berkeley and Los Angeles: University of California Press.

Dunsheath, Percy. 1962. *A History of Electrical Power Engineering.* Cambridge: MIT Press.

Dyos, H. J., & D. H. Aldcroft. [1969] 1974. *British Transport. An Economic Survey from the Seventeenth Century to the Twentieth.* Baltimore: Penguin.

Earle, Carville V. 1978. "A Staple Interpretation of Slavery and Free Labor." *Geographical Review* 68:51–65.

References   Earle, Carville V., & Ronald Hoffman. 1980. "The Foundation of the Modern Economy: Agriculture and the Costs of Labor in the United States and England, 1800–1860." In *American Historical Review* 85:1055–96.

East, W. Gordon. 1950. *An Historical Geography of Europe,* 4th ed. London: Methuen.

Edwards, Dennis, & Ron Pigram. 1986. *London's Underground Suburbs.* London: Baton.

Edwards, Harry. 1983. *The Morris Motor Car, 1913–1983.* Ashbourne, Derbyshire: Moorland.

Ekwall, Eilert. 1936. "The Scandinavian Settlement." In Darby, 133–64.

Ellis, David M., et al. 1967. *A History of New York State.* Ithaca: Cornell University Press.

Ellis, John. 1986. *The Social History of the Machine Gun.* Baltimore: Johns Hopkins University Press.

Emmerson, George S. 1977. *John Scott Russell. A Great Victorian Engineer and Naval Architect.* London: Murray.

————. c. 1980. *S.S. Great Eastern: The Greatest Iron Ship.* London: David & Charles.

Engels, Friedrich. [1845] 1968. *The Condition of the Working Class in England.* Stanford: Stanford University Press.

Eyre, Samuel R. 1978. *The Real Wealth of Nations.* London: Edward Arnold.

Farnie, D. A. 1979. *The English Cotton Industry and the World Market, 1815–1896.* Oxford: Clarendon.

Ferneyhough, Frank. 1975. *The History of Railways in Britain.* Reading: Osprey.

————. 1980. *Liverpool & Manchester Railway 1830–1980.* London: Hale.

Fischer, Fritz. 1967. *Germany's Aims in the First World War.* New York: Norton.

Fite, Gilbert C. 1984. *Cotton Fields No More. Southern Agriculture 1865–1980.* Lexington: University Press of Kentucky.

Flandrin, Jean-Louis. 1979. *Families in Former Times. Kinship, Household and Sexuality in Early Modern France.* Cambridge: Cambridge University Press.

Fleming, Lamar, Jr., & James A. Tinsley. 1966. *Growth of the Business of Anderson, Clayton and Co.* Texas Gulf Coast Historical Association, Publication Series, Vol. 10, September.

Fleure, H. J., & M. Davies. 1970. *A Natural History of Man in Britain. Conceived as a Study of Changing Relations between Men and Environments.* London: Collins.

Flexner, James Thomas. 1978. *Steamboats Come True. American Inventors in Action.* Boston: Little, Brown.

Flink, James J. 1988. *The Automobile Age.* Cambridge: MIT Press.

Fogel, Robert W. 1964. *Railroads and American Economic Growth: Essays in Econometric History.* Baltimore: Johns Hopkins Press.

Forer, Pip. 1978. "Time-Space and Area in the City of the Plains." In Carlstein, Parkes, & Thrift 1:99–118.

Francillon, R. J. 1970. *Japanese Aircraft of the Pacific War.* New York: Funk & Wagnalls.

Frazer, John Foster. 1899. *Round the World on a Wheel.* New York: Frederick Stokes.

Fridenson, Patrick. 1972. *Histoire des Usines Renault: 1. Naissance de la Grande Enterprise, 1898–1939.* Paris: Editions du Seuil.

————. 1982. "The Spread of the Automobile Revolution." In Bardou et al., 77–167.

Fryxell, Paul A. n.d. *The Natural History of the Cotton Tribe.* College Station: Texas A&M University Press.

Galloway, J. H. 1977. "The Mediterranean Sugar Industry." *Geographical Review* 67:177–94.

Garrett, Richard. 1977. *Submarines.* Boston: Little, Brown.

Giacosa, Dante. 1979. *Forty Years of Design with Fiat.* Milano: Automobilia.

Gibbs-Smith, Charles H. 1970. *Aviation. An Historical Survey from Its Origins to the End of World War II.* London: Science Museum, H.M.S.O.

Gibson, James R. 1969. *Feeding the Russia Fur Trade. Provisionment of the Okhotsk Seaboard and the Kamchatka Peninsula, 1634–1856.* Madison: University of Wisconsin Press.

Giddens, Anthony. 1981. *A Contemporary Critique of Historical Materialism.* Vol. 1, *Power, prosperity and the state.* Berkeley and Los Angeles: University of California Press.

Gilbreth, Frank B. 1911. *Motion Study: A Method for Increasing the Efficiency of the Workman.* New York: Van Nostrand.

Gillette, King Camp. 1910. *World Corporation.* Boston: New England News.

Gimpel, Jean. 1976. *The Medieval Machine. The Industrial Revolution of the Middle Ages.* New York: Holt, Rinehart & Winston.

Glass, D. V., & D. E. C. Eversley, eds. 1965. *Population in History. Essays in Historical Demography.* Chicago: Aldine.

Goldenweiser, A. A. 1913. "The Principle of Limited Possibilities in the Development of Culture." *Journal of American Folklore* 26:259–90.

Goldstein, Joshua S. 1988. *Long Cycles. Prosperity and War in the Modern Age.* New Haven: Yale University Press.

Gollin, Alfred. 1984. *No Longer an Island. Britain and the Wright Brothers, 1902–1909.* London: Heinemann.

Gray, R. B., ed. 1975. *The Agricultural Tractor: 1855–1950.* St. Joseph, Mich.: American Society of Agricultural Engineers.

Grayson, Stan. 1976. "The Front-Wheel-Drives of John Walter Christie, Inventor." *Automobile Quarterly* 14:256–73.

———. 1978. "The All-Steel World of Edward Budd." *Automobile Quarterly* 16:352–67.

Greenhill, Basil. 1976. *Archeology of the Boat.* Middletown, Conn.: Wesleyan University Press.

Haddow, G. W., & Peter M. Grosz. 1969. *The German Giants. The German R-Planes, 1914–1918,* 2d ed. New York: Funk & Wagnalls.

Hadfield, Charles. 1968. *The Canal Age.* Newton Abbot, Devon: David & Charles.

———. 1986. *World Canals. Inland Navigation Past & Present.* New York: Facts on File.

Hägerstrand, Torsten. 1978. "Survival and Arena." In Carlstein, Parkes, & Thrift 2:122–145.

———. 1988. "Some Unexplored Problems in the Modeling of Culture Transfer and Transformation." In Hugill & Dickson, 217–32.

Haigh, K. R. 1968. *Cableships and Submarine Cables.* Washington, D.C.: United States Underseas Cable Corporation.

Hallion, Richard P. 1977. *Legacy of Flight. The Guggenheim Contribution to American Aviation.* Seattle: University of Washington Press.

Handy, Robert T. 1977. *A History of the Churches in the United States and Canada.* New York: Oxford University Press.

Hanle, Paul A. 1982. *Bringing Aerodynamics to America.* Cambridge: MIT Press.

References

Hanley, David H., & Raymond F. Corley. 1973. "About the Railcars Which (Unintentionally) Forecast Dieseldom." *Trains* 34(1):36–49.

Hansen, Hans Jürgen. 1975. *The Ships of the German Fleet, 1848–1945*. New York: Arco.

Hayden, Dolores. 1981. *The Grand Domestic Revolution: A History of Feminist Designs for American Homes, Neighborhoods, and Cities*. Cambridge: MIT Press.

Heinze, Edwin P. A. 1926. "Germany Best Potential Automobile Mass Market in Europe." *Automotive Industries* 55:1082–83.

————. 1938. "Berlin Automobile Show. Economy of Materials and Substitute Materials at the . . ." *Automotive Industries* 78:442–47, 457–58.

Hershberg, Theodore, ed. 1981. *Philadelphia. Work, Space, Family, and Group Experience in the Nineteenth Century. Essays Toward an Interdisciplinary History of the City*. New York: Oxford University Press.

Herwig, Holger H. 1976. *Politics of Frustration: The United States in German Naval Planning, 1889–1941*. Boston: Little, Brown.

Hindle, Brooke, & Stephen Lubar. 1986. *Engines of Change. The American Industrial Revolution, 1790–1860*. Washington, D.C.: Smithsonian.

Hoare, Sir Samuel. See Templewood, Viscount.

Hobsbawm, Eric. 1987. *The Age of Empire, 1875–1914*. New York: Beacon.

Hobson, J. A. [1902] 1938. *Imperialism: A Study*, 3d ed. London: Allen & Unwin.

Hogg, Ian, & John Batchelor. 1978. *Naval Gun*. Poole, Dorset: Blandford.

Hough, Richard. 1979. *Man O'War. The Fighting Ship in History*. London: Dent.

————. 1983. *The Great War at Sea, 1914–1918*. New York: Oxford University Press.

Hounshell, David. 1984. *From the American System to Mass Production*. Baltimore: Johns Hopkins University Press.

Howard, Frank. 1979. *Sailing Ships of War, 1400–1800*. New York: Mayflower.

Howarth, David. 1981. *The Voyage of the Armada: The Spanish Story*. New York: Viking.

Howson, H. F. 1981. *London's Underground*, 5th ed. London: Ian Allan.

Hughes, Thomas P. 1983. *Networks of Power: Electrification in Western Society, 1880–1930*. Baltimore: John Hopkins University Press.

Hughes, Tom. 1973. *The Blue Riband of the Atlantic*. Cambridge: Patrick Stephens.

Hugill, Peter J. 1977. "A Small Town Landscape as Sustained Gesture on the Part of a Dominant Social Group: Cazenovia, New York, 1794–1976." Ph.D. diss., Syracuse University.

————. 1982. "Good Roads and the Automobile in the United States, 1880–1929." *Geographical Review* 72:327–49.

————. 1983. "The Commuters Who Got on their Bikes." *Geographical Magazine* 55(7):371–74.

————. 1986. "English Landscape Tastes in the United States." *Geographical Review* 76:408–23.

————. 1988a. "Technology Diffusion in the World Automobile Industry, 1885–1985." In Hugill & Dickson, 110–42.

————. 1988b. "Structural Changes in the Core Regions of the World-Economy, 1830–1945." *Journal of Historical Geography* 14:111–27.

————. 1988c. "Transport of Desires: Trade as the Engine of History." Paper presented at conference "What Is the Engine of History?" Texas A&M University.

———. 1988d. "The Macro-Landscape of the Wallersteinian World-Economy." In *Geoscience and Man*. Vol. 25, *The American South*, edited by Richard L. Nostrand & Sam B. Hilliard. Baton Rouge: Department of Geography, Louisiana State University, 77–84.

———. 1990. "Technology and Geography in the Emergence of the American Automobile Industry, 1895–1915." In *Roadside America. The Automobile in Design and Culture*, edited by Jan Jennings. Ames: Iowa State University Press, 29–39.

Hugill, Peter J., & D. Bruce Dickson, eds. 1988. *The Transfer and Transformation of Ideas and Material Culture*. College Station: Texas A&M University Press.

Hugill, Peter J., & John C. Everitt. 1992. "Macro-Landscapes: The Cultural Landscape Revised by World-System Theory." In *Geoscience and Man*. Vol. 32, *Person, Place and Thing: Interpretive and Empirical Essays in Cultural Geography*, edited by Shue Tuck Wong. Baton Rouge: Department of Geography, Louisiana State University, 177–94.

Israel, Jonathan I. 1989. *Dutch Primacy in World Trade, 1575–1740*. Oxford: Clarendon.

Jackson, Alan A. 1973. *Semi-Detached London: Suburban Development, Life & Transport, 1900–1939*. London: Allen & Unwin.

Jeremy, David J. 1981. *Transatlantic Industrial Revolution: The Diffusion of Textile Technologies between Britain and America, 1770–1830*. Cambridge: MIT Press.

Jones, Eric L. 1981. *The European Miracle: Environments, Economies and Geopolitics in the History of Europe and Asia*. New York: Cambridge University Press.

Keegan, John. 1989. *The Price of Admiralty. The Evolution of Naval Warfare*. New York: Viking.

Keith, Donald H. 1987. "The Molasses Reef Wreck." Ph.D. diss., Texas A & M University.

Kemp, Peter. 1978. *The History of Ships*. London: Orbis.

Kennedy, Paul. 1987. *The Rise and Fall of the Great Powers. Economic Change and Military Conflict from 1500 to 2000*. New York: Random House.

Kirsch, Peter. 1990. *The Galleon: The Great Ships of the Armada Era*. London: Conway.

Knight, Frank W. 1970. *Slave Society in Cuba during the Nineteenth Century*. Madison: University of Wisconsin Press.

Kniseley (Marshall), Mary Ann. 1987. "Bryan and College Interurban Railway: Historical Perspective on Its Development, Ownership, and Disposition." Unpublished MS project, Department of Geography, Texas A&M University.

Knott, Richard C. 1979. *The American Flying-boat: An Illustrated History*. Greenwich, Conn.: Conway.

Kropotkin, Petr. [1899] 1974. *Fields, Factories and Workshops Tomorrow*. London: Allen & Unwin.

Ladurie, Emmanuel LeRoy. 1971. *Times of Feast, Times of Famine. A History of Climate since the Year 1000*. New York: Doubleday.

Landes, David S. 1983. *Revolution in Time. Clocks and the Making of the Modern World*. Cambridge: Harvard University Press.

Laslett, Peter. 1984. *The World We Have Lost. England before the Industrial Age*. New York: Scribner's.

References

Laslett, Peter, & Richard Wall, eds. 1972. *Household and Family in Past Time.* Cambridge: Cambridge University Press.

Laux, James M. 1976. *In First Gear. The French Automobile Industry to 1914.* Liverpool: Liverpool University Press.

Lawton, Richard. 1986. "Population." In Langton & Morris (see Atlases), 10–29.

Leary, William M. 1985. *Aerial Pioneers. The U.S. Air Mail Service, 1918–1927.* Washington, D.C.: Smithsonian.

Lemon, James T. 1987. "Colonial America in the Eighteenth Century." In Mitchell & Groves, 121–46.

Lenin, Vladimir Ilyich. [1917] 1939. *Imperialism, the Highest Stage of Capitalism.* New York: International.

———. [1920] 1972. *On the Development of Heavy Industry and Electrification.* Moscow: Progress.

Lofland, Lyn H. 1973. *A World of Strangers. Order and Action in Urban Public Space.* New York: Basic Books.

Lopez, Robert S. 1971. *The Commercial Revolution of the Middle Ages, 950–1350.* Englewood Cliffs, N.J.: Prentice-Hall.

Lowenthal, David. 1972. *West Indian Societies.* New York: Oxford University Press.

Lyon, David. 1980. *The Ship.* Vol. 8, *Steam, Steel and Torpedoes. The Warship in the Nineteenth Century.* London: National Maritime Museum, H.M.S.O.

McAdam, Roger Williams. 1959. *Commonwealth: Giantess of the Sound.* New York: Roger Daye.

MacBride, Robert. 1962. *Civil War Ironclads: The Dawn of Naval Armor.* Philadelphia: Chilton.

McCall, Edith S. 1984. *Conquering the Rivers: Henry Miller Shreve and the Navigation of America's Inland Waterways.* Baton Rouge: Louisiana State University Press.

McGowan, Alan. 1980. *The Ship.* Vol. 4, *The Century before Steam. The Development of the Sailing Ship, 1700–1820.* London: National Maritime Museum, H.M.S.O.

———. 1981. *The Ship.* Vol. 3, *Tiller and Whipstaff: The Development of the Sailing Ship, 1400–1700.* London: National Maritime Museum, H.M.S.O.

McGrail, Sean. 1981. *The Ship.* Vol. 1, *Rafts, Boats and Ships from Prehistoric Times to the Medieval Era.* London: National Maritime Museum, H.M.S.O.

McGurn, James. 1987. *On Your Bicycle: An Illustrated History of Cycling.* New York: Facts on File.

Mackinder, Halford J. 1900. "The Great Trade Routes (Their Connection with the Organization of Industry, Commerce, and Finance)." *Journal of the Institute of Bankers* 21: 1–6, 137–46, 147–55, 266–73.

———. 1904. "The Geographical Pivot of History." *Geographical Journal* 23:421–37.

McKinley, Marvin. 1980. *Wheels of Farm Progress.* St. Joseph, Mich.: American Society of Agricultural Engineers.

Macksey, Kenneth, & John H. Batchelor. 1970. *Tank. A History of the Armoured Fighting Vehicle.* New York: Scribner's

McLuhan, Marshall. 1964. *Understanding Media: The Extensions of Man.* Toronto: Signet.

MacMechen, T. R., & Carl Dienstbach. [1910] 1987. "Over Sea by Air-Ship. Surprising Progress of German Plans for a Transatlantic Service." In Oppel, 325–41.

————. [1912] 1987. "The Greyhounds of the Air." In Oppel, 278–92.

McNeill, William H. 1963. *The Rise of the West. A History of the Human Community.* Chicago: University of Chicago Press.

————. 1982. *The Pursuit of Power. Technology, Armed Force, and Society since A.D. 1000.* Chicago: University of Chicago Press.

————. 1988. "Diffusion in History." In Hugill & Dickson, 75–90.

MacPherson, C. B. 1962. *The Political Theory of Possessive Individualism: Hobbes to Locke.* New York: Oxford University Press.

Mahan, Alfred Thayer. [1892] 1957. *The Influence of Seapower upon History, 1660–1783.* New York: Hill & Wang.

————. [1890] 1957. *The Influence of Sea Power upon the French Revolution and Empire, 1793–1812.* 2 vols. Boston: Little, Brown.

Majumdar, R. C. 1963. *Hindu Colonies in the Far East.* Calcutta: Mukhopadhyay.

Marshall, T. H. 1929. "The Population Problem during the Industrial Revolution: A Note on the Present State of the Controversy" (*Economic History*, vol. 1). In Glass & Eversley, 247–68.

Martin, Robert E. 1969. "Republic's 'Rainbow'." *Air Classics* 5(6):38–42.

Marx, Karl. 1954. *Capital.* Vol. 1, *The Process of the Production of Capital.* Moscow: Progress.

————. 1959. *Capital.* Vol. 3, *The Process of Capitalist Production as a Whole.* Moscow: Progress.

Maxcy, George. 1981. *The Multinational Motor Industry.* London: Croom Helm.

Mayer, Harold M., & Richard C. Wade. 1969. *Chicago: Growth of a Metropolis.* Chicago: University of Chicago Press.

Mayer, S. L., ed. 1984. *The Rise and Fall of Imperial Japan.* New York: Military Press.

Mayr, Otto. 1986. *Authority, Liberty & Automatic Machinery in Early Modern Europe.* Baltimore: Johns Hopkins University Press.

Meinig, Donald W. 1962. *On the Margins of the Good Earth. The South Australian Wheat Frontier 1869–1884.* Washington, D.C.: Association of American Geographers.

————. 1965. "The Mormon Culture Region: Strategies and Patterns in the Geography of the American West, 1847–1964." In Association of American Geographers, *Annals* 55:191–220.

————. 1966. "Geography of Expansion, 1785–1855." In Thompson, 140–71.

————. 1986. *The Shaping of America. A Geographical Perspective on Five Hundred Years of History.* Vol. 1, *Atlantic America, 1492–1800.* New Haven: Yale University Press.

Melbin, Murray. 1978. "The Colonization of Time." In Carlstein, Parkes, & Thrift 2:110–13.

Mensch, Gerhard. 1979. *Stalemate in Technology. Innovations Overcome the Depression.* Cambridge, Mass.: Ballinger.

Meyer, Balthasar H. [1910] 1948. *History of Transportation in the United States before 1860.* Magnolia, Mass.: Peter Smith.

Middlebrook, Martin, & Patrick Mahoney. 1979. *Battleship. The Sinking of the "Prince of Wales" and the "Repulse."* New York: Scribner's.

Middleton, William D. 1961. *The Interurban Era.* Milwaukee: Kalmbach.

Mikesh, Robert C., & Shorzoe Abe. 1990. *Japanese Aircraft, 1910–1941.* London: Putnam.

References

Mintz, Sidney W. 1985. *Sweetness and Power. The Place of Sugar in Modern History.* New York: Penguin.

Mirsky, Jeannette, ed. 1965. *The Great Chinese Travellers.* London: Allen & Unwin.

Mitchell, J. Clyde. 1969. "The Concept and Use of Social Networks." In *Social Networks in Urban Situations,* edited by J. Clyde Mitchell. Manchester: Manchester University Press, 1–50.

Mitchell, Robert A., & Paul A. Groves, eds. 1987. *North America. The Historical Geography of a Changing Continent.* New York: Rowman & Littlefield.

Modelski, George, & William R. Thompson. 1988. *Seapower in Global Politics, 1494–1993.* Seattle: University of Washington Press.

Mohr, Anton. 1926. *The Oil War.* New York: Harcourt, Brace.

Moody, G. T. 1979. *Southern Electric, 1909–1979. The History of the World's Largest Suburban Electrified System,* 5th ed. London: Ian Allan.

Morison, Samuel Eliot. 1965. *The Oxford History of the American People.* New York: Oxford University Press.

Morpugo, J. E. 1972. *Barnes Wallis. A Biography.* New York: St. Martin's Press.

Morris, John. 1973. *The Age of Arthur. A History of the British Isles from 350 to 650.* New York: Scribner's

Morris, R. J. 1986. "Urbanization." In Langton & Morris (See Atlases), 164–79.

Morrison, John. 1980. *The Ship.* Vol. 2, *Long Ships and Round Ships. Warfare and Trade in the Mediterranean, 3000 B.C.–500 A.D.* London: National Maritime Museum, H.M.S.O.

Morrow, John Howard, Jr. 1976. *Building German Airpower, 1909–1914.* Knoxville: University of Tennessee Press.

Mullin, John R. 1981. "The Impact of National Socialist Policies upon Local City Planning in Pre-war Germany." *Journal of the American Planning Association* 47:35–47.

———. 1982. "Henry Ford and Field and Factory: An Analysis of the Ford Sponsored Village Industries Experiment in Michigan." *Journal of the American Planning Association* 48:419–31.

Mullineux, Frank. 1959. *The Duke of Bridgewater's Canal.* Eccles, Lancashire: Eccles & District History Society.

Mumford, Lewis. 1934. *Technics and Civilization.* New York: Harcourt Brace.

———. 1967. *The Myth of the Machine.* Vol. 1, *Technics and Human Development.* New York: Harvest.

———. 1970. *The Myth of the Machine.* Vol. 2, *The Pentagon of Power.* New York: Harvest.

Munro (Saki), Hector Hugh. 1914. *When William Came: A Story of London under the Hohenzollerns.* London: John Lane.

Nader, Ralph. 1972. *Unsafe at Any Speed: The Designed-in Dangers of the American Automobile,* 2d ed. New York: Grossman.

Needham, Joseph. 1954. *Science and Civilization in China.* Vol. 1, *Introductory Orientations.* Cambridge: Cambridge University Press.

Nelson, Walter Henry. 1965. *Small Wonder. The Amazing Story of the Volkswagen.* Boston: Little, Brown.

Nock, O. S. 1969. *LNER Steam.* Newton Abbot, Devon: David & Charles.

———. 1980. *Two Miles a Minute. The Story Behind the Conception and Operation of Britain's High Speed and Advanced Passenger Trains.* Cambridge: Patrick Stephens.

Norris, Frank. [1903] 1969. *The Pit. A Story of Chicago.* Gloucester, Mass.: Peter Smith.

North, Douglass C., & Robert P. Thomas, eds. 1968. *The Growth of the American Economy to 1860.* New York: Harper & Row.

North, John D. 1923. "The Case for Metal Construction." *Journal of the Royal Aeronautical Society* 28:3–25.

Nowarra, Heinz. 1980. *Heinkel He 111. A Documentary History.* London: Jane's.

Offer, Avner. 1989. *The First World War: An Agrarian Interpretation.* Oxford: Clarendon.

Oppel, Frank, ed. 1987. *Early Flight: From Balloons to Biplanes.* Secaucus, N.J.: Castle.

Overy, R. J. 1975. "Cars, Roads, and Economic Recovery in Germany, 1932–1938." *Economic History Review* 28:466–83.

Owen, David. 1975. "Gear Maker, Arms Maker, Car Maker—The Legacy of Andre Citroen." *Automobile Quarterly* 13:192–217.

Padfield, Peter. 1979. *Tide of Empires: Decisive Naval Campaigns in the Rise of the West.* Vol. 1, *1481–1654.* London: Routledge & Kegan Paul.

———. 1982. *Tide of Empires. Decisive Naval Campaigns in the Rise of the West.* Vol. 2, *1654–1763.* London: Routledge & Kegan Paul.

Parkin, George R. 1894. "The Geographical Unity of the British Empire." *Scottish Geographical Magazine* 10:225–42.

Parry, J. H. 1966. *The Establishment of the European Hegemony: 1415–1715. Trade and Exploration in the Age of the Renaissance,* 3d ed. New York: Harper Torchbooks.

Paterson, Alan J. S. 1969. *The Golden Years of the Clyde Steamers (1889–1914).* New York: Kelley.

Pearson, Drew. 1924. "Henry Ford Says—Farmer-Workmen Will Build Automobile of the Future: An Interview with Henry Ford." *Automotive Industries* 51(9):389–92.

Penrose, Boies. 1952. *Travel and Discovery in the Renaissance, 1420–1620.* New York: Atheneum.

Penrose, Harald. 1969. *British Aviation. The Great War and Armistice.* New York: Funk & Wagnalls.

Phillips, Carla Rahn. 1986. *Six Galleons for the King of Spain: Imperial Defense in the Early Seventeenth Century.* Baltimore: Johns Hopkins University Press.

Platt, Maurice. 1980. *An Addiction to Automobiles: The Occupational Biography of an Engineer and Journalist.* London: Warne.

Polmar, Norman, & Norman Friedman. 1981. *Warships: From Early Steam to Nuclear Power.* London: Octopus.

Polo, Marco. 1926. *The Travels of Marco Polo,* revised and edited by Manuel Komroff. New York: Boni & Liveright.

Porteous, J. Douglas. 1977. *Canal Ports: The Urban Achievement of the Canal Age.* New York: Academic.

Pounds, Norman J. G. 1979. *An Historical Geography of Europe, 1500–1840.* Cambridge: Cambridge University Press.

———. 1985. *An Historical Geography of Europe, 1800–1914.* Cambridge: Cambridge University Press.

Prager, Hans Georg. 1977. *Blohm and Voss. Ships and Machinery for the World.* London: Brassey's.

Preston, Antony. 1979. *Aircraft Carriers.* New York: Grosset & Dunlap.

———. 1980. *The Ship.* Vol. 9, *Dreadnought to Nuclear Submarine.* London: National Maritime Museum, H.M.S.O.

Puryear, Pamela A., & Nath Winfield, Jr. 1976. *Sandbars and Sternwheelers.*

References
    Steam Navigation on the Brazos. College Station: Texas A&M University Press.

Reed, Brian, ed. 1972. *Locomotives in Profile*, vol. 2. Garden City, N.Y.: Doubleday.

Renshaw, Patrick. 1968. *The Wobblies. The Story of Syndicalism in the United States.* Garden City, N.Y.: Anchor.

Rhys, D. G. 1976. "Concentration in the Inter-War [British] Motor Industry." *Journal of Transport History* 3:241–64.

Ritchie, Andrew. 1975. *King of the Road. An Illustrated History of Cycling.* Berkeley: Ten Speed.

Roscoe, Theodore. 1949. *U.S. Submarine Operations in World War II.* Annapolis, Md.: Naval Institute.

Rose, Mary B. 1986. *The Greggs of Quarry Bank Mill. The Rise and Decline of a Family Firm, 1750–1914.* Cambridge: Cambridge University Press.

Rosenberg, Nathan, & L. E. Birdzell. 1986. *How the West Grew Rich. The Economic Transformation of the Industrial World.* London: I. B. Tauris.

Rostow, W. W. 1962. *The Process of Economic Growth.* New York: Norton.

Rowe, Vivian. 1959. *The Great Wall of France: The Triumph of the Maginot Line.* London: Putnam.

Rowland, K. T. 1970. *Steam at Sea. A History of Steam Navigation.* New York: Praeger.

Russell, J. C. 1972. "Population in Europe, 500–1500." In Cipolla, 1972, 25–70.

Rutman, Darrett B. 1965. *Winthrop's Boston. A Portrait of a Puritan Town, 1630–1649.* Chapel Hill: University of North Carolina Press.

Saki. See Munro, Hector Hugh.

Sandler, Stanley. 1979. *The Emergence of the Modern Capital Ship.* Newark: University of Delaware Press.

Sauer, Carl O. 1939. *Man in Nature.* New York: Scribner's.

———. 1952. *Agricultural Origins and Dispersals.* New York: American Geographical Society.

———. 1962. "Middle America as a Culture Historical Location." In Wagner & Mikesell, 195–202.

———. 1963. *Land and Life.* Berkeley and Los Angeles: University of California Press.

———. 1966. *The Early Spanish Main.* Berkeley and Los Angeles: University of California Press.

Scammell, G. V. 1981. *The World Encompassed. The First European Maritime Empires, c. 800–1650.* Berkeley and Los Angeles: University of California Press.

Schlucter, Wolfgang. 1981. *The Rise of Western Rationalism. Max Weber's Developmental History.* Berkeley and Los Angeles: University of California Press.

Schmitt, Gunter. 1988. *Hugo Junkers and His Aircraft.* Berlin: Transpress VEB Verlag.

Schumpeter, Joseph A. 1950. *Capitalism, Socialism and Democracy,* 3d ed. New York: Harper & Row.

———. 1955. *Imperialism: Social Classes.* New York: Meridian.

Sedgwick, Michael. 1974. "The Fiat Tipo 508s." In *Classic Cars in Profile,* edited by Anthony Harding. Vol. 1, *Profile Nos. 1–24.* Garden City, N.Y.: Doubleday, 265–76.

Seeley, Bruce. 1988. "The Diffusion of Science into Engineering: Highway Research at the Bureau of Public Roads, 1900–1940." In Hugill & Dickson, 145–62.

Semple, Ellen Churchill. 1903. *American History and Its Geographic Conditions*. Boston: Houghton Mifflin.

Senior, Clive. 1976. *A Nation of Pirates. English Piracy in Its Heyday.* Newton Abbot, Devon: David & Charles.

Setright, L. J. K. 1975. *Some Unusual Engines*. London: Institution of Mechanical Engineers.

Sharp, Archibald. [1896] 1979. *Bicycles and Tricycles. An Elementary Treatise on Their Design and Construction.* Cambridge: MIT Press.

Sloan, Alfred P. 1964. *My Years with General Motors.* Garden City, N.Y.: Doubleday.

Sloniger, Jerry. 1980. *The VW Story.* Cambridge: Patrick Stephens.

Smith, Adam. [1776] n.d. *The Wealth of Nations,* vol. 1. New Rochelle, N.Y. Arlington House.

Smith, Edgar C. 1938. *A Short History of Naval and Marine Engineering.* Cambridge: Cambridge University Press.

Smith, Herschel. 1981. *A History of Aircraft Piston Engines.* Manhattan, Kans.: Sunflower.

Solberg, Carl. 1979. *Conquest of the Skies. A History of Commercial Aviation in America.* Boston: Little, Brown.

Sraffa, Piero, & Maurice H. Dobb, eds. 1951. *The Works and Correspondence of David Ricardo,* vol. 1. Cambridge: Cambridge University Press.

Stedman, Edmund Clarence. [1908] 1987. "The Prince of the Power of the Air: Aerial Navigation a Menace to British Supremacy." In Oppel, 375–83.

Steiner, John E. 1979. "Jet Aviation Development: A Company Perspective." In Boyne & Lopez, 141–84.

Sternlicht, Sanford, & Edwin M. Jameson. 1981. *U.S.F. Constellation: "Yankee Racehorse."* Cockeysville, Md.: Liberty.

Stevens, Thomas. 1889. *Around the World on a Bicycle.* 2 vols. New York: Scribner's.

Stilgoe, John R. 1983. *Metropolitan Corridor. Railroads and the American Scene.* New Haven: Yale University Press.

Stinchcombe, William C. 1969. *The American Revolution and the French Alliance.* Syracuse, N.Y.: Syracuse University Press.

Stover, Leon, & Takeko Kawai Stover. 1976. *China: An Anthropological Perspective.* Pacific Palisades, Calif.: Goodyear.

Strauss, Anselm H. 1961. *Images of the American City.* New York: Free Press.

Taylor, C. Fayette. 1971. *Aircraft Propulsion. A Review of the Evolution of Aircraft Piston Engines. Smithsonian Annals of Flight* 1(4). Washington, D.C.

Taylor, Frederick W. 1911. *The Principles of Scientific Management.* New York: Harper.

Templewood, Viscount. 1957. *Empire of the Air. The Advent of the Air Age, 1922–1929.* London: Collins.

Thirring, Hans. 1958. *Energy for Men, Windmills to Nuclear Power.* Bloomington: Indiana University Press.

Thompson, John H., ed. 1966. *Geography of New York State.* Syracuse, N.Y.: Syracuse University Press.

Thørdarson, Matthias. 1930. *The Vinland Voyages.* New York: American Geographical Society, Research Series 18.

Throckmorton, Peter. 1962. "Oldest Known Shipwreck Yields Bronze Age Cargo." *National Geographic* 121(5):676–711.

References

Toyota Motor Company. 1967. *Thirty-Year History of Toyota Motor Company.* Toyota City, Japan: Toyota Motor Company.

Tracey, Michael. 1985. "The Poisoned Chalice? International Television and the Idea of Dominance." *Daedalus* 114(4):17–56.

Trebilcock, Clive. 1981. *The Industrialization of the Continental Powers 1780–1914.* London: Longman.

Tubbs, D. B. 1969. "The Lancia Lambda." In *Classic Cars in Profile,* edited by Anthony Harding. Vol. 2, *Profile Nos. 25–48,* 229–40. Garden City, N.Y.: Doubleday.

Turner, Frederick Jackson. 1893. "The Significance of the Frontier in American History." In *Annual Report of the American Historical Association,* 190–227.

Unger, Richard W. 1978. *Dutch Shipbuilding before 1800.* Amsterdam: Van Gorcum.

Usher, Abbott Payson. 1954. *A History of Mechanical Inventions,* 2d ed. Cambridge: Harvard University Press.

Vance, James E., Jr. 1970. *The Merchant's World: The Geography of Wholesaling.* Englewood Cliffs, N.J.: Prentice-Hall.

———. [1986] 1990. *Capturing the Horizon. The Historical Geography of Transportation since the Transportation Revolution of the Sixteenth Century.* Baltimore: Johns Hopkins University Press.

van der Vat, Dan. 1988. *The Atlantic Campaign. World War II's Great Struggle at Sea.* New York: Harper & Row.

Verlinden, Charles. 1970. *The Beginnings of Modern Colonization.* Ithaca: Cornell University Press.

Vissering, Harry. 1922. *Zeppelin. The Story of a Great Achievement.* Chicago: the author.

von Karman, Theodore, & Lee Edson. 1967. *The Wind and Beyond. Theodore von Karman: Pioneer in Aviation and Pathfinder in Space.* Boston: Little, Brown.

von Ohain, Hans. 1979. "The Evolution and Future of Aeropropulsion Systems." In Boyne & Lopez, 25–46.

von Thünen, Johan Heinrich. 1966. *Von Thünen's Isolated State: An English Edition of* Der Isolierte Staat. Oxford: Pergamon.

Von Tunzelmann, G. N. 1978. *Steam Power and British Industrialization to 1860.* Oxford: Oxford University Press.

Wagner, Hermann. 1903. *Lehrbuch der Geographie.* Hannover: Hahn'sche.

Wagner, Philip L., & Marvin W. Mikesell, eds. 1962. *Readings in Cultural Geography.* Chicago: University of Chicago Press.

Walden, Don. 1990. "Raising Galveston." *American Heritage of Science and Technology* 5(3):8–19.

Wall, Richard, Jean Robin, & Peter Laslett, eds. 1983. *Family Forms in Historic Europe.* Cambridge: Cambridge University Press.

Wall, Robert. 1980. *Airliners.* Englewood Cliffs, N.J.: Prentice-Hall.

Wallerstein, Immanuel. 1974. *The Modern World-System.* Vol. 1, *Capitalist Agriculture and the Origins of the European World-Economy in the Sixteenth Century.* New York: Academic Press.

———. 1980. *The Modern World-System.* Vol. 2, *Mercantilism and the Consolidation of the European World-Economy, 1600–1750.* New York: Academic Press.

———. 1984. *The Politics of the World-Economy. The States, the Movements, and the Civilizations.* Cambridge: Cambridge University Press.

————. 1989. *The Modern World-System*. Vol. 3, *The Second Era of Great Expansion of the Capitalist World-Economy, 1730–1840s*. New York: Academic Press.

Walzer, Michael. 1971. *The Revolution of the Saints. A Study in the Origins of Radical Politics*. New York: Atheneum.

Ward, David. 1971. *Cities and Immigrants. A Geography of Change in Nineteenth-Century America*. New York: Oxford University Press.

Warner, Sam Bass, Jr. 1968. *Streetcar Suburbs. The Process of Growth in Boston, 1870–1900*. New York: Atheneum.

Watson, Ken. 1985. *Paddle Steamers: An Illustrated History of Steamboats on the Mississippi and Its Tributaries*. New York: Norton.

Webb, Walter Prescott. 1952. *The Great Frontier*. Boston: Houghton Mifflin.

Weber, Max. [1904–5] 1958. *The Protestant Ethic and the Spirit of Capitalism*. New York: Scribner's.

Wells, H. G. [1896] 1984. *The Wheels of Chance*. London: Dent.

Westwood, J. N. 1977. *Locomotive Designers in the Age of Steam*. Rutherford, N.J.: Fairleigh Dickinson University Press.

————. 1982. *Soviet Locomotive Technology during Industrialization, 1928–1952*. London: Macmillan.

Wheatley, Paul. 1961. *The Golden Khersonese: Studies in the Historical Geography of the Malay Peninsula*. Kuala Lumpur: University of Malaya Press.

White, H. P. 1963. *A Regional History of the Railways of Great Britain*. Vol. 3, *Greater London*. London: Phoenix House.

White, Lynn, Jr. 1962. *Medieval Technology and Social Change*. New York: Oxford University Press.

Whitehead, R. A. 1977. *Steam in the Village*. Newton Abbot, Devon: David & Charles.

Whitehurst, Clinton H., Jr. 1986. *The U.S. Shipbuilding Industry: Past, Present, and Future*. Annapolis, Md.: Naval Institute.

Whittle, Sir Frank. 1979. "The Birth of the Jet Engine in Britain." In Boyne & Lopez, 3–24.

Wilkins, Mira, & Frank E. Hill. 1964. *American Business Abroad. Ford on Six Continents*. Detroit: Wayne State University Press.

Williams, John. 1972. *The Other Battleground. The Home Fronts: Britain, France and Germany, 1914–1918*. Chicago: Regnery.

Williams, Michael. 1980. *British Tractors for World Farming*. Poole, Dorset: Blandford.

Williamson, Reg. 1988. "From Black to Silver: A Short History of the CD." *The Absolute Sound* 14 (March/April): 45–53.

Wilson, Geoffrey. 1976. *The Old Telegraphs*. London: Phillimore.

Wilson, S. S. 1973. "Bicycle Technology." *Scientific American* 228:81–91.

Wise, William. 1968. *Killer Smog. The World's Worst Air Pollution Disaster*. New York: Ballantine.

Wixom, Charles W. 1975. *ARBA Pictorial History of Roadbuilding*. Washington, D.C.: American Road Builders' Association.

Wolf, Eric R. 1982. *Europe and the People without History*. Berkeley and Los Angeles: University of California Press.

Wolko, Howard S. 1981. *In the Cause of Flight. Technologists of Aeronautics and Astronautics. Smithsonian Studies in Air and Space No. 4*. Washington, D.C.

Wooldridge, E. T. 1983. *Winged Wonders. The Story of the Flying Wings*. Washington, D.C.: Smithsonian.

References

Wright, Gavin. 1978. *The Political Economy of the Cotton South. Households, Markets and Wealth in the Nineteenth Century.* New York: Norton.

Wrigley, E. A. 1969. *Population and History.* New York: McGraw-Hill.

———. 1983. "The Growth of Population in Eighteenth-Century England: A Conundrum Resolved." *Past and Present* 98:121–50.

Wrigley, E. A., & R. S. Schofield. 1981. *The Population History of England, 1541–1871. A Reconstruction.* London: Arnold.

Young, David, & Neal Callahan. 1981. *Fill the Heavens with Commerce. Chicago Aviation, 1855–1926.* Chicago: Chicago Review.

Young, Peter. 1983. *Power of Speech. A History of Standard Telegraphs and Cables, 1883–1983.* London: Allen & Unwin.

World Maps

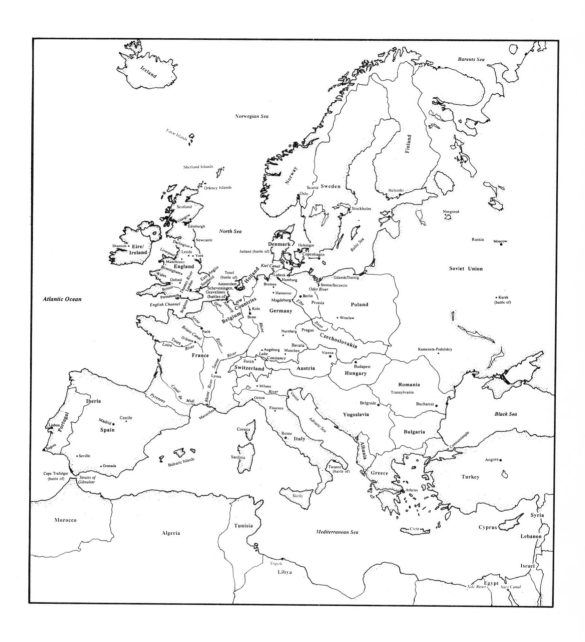

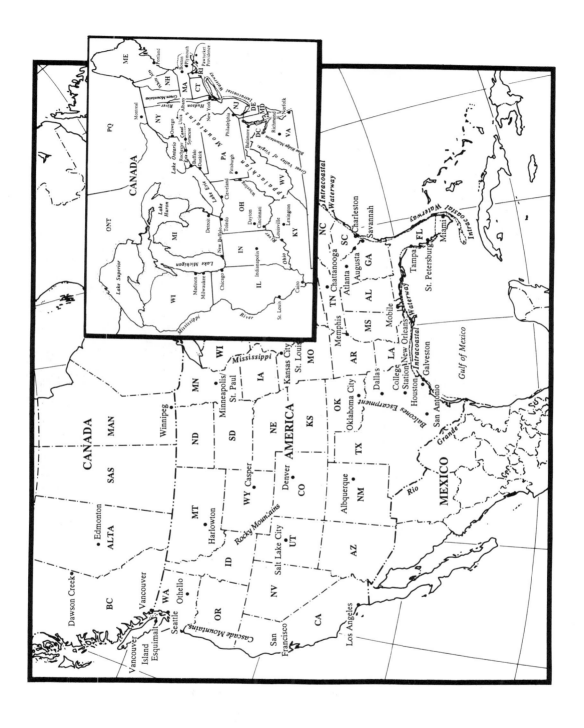

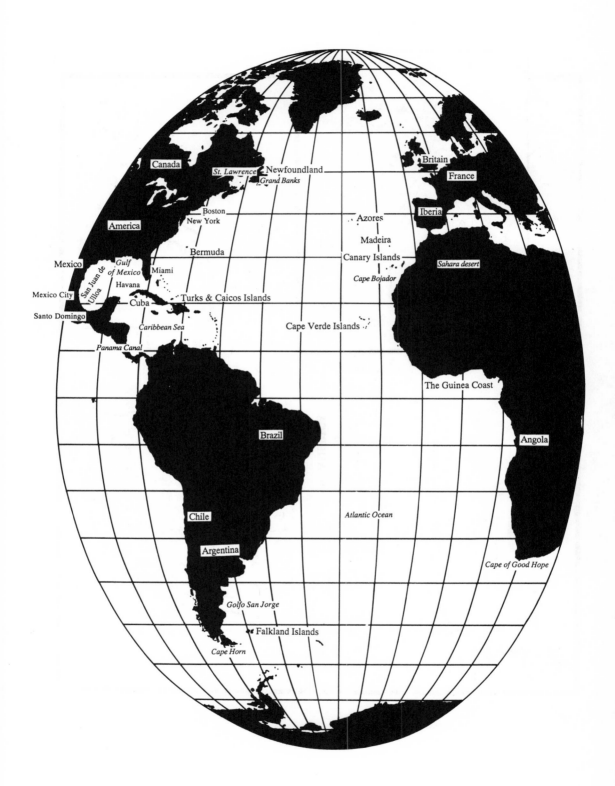

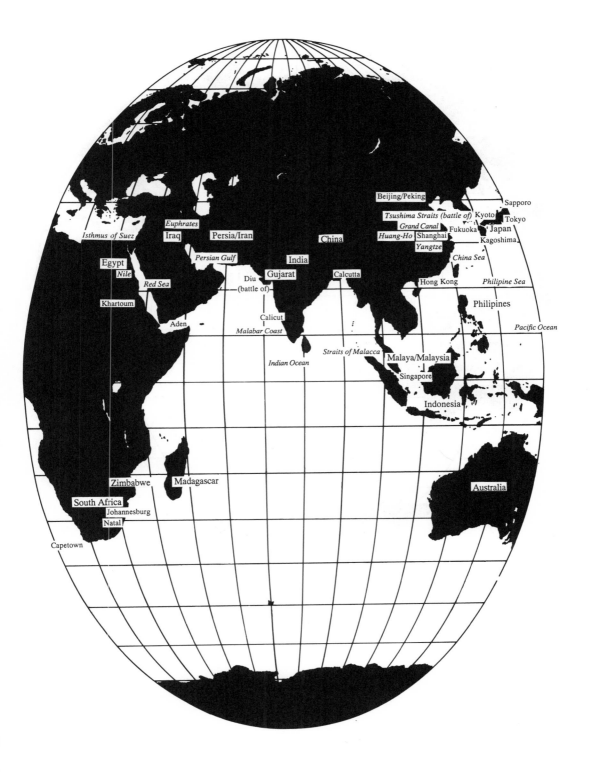

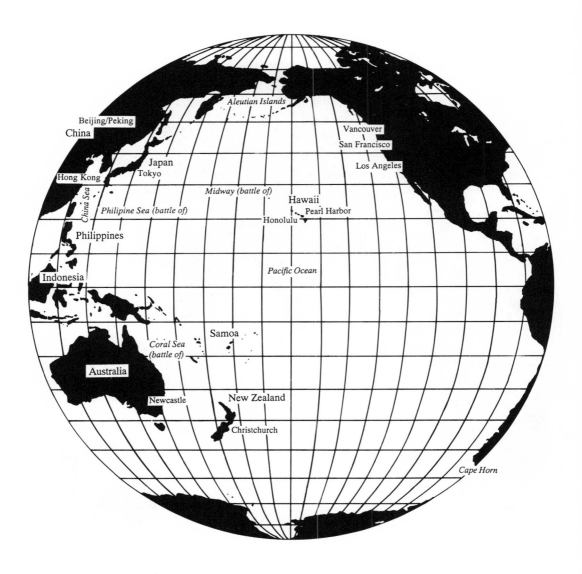

Aleutian Islands

Beijing/Peking
China

Vancouver
San Francisco

Japan
Tokyo

Los Angeles

Hong Kong

Midway (battle of)

China Sea

Hawaii

Philipine Sea (battle of)

Pearl Harbor
Honolulu

Philippines

Indonesia

Pacific Ocean

Samoa

Coral Sea
(battle of)

Australia

New Zealand

Newcastle

Christchurch

Cape Horn

steam, 241–42, two- and three-point hitches, 244; unit-construction, 243
Trafalgar, Battle of, 1805, 121, 123–24, 167
Trans Europe Express (TEE), 205
transatlantic air service, 282, 287, 289, 291–92
transfer costs, Portuguese reduction of, 109
transoceanic polities made possible, 8
Trent (river), England, 161, 163
trolley car, derivation of, 190
Troy, N.Y., 79
trucks, 8, 237–41, 309–10, 314
Tsushima Straits, Battle of, 1905, 137–38, 142
Tupoley. *See* airplane
"turbojet conclusion," 288
Turks & Caicos Islands, 111
"turnstile of trade," 58
TWA (Transcontinental & Western Airlines), 279, 280
two-field crop rotation, 44–45
two-tier society, 317, 318

Union Pacific Railroad, 175, 204
unions, truckers, 239
unit-construction of automobiles, 221, 228, 230
United Airlines, 280
United States: as hegemon, 14, 31–41; as hegemonic challenger, 14
university, role in research and development, 327
Unsafe at Any Speed (Nader), 222
unstable ecologies, 8
Ural Mountains, Russia, 247, 310
urban size and shape by mode of transport, 213–14
Utica, N.Y., 165

Vauxhall. *See* automobile; automobile companies
Venetian capital, 60
Venetian support of Islam, 112
Versailles, Treaty of, 1919, 271, 288
vertical frontier. *See* frontier, energy
vertical integration, 217
Vickers. *See* airplane; airplane companies
Vienna, Austria, 279
Vikings, 106, 109
Virginia, Great Valley of, 27
virtue, role of, in republic of North, 77
Vladivostok, Russia, 137
Volkswagen. *See* automobile; automobile companies
von Thunen model, world system as global, 29

Waco, Tex., 171, 194
War: American Revolution, 1776–83, 14, 26, 27, 58, 121, 123; Anglo-Dutch wars (1652–54, 1664–67, 1672–78), 22–23, 51, 57, 119–20, 121, 152, 308; Anglo-French wars of eighteenth

century, 161; of Austrian Succession, 1740–1748, 123; *blitzkrieg*, 100, 245–47, 250, 255, 274, 301; Boer, 1899–1902, 155, 210; cold, 304; economic, 12, 325, 328; of 1812, 121, 124, 152, 165; Falklands, 1982, 129, 149; Franco-Prussian, 1870, 32, 154, 252; French Revolution, 1793–1802, 14, 123; *guerre de course*, 314; Gulf War, 1990, 314, 328; Iran-Iraq, 1980–1988, 314; Iraqi uprising, 1920, 268; King William's, 1688–97, 121; Korean, 1950–1953, 291; Napoleonic Wars, 1803–15, 14, 97, 123–25, 152, 153, 313; Russian civil, 1917–22, 187; Russo-Japanese, 1904–5, 137, 142; Seven Years' (French and Indian), 1756–63, 123; of Spanish Succession, 1702–13, 121; U.S. Civil, 1861–65, 29, 141, 143, 152, 153; Vietnam, 1964–73, 148, 157; World War I, 1914–1918, 8, 14, 33, 38–39, 74, 86, 99–100, 135, 138, 140, 144–47, 155–56, 188–89, 204, 207, 241, 243–44, 246, 253, 258, 277, 301, 302, 312, 317, 329; World War II, 1939–1945, 37–39, 86, 96, 100, 144, 146, 151, 156, 189, 246–47, 279, 282, 284, 287, 289, 304, 313, 317, 323, 330
warships: dreadnought battleships, 133; guns, 155; iron-hulled, 152; monitors, 152. *See also* ships; wind-driven ships
Wartburg. *See* automobile companies
Washington, D.C., 302
Washington Naval Treaty, 1922, 147, 155
Washington-on-the-Brazos, Tex., 170–71
water-power, ubiquity of, in Europe by 1086, 52
water-powered mills, 7, 52–53
Waterloo, Battle of, 1815, 124
waterways, as arteries of empire, 30
"wave-line" principle of hull design, 140
Weser (river), Germany, 160
Western Union, 320–21
wheat frontier, 312
wheat productivity, 33
Wheeling, W.Va., 79
Wheels of Chance (Wells), 211
When William Came (Saki), 324
"white man's graveyard," 113
White Mountains, U.S., 27
wind-driven ships: armament, importance of, 16, 19, 50, 107, 111–13; caravel, or three-masted, 16, 107, 111–13; carrack, 114, 116; cogs, 106; defensible, 15, 50, 106; as energy converters, 52; first-rate, three-deck, ship of the line, 121, 152; fluyt, 21, 116, 151; frigate, 27, 121, 152; galleon, 20, 51, 114, 116; Iberian, inferiority of, by late 1500s, 21; skeleton construction of, 107; third-rate, 74 gun, ship of the line, 27, 121, 152; wooden, 6
wind-powered mills, 52
Winnipeg, Manitoba, 78
Winton 201 diesel engine, 202

Index to Proper Names